AF595083

Celebrating Applied Sciences Reaches 20,000 Articles Milestone: Feature Papers in Applied Biosciences and Bioengineering Section

Celebrating Applied Sciences Reaches 20,000 Articles Milestone: Feature Papers in Applied Biosciences and Bioengineering Section

Editors

Francesco Cappello
Magdalena Gorska-Ponikowska

MDPI • Basel • Beijing • Wuhan • Barcelona • Belgrade • Manchester • Tokyo • Cluj • Tianjin

Editors

Francesco Cappello
University of Palermo
Italy

Magdalena Gorska-Ponikowska
Medical University of Gdansk
Poland

Editorial Office
MDPI
St. Alban-Anlage 66
4052 Basel, Switzerland

This is a reprint of articles from the Special Issue published online in the open access journal *Applied Sciences* (ISSN 2076-3417) (available at: https://www.mdpi.com/journal/applsci/special_issues/Celebrating_Biosciences_Section).

For citation purposes, cite each article independently as indicated on the article page online and as indicated below:

LastName, A.A.; LastName, B.B.; LastName, C.C. Article Title. *Journal Name* **Year**, *Volume Number*, Page Range.

ISBN 978-3-0365-4427-4 (Hbk)
ISBN 978-3-0365-4428-1 (PDF)

Contents

About the Editors

Francesco Cappello

Francesco Cappello was born in Palermo, Italy, in 1973. In 1997, he completed his degree in Medicine and Surgery with honors and was awarded with Best Thesis prize. In 2002, he became a specialist in Pathological Anatomy with the maximum score. Since 2016, he has served as Full Professor at the University of Palermo, Palermo, Italy, where he teaches Human Anatomy and Histology. His current honorific appointments include the following: Scientific Director of the Euro-Mediterranean Institute of Science and Technology, Palermo, Italy; Associate Member of the Faculty at the Neuroscience Graduate Program, University of Texas Medical Branch, Galveston (TX), USA; Adjunct Associate Professor at the Department of Biology, Temple University, Philadelphia (PA), USA; President of the Italian Society of Experimental Biology; Senior Fellow of the Cell Stress Society International. His research interests include cell differentiation, tissue homeostasis, organ regeneration, cell stress and chaperones, and nanovesicles.

Magdalena Gorska-Ponikowska

Magdalena Gorska-Ponikowska (born in 1984), Associate Professor, is the Head of the Department of Medical Chemistry, Medical University of Gdansk, Poland. She has been a Visiting Researcher at the University of Stuttgart, Germany, since 2017 and has also served at the Institute 'Istituto Euro Mediterraneo di Scienza' (Palermo, Italy) since her appointment in 2018. She is a creator and co-owner of the cosmetic brand Skin Science, contributing with her expert authorship. Prof. Magdalena Górska-Ponikowska graduated from the Faculty of Pharmacy, Medical University of Gdansk, Poland. She subsequently obtained her habilitation degree in 2018. She has participated in numerous international scientific research internships, including at University of Palermo (Italy), University of Stuttgart (Germany), and University of Alberta (Edmonton, Canada). Prof. Magdalena Górska-Ponikowska was awarded a scholarship from the Minister of Science and Higher Education (2017) and granted the title 'Muse of the City of Sopot' (Poland) in the field of regional scientific activity (2017). Her research areas involve medical chemistry, biochemistry, molecular biology, oncology, cosmetology, nanotechnology, and neurobiology. She is an author and co-author of 74 peer-reviewed articles in international journals. ORCID: 0000-0002-7366-8429.

MDPI

Editorial

Special Issue "Celebrating Applied Sciences Reaches 20,000 Articles Milestone: Feature Papers in Applied Biosciences and Bioengineering Section"

Magdalena Gorska-Ponikowska [1,2,3,*] and Francesco Cappello [3,4,*]

1 Department of Medical Chemistry, Medical University of Gdansk, 1 Debinki Street, 80-211 Gdansk, Poland
2 Department of Biophysics, University of Stuttgart, 71254 Sttutgart, Germany
3 IEMEST Istituto Euro-Mediterraneo di Scienza e Tecnologia, 90127 Palermo, Italy
4 Department of Biomedicine, Neuroscience and Advanced Diagnostics, University of Palermo, 90127 Palermo, Italy
* Correspondence: magdalena.gorska-ponikowska@gumed.edu.pl (M.G.-P.); francesco.cappello@unipa.it (F.C.)

Citation: Gorska-Ponikowska, M.; Cappello, F. Special Issue "Celebrating Applied Sciences Reaches 20,000 Articles Milestone: Feature Papers in Applied Biosciences and Bioengineering Section". *Appl. Sci.* **2022**, *12*, 3978. https://doi.org/10.3390/app12083978

Received: 6 April 2022
Accepted: 11 April 2022
Published: 14 April 2022

This Special Issue celebrates the publication of 20,000 articles in *Applied Sciences*. This Special Issue intended to collect papers featuring important and recent developments or achievements in biosciences and bioengineering, with a special emphasis on recently discovered techniques or applications. Thanks to it, we managed to gather the interdisciplinary papers, including the broad spectrum of the following topics: clinical studies, advanced diagnostics, biochemistry, bioinformatics, biomaterials, biomechanics, biomedicine, biotechnology, clinical engineering, drug delivery, microbiology, life science, neuroscience, oncobiology, physiology, radiology, and tissue engineering.

In vivo clinical studies are the most important determination of drugs' pharmacokinetics and efficacy [1–4]. In an intensive care unit clinical study, the pharmacokinetics of levetiracetam, a second-generation antiepileptic drug has been evaluated [1]. Interestingly, a controlled clinical trial supervised by Lai et al. proved that combining yoga with rehabilitation has the potential to improve depressive disorders [2], while Khan and Reilly proposed suggestions for enhanced confidence for randomized controlled trials in protective treatments against endothelial glycocalyx degradation in surgery [3]. The effects of essential oils and other substances derived from the Lamiaceae family plants as adjuvants for the treatment of periodontitis have also been discussed [4].

A number of works concerned the very important topic of searching for biomarkers of pathogenesis and the progression of pathologies, such as cancer. Heat shock proteins (HSPs) are ubiquitously expressed housekeeping chaperones responsible for maintaining homeostasis of the organisms and can be considered as physiologically expressed biomarkers of cancer, e.g., leukemia, or different pathologies including gastric diseases [5–9]. The biomarkers may be localized extracellular or extracellularly excreted [8,9]. The molecular mechanism of asthma and COPD based on extracellular nanovesicles and their putative use in therapy is discussed by Fucarino et al. [8], while Alberti et al. describe tumor-secreted extracellular vesicles as the main mediators of cell–cell communication, permitting cells to exchange proteins, lipids, and metabolites under varying pathophysiological conditions [9].

When it is not possible to perform in vivo tests, establishing an in vitro-based system that can realistically simulate in vivo conditions is desirable [10–14]. A new cell culture method by combining fluoropolymers and dot-patterned extracellular matrix substrates to achieve spheroids has thus been successfully developed [10]. A new low-cost and simple-to-use method for the determination of free biothiols in biological fluids has been also proposed [11]. Moreover, a bioinformatics-based method, which introduces thermodynamic measures and topological characteristics aimed to identify potential drug targets for pharmacoresistant epileptic patients has been established [13]. Di Bella et al. show a relative accuracy, sensitivity, and specificity of 100% for *Salmonella* spp. detection and identification in comparison with the reference method ISO 6579-1:20 [14]. Interestingly, it was determined that the temperature of storage up to 7 months does not significantly affect the antioxidant properties of elderberry (*Sambucus nigra* L.) juice, which is highly rich in polyphenols, particularly flavonoids [15]. The nutritional characteristics of *Halimione portulacoides* (L.) has been widely described [16].

Author Contributions: Conceptualization M.G.-P. & F.C.; writing—original draft preparation, M.G.-P. & F.C. writing—review and editing, M.G.-P. & F.C.; supervision, M.G.-P. & F.C. All authors have read and agreed to the published version of the manuscript.

Funding: This research received no external funding.

Acknowledgments: M.G.-P. kindly acknowledge ST46 funding from Medical University of Gdansk.

Conflicts of Interest: The authors declare no conflict of interest.

References

1. Markantonis, S.; Markou, N.; Karagkounis, A.; Koutrafouri, D.; Stefanatou, H.; Kousovista, R.; Karalis, V. The Pharmacokinetics of Levetiracetam in Critically Ill Adult Patients: An Intensive Care Unit Clinical Study. *Appl. Sci.* **2022**, *12*, 1208. [CrossRef]
2. Lai, Y.; Lin, C.; Hsieh, C.; Yang, J.; Tsou, H.; Lin, C.; Li, S.; Chan, H.; Liu, W. Combining Yoga Exercise with Rehabilitation Improves Balance and Depression in Patients with Chronic Stroke: A Controlled Trial. *Appl. Sci.* **2022**, *12*, 922. [CrossRef]
3. Khan, H.; Reilly, G. Protective Treatments against Endothelial Glycocalyx Degradation in Surgery: A Systematic Review and Meta-Analysis. *Appl. Sci.* **2021**, *11*, 6994. [CrossRef]
4. Castellino, G.; Mesa, F.; Cappello, F.; Benavides-Reyes, C.; Malfa, G.; Cabello, I.; Magan-Fernandez, A. Effects of Essential Oils and Selected Compounds from Lamiaceae Family as Adjutants on the Treatment of Subjects with Periodontitis and Cardiovascular Risk. *Appl. Sci.* **2021**, *11*, 9563. [CrossRef]
5. Pawlik-Gwozdecka, D.; Sakowska, J.; Zieliński, M.; Górska-Ponikowska, M.; Cappello, F.; Trzonkowski, P.; Niedźwiecki, M. Association between Serum Heat Shock Proteins and Gamma-Delta T Cells—An Outdated Clue or a New Direction in Searching for an Anticancer Strategy? A Short Report. *Appl. Sci.* **2021**, *11*, 7325. [CrossRef]
6. Gorska-Ponikowska, M.; Kuban-Jankowska, A.; Marino Gammazza, A.; Daca, A.; Wierzbicka, J.M.; Zmijewski, M.A.; Luu, H.H.; Wozniak, M.; Cappello, F. The Major Heat Shock Proteins, Hsp70 and Hsp90, in 2-Methoxyestradiol-Mediated Osteosarcoma Cell Death Model. *Int. J. Mol. Sci.* **2020**, *21*, 616. [CrossRef] [PubMed]
7. Pitruzzella, A.; Burgio, S.; Lo Presti, P.; Ingrao, S.; Fucarino, A.; Bucchieri, F.; Cabibi, D.; Cappello, F.; Conway de Macario, E.; Macario, A.; et al. Hsp60 Quantification in Human Gastric Mucosa Shows Differences between Pathologies with Various Degrees of Proliferation and Malignancy Grade. *Appl. Sci.* **2021**, *11*, 3582. [CrossRef]
8. Fucarino, A.; Pitruzzella, A.; Burgio, S.; Zarcone, M.; Modica, D.; Cappello, F.; Bucchieri, F. Extracellular Vesicles in Airway Homeostasis and Pathophysiology. *Appl. Sci.* **2021**, *11*, 9933. [CrossRef]
9. Alberti, G.; Sánchez-López, C.; Andres, A.; Santonocito, R.; Campanella, C.; Cappello, F.; Marcilla, A. Molecular Profile Study of Extracellular Vesicles for the Identification of Useful Small "Hit" in Cancer Diagnosis. *Appl. Sci.* **2021**, *11*, 10787. [CrossRef]
10. Togo, H.; Yoshikawa-Terada, K.; Hirose, Y.; Nakagawa, H.; Takeuchi, H.; Kusunoki, M. Development of a Simple Spheroid Production Method Using Fluoropolymers with Reduced Chemical and Physical Damage. *Appl. Sci.* **2021**, *11*, 10495. [CrossRef]
11. Akrivi, E.; Vlessidis, A.; Giokas, D.; Kourkoumelis, N. Gold-Modified Micellar Composites as Colorimetric Probes for the Determination of Low Molecular Weight Thiols in Biological Fluids Using Consumer Electronic Devices. *Appl. Sci.* **2021**, *11*, 2705. [CrossRef]
12. Mobaraki, M.; Karnik, S.; Li, Y.; Mills, D. Therapeutic Applications of Halloysite. *Appl. Sci.* **2022**, *12*, 87. [CrossRef]
13. Yu, C.; Rietman, E.; Siegelmann, H.; Cavaglia, M.; Tuszynski, J. Application of Thermodynamics and Protein–Protein Interaction Network Topology for Discovery of Potential New Treatments for Temporal Lobe Epilepsy. *Appl. Sci.* **2021**, *11*, 8059. [CrossRef]
14. Di Bella, C.; Costa, A.; Sciortino, S.; Oliveri, G.; Cammilleri, G.; Geraci, F.; Lo Monaco, D.; Carpintieri, D.; Lo Bue, G.; Bongiorno, C.; et al. Validation of a Commercial Loop-Mediated Isothermal Amplification (LAMP)-Based Kit for the Detection of Salmonella spp. According to ISO 16140:2016. *Appl. Sci.* **2021**, *11*, 6669. [CrossRef]
15. Neves, C.; Pinto, A.; Gonçalves, F.; Wessel, D. Changes in Elderberry (*Sambucus nigra* L.) Juice Concentrate Polyphenols during Storage. *Appl. Sci.* **2021**, *11*, 6941. [CrossRef]
16. Pires, A.; Agreira, S.; Ressurreição, S.; Marques, J.; Guiné, R.; Barroca, M.; Moreira da Silva, A. Sea Purslane as an Emerging Food Crop: Nutritional and Biological Studies. *Appl. Sci.* **2021**, *11*, 7860. [CrossRef]

Article

The Pharmacokinetics of Levetiracetam in Critically Ill Adult Patients: An Intensive Care Unit Clinical Study

Sophia-Liberty Markantonis [1], Nikolaos Markou [2], Apostolos Karagkounis [1], Dionysia Koutrafouri [2], Helen Stefanatou [2], Rania Kousovista [3] and Vangelis Karalis [1,*]

1 Department of Pharmacy, School of Health Sciences, National and Kapodistrian University of Athens, 15784 Athens, Greece; kyroudi@pharm.uoa.gr (S.-L.M.); apos.karagkounis@gmail.com (A.K.)
2 ICU, Latsio Burn Center, Thriassio General Hospital of Elefsina, 19600 Attica, Greece; nikolaos_markou@hotmail.com (N.M.); denouk@gmail.com (D.K.); helenstefanatou@yahoo.com (H.S.)
3 Department of Mathematics, University of Crete, Heraklion, 70013 Crete, Greece; mathp359@math.uoc.gr
* Correspondence: vkaralis@pharm.uoa.gr; Tel.: +30-210-7274267

Abstract: The aim of this study was to investigate levetiracetam pharmacokinetics in critically ill adult intensive care patients and to identify pathophysiological factors affecting its kinetics. Fourteen critically ill patients in an intensive care unit were enrolled in the study and received intravenous levetiracetam. Blood samples were collected at specific time points to determine the levetiracetam pharmacokinetics. Patient characteristics such as renal function, demographics, disease severity, organ dysfunction, and biochemical laboratory tests were evaluated for their influence on the kinetics of levetiracetam. Estimated glomerular filtration rate (eGFR) had a statistically significant ($p = 0.001$) effect on levetiracetam clearance. None of the other patient characteristics had a statistically significant effect on the pharmacokinetics. Simulations of dosing regimens revealed that even typically administered doses of levetiracetam may result in significantly increased concentrations and risk of drug toxicity in patients with impaired renal function. The Acute Physiology and Chronic Health Evaluation II (APACHE II) score differed significantly among the three groups with different epileptic activity ($p = 0.034$). The same groups also differed in terms of renal function ($p = 0.031$). Renal dysfunction should be considered when designing levetiracetam dosage. Patients with a low APACHE II score had the lowest risk of experiencing epileptic seizures.

Keywords: levetiracetam; critically ill patients; population pharmacokinetic modeling; simulated dosage schemes; intensive care unit

Citation: Markantonis, S.-L.; Markou, N.; Karagkounis, A.; Koutrafouri, D.; Stefanatou, H.; Kousovista, R.; Karalis, V. The Pharmacokinetics of Levetiracetam in Critically Ill Adult Patients: An Intensive Care Unit Clinical Study. *Appl. Sci.* **2022**, *12*, 1208. https://doi.org/10.3390/app12031208

Academic Editors: Francesco Cappello and Magdalena Gorska-Ponikowska

Received: 30 November 2021
Accepted: 23 January 2022
Published: 24 January 2022

1. Introduction

Levetiracetam, a second-generation antiepileptic drug (AED) with a unique mechanism of action, is approved by both the U.S. Food and Drug Administration and the European Medicines Agency (EMA) as an add-on therapy for the treatment of partial seizures, myoclonic seizures, and primary generalized tonic-clonic seizures. In Europe, it is also the only AED approved for the treatment of partial seizures in adults and adolescents aged 16 years and older [1].

In critical care, levetiracetam has been studied for the treatment of status epilepticus, tumor-related seizures, and seizures following subarachnoid or intracerebral hemorrhage, trauma, and stroke [2–7]. The pharmacokinetic and pharmacodynamic properties of this antiepileptic drug make it an advantageous option in the care of patients in the ICU. Compared to other AEDs, the pharmacokinetic profile of levetiracetam is favorable after both oral and intravenous administration (IV) [8,9]. After oral administration, there is rapid and nearly complete absorption (95%) with dose-proportional pharmacokinetics, low protein binding (10%), and low within-subject variability. It is excreted primarily renally by glomerular filtration with partial tubular reabsorption and has a plasma elimination half-life of approximately 6–8 h in adults. There are no known clinically significant drug interactions, and it is not associated with the common hemodynamic or cardiovascular side effects of other AEDs [8,10–12].

A clear association between levetiracetam serum levels and therapeutic or toxic effects has not been demonstrated [10,13]. In accordance with the International League Against Epilepsy (ILAE) guidelines for medication adherence monitoring, overdose and dose adjustment, a therapeutic range of 12–46 mg/L has been recommended [10,14,15]. However, different levetiracetam levels outside the reference range in different populations have been reported in the literature. Pregnant women, elderly and pediatric patients, and patients with end-stage renal failure or those requiring renal replacement therapy exhibit altered pharmacokinetics [16–27].

Relatively few studies in critically ill ICU patients have reported increased drug clearance (greater than 120–160 mL/min/1.73 m^2) and treatment failure for seizures [21,28–31]. Based on empirical evidence, therapeutic drug monitoring for levetiracetam has been classified as 'potentially useful' but increasing evidence that altered pharmacokinetics of levetiracetam may be observed in critically ill patients has increased the need for therapeutic drug monitoring and pharmacokinetic modeling to determine optimal dosing regimens to maximize therapeutic effect and limit toxicity [22].

The aim of this study was to investigate the pharmacokinetics of levetiracetam in critically ill intensive care patients using nonlinear mixed-effect modeling approaches. The latter allow the identification of important pathophysiological factors (such as renal function) that could influence the kinetics of levetiracetam. Special emphasis was placed on the selection of the most suitable index for renal function and for this purpose several indices have been measured and investigated. Most of the patients participating in the study had from mild loss to severe loss of kidney function which allowed this exploration. Simulations were performed to assess the impact of important patient characteristics on the kinetics of levetiracetam. In addition, the potential influence of disease severity and renal function on epileptic status was investigated.

2. Materials and Methods

2.1. Patients and Data Collection

The study was conducted in the intensive care unit of a tertiary hospital (Latsio Burn Center, Thriassio General Hospital of Elefsina, Attica, Greece). Fourteen critically ill patients were enrolled in the study and all of them received intravenous (IV) levetiracetam (Keppra®) either for prophylaxis or treatment of epilepsy. The study protocol was reviewed by the hospital Scientific Committee and written informed consent was obtained from all patients or their legal representatives before enrolment in the study. The study was conducted in accordance with the International Conference on Harmonization guidelines for good clinical practice and the Declaration of Helsinki, while all patient data were collected and processed in full compliance with the EU General Data Protection Regulation 2016/679 and EU Directive 2016/680 of the European Parliament [32–35].

Patient data included information on levetiracetam administration and ICU admission, demographics, biochemical parameters (e.g., liver enzymes, urea, albumin), fluid balance (in mL) and organ function expressed by the Sequential Organ Failure Assessment Score (SOFA) and the Acute Physiology and Chronic Health Evaluation II (APACHE II) score [36]. From patients' weight and height, the body mass index (BMI) and body surface area (BSA) were also calculated [37]. The SOFA score is used to track the status of an ICU patient and determine the extent of their organ function or rate of failure. In this study, several expressions of this score were calculated, such as SOFA hemodynamic, SOFA liver, SOFA kidneys, and overall SOFA. The APACHE II is calculated within 24 h of a patient's admission to the ICU using various measurements such as heart rate, respiratory rate, blood pH, and hematocrit. Taking all these factors into account, an integer value from 0 to 71 is calculated, with higher values indicating a more severe condition and a higher risk of death. Epileptic activity was also monitored and classified as negative, possible, and positive. The "positive" group included patients who had epileptic seizures, while the "negative" group included those who had no epileptic activity and levetiracetam was administered for prophylactic reasons only. The "possible" category included patients who had a clinical profile of epileptic activity, but this was not verified by electroencephalogram.

In this study, particular attention was paid to the role of renal function, as many of the patients in the ICU had impaired renal function. For this reason, various measures of renal function were calculated, such as creatinine clearance over time, Cockcroft–Gault creatinine clearance, and measures of estimated renal function (eGFR) such as MDRD eGFR (Modification of Diet in Renal Disease study group), CKD-EPI eGFR (Chronic Kidney Disease epidemiology collaboration), Jelliffe eGFR in mL/min/1.73 m^2 [38–40]. Patients in this study were divided into two subgroups according to their renal function; patients with stable renal function (group A) and patients with acute renal impairment (group B) (Table 1). Levetiracetam treatment data included dosing regimen, cause of administration, chronic treatment, and loading dose administration.

Table 1. Patients' demographic characteristics, levetiracetam dosing regimen, epileptic activity, and subgroup classification.

ID	Subgroup	Renal Function	Sex	Age (Years)	Weight (Kg)	Height (cm)	Dosing Regimen	Epileptic Activity
1	A	Stable renal function	Female	62	84	165	1000 mg BID	±
2			Male	26	70	180	1000 mg BID	±
3			Female	72	68	160	1500 mg BID	+
4			Female	59	75	165	1500 mg BID	+
5			Male	47	80	170	1000 mg BID	-
6			Male	75	110	170	1000 mg BID	-
7			Female	48	58	160	1500 mg BID	±
8			Male	33	80	170	1000 mg BID	+
9			Male	33	80	170	1500 mg BID	+
10			Male	61	65	160	1000 mg BID	-
11			Male	83	65	170	1000 mg BID	-
12	B	Acute kidney injury	Female	45	75	170	1000 mg BID	-
13			Male	73	65	170	1500 mg BID	-
14			Female	82	50	150	1000 mg BID	-

Key: Epileptic activity: - no activity, ± possible, + epileptic activity; BID, twice daily; eGFR, estimated glomerular filtration rate; Stable renal function, when eGFR > 60 mL/min/1.73 m^2.

Levetiracetam was administered diluted in 100 mL N/S 0.9% *w/v* by IV infusion over 15 min, while the loading dose was administered in 250 mL N/S 0.9% *w/v* by IV infusion over 30 min. Blood samples for all participants were taken two days after the start of dosing to ascertain whether the serum levels had achieved a steady state. The specific time points for blood collection were determined according to the pharmacokinetics of levetiracetam to obtain as much information as possible. The schedule for blood collection was 0 h (trough level), 20 min, and 3 h after steady-state administration of levetiracetam. The actual blood sampling times were recorded.

For blood collection and storage of serum or plasma samples the following procedure was applied: collection of 5 mL of whole blood in a coated Vacutainer tube with a red tip (Becton Dickinson). After blood collection, it was allowed to clot undisturbed at room temperature (usually 15–30 min). The clot was removed by centrifugation at 1000–2000 rpm for 10 min. After centrifugation, the serum was immediately divided into two fractions >200 μL and immediately transferred to polypropylene tubes (not gel tubes) where they were stored in a freezer −70 °C. For pharmacokinetic analysis, serum concentrations of levetiracetam were determined by high performance liquid chromatography using a UV detector at 205 nm after sample preparation by solid phase extraction using ClinRep® HPLC Complete Kit, levetiracetam (Keppra®) in serum/plasma (order number 15500) (RECIPE Chemicals & Instruments GmbH, Munich, Germany). The lower limit of detection was 0.14 mg/L, the lower limit of quantification was 0.46 mg/L, the upper quantification limit was 104 mg/L, and the recovery was 97–105%. The chromatographic test was performed with the ClinTest® standard solution, calibration with the ClinCal® serum calibrator and quality control with the ClinChek® serum controls level I & II test solutions (order number 15582). The precision of the method was determined by calculating the relative standard deviation at three plasma concentrations during the same analysis (within-day precision) and in triplicate over six analyses (between-day precision). For the entire concentration range, the relative standard errors for both within- and between-assay precision were less than 7.2%.

2.2. Statistical Analysis

All statistical analyses of dosing schedules and clinical data were performed using IBM SPSS Statistics version 25 (IBM Corporation, Armonk, NY, USA). Statistical comparisons were performed at 5% significance level. For scale variables such as SOFA and eGFR measurements, a normality test was first performed (using the Shapiro–Wilk test) to determine whether the variables followed a normal distribution. In the case of normally distributed data, the independent t-test was performed to compare these variables between the two groups of patients. For comparison of more than two groups, the one-way method ANOVA was used with the post-hoc criteria of least significant difference and Tukey. The non-parametric analogue, Mann–Whitney, was used for variables that deviated from the normal distribution. The relationship between two nominal (or ordinal) measures (e.g., patient group and epilepsy status) was examined using the chi-square test. APACHE II score is actually an ordinal scale, but with many values. In other words, the APACHE II score can be considered as a Likert scale and therefore parametric statistical methods can be applied.

2.3. Population Pharmacokinetic Analysis

2.3.1. Non-Linear Mixed Effect Modeling

Population pharmacokinetic analysis was implemented in Monolix™ 2020R1 (Lixoft, Orsay, France, Simulation Plus) where individual serum concentration profile data were analyzed using the expectation maximization algorithm of stochastic approximation for nonlinear mixed effects followed by importance test methods. The value of the objective function was calculated using the Monte Carlo method for the final population parameter values.

2.3.2. Structural Models, Error Models, and Covariates

One- and two-compartment models for intravenous infusion with first-order elimination and initial estimates for the parameters were examined [23,25,26,32–34]. A lognormal distribution of pharmacokinetic parameters was assumed, while several residual error models were tested, including constant, proportional, and combined. The parameters were estimated using the stochastic approximation estimation method (SAEM), and the objective function value was determined using the importance sampling Monte Carlo approach at the final population parameter values.

Once the structural model was determined, several covariates were tested. The covariates examined in this study were related to subject-specific characteristics, including sex, age, body weight, height, APACHE II score, timed creatinine clearance, Cockcroft–Gault eGFR, Jelliffe eGFR, CKD-EPI eGFR, MDRD eGFR, serum creatinine, urea, SGOT, and total SOFA. Three expressions of body weight were used: typical body weight, ideal body weight, and adjusted body weight. Linear and lognormal (allometric) models were assessed. Analyses of covariates were performed using stepwise forward selection and backward elimination. Continuous covariates were examined either untransformed or centered around the mean or median value of the covariate. The categorical covariate examined was sex, while SOFA hemodynamic, SOFA renal, and SOFA liver were ordinal variables. The total SOFA score, which comes from the sum of all individual SOFA scores, can be considered a continuous variable

The Pearson correlation test and one-sided ANOVA were used for continuous and categorical covariates, respectively. The Wald test was used to check whether the covariates could explain the variability of the parameters in the final model. For all pharmacokinetic analyses, the significance threshold was set at 5%.

2.3.3. Model Evaluation

Model selection was based on goodness-of-fit criteria, visual inspection of diagnostic plots, comparison of relative standard deviations of estimated parameters, precision of estimates, and changes in Akaike and Bayesian information criteria and log-likelihood. Visual inspection of goodness-of-fit was performed using plots of observed values versus predicted values for the population and individual weighted residuals versus time. Visual predictive checks (VPCs) were used to assess the predictive performance, stability, and robustness of the model. VPCs were generated from 1000 Monte Carlo simulations and 90% prediction intervals. Normalized prediction distribution errors versus concentration were also used.

2.3.4. Simulations

The final pharmacokinetic model was used to simulate the concentration–time profiles of levetiracetam at different levels of renal function. Three simulated subject groups of 50 subjects each were formed: (a) subjects with normal renal function using the model parameters reported in the literature [23], (b) subjects with 50% clearance limitation, and (c) subjects with the pharmacokinetic characteristics determined in the fourteen patients in this study. The dosing regimen for each subject was identical to that actually used in the ICU (see Table 1). All simulations were performed in Simulx® (Monolix™ 2020R1).

3. Results

Eleven patients with stable renal function in the ICU had an eGFR estimate greater than 60 mL/min/1.73 m^2 and were classified in subgroup "A". The other three patients had acute renal failure and were classified in subgroup "B" (Table 1).

Regarding the reasons for administration of levetiracetam, most patients (71.4%) received levetiracetam for prophylactic purposes and only 28.6% for therapy. Only one patient (i.e., 7.1%) received levetiracetam for chronic treatment. The proportion of patients receiving a loading dose of levetiracetam was 42.9%, while the higher proportion (57.1%) initiated levetiracetam without a loading dose. The reason for levetiracetam administration was either prophylaxis (71.4% of patients) or treatment of seizures (28.6%). The mean eGFR of patients was 94.3 mL/min/1.73 m^2 (or equivalent to 5.66 L/h/1.73 m^2) and the mean APACHE II score was 17.4.

3.1. *Epileptic Activity*

The possible influence of disease severity (expressed by the APACHE II score or the SOFA score) and renal function (expressed by the CKD-EPI eGFR) on epileptic status was investigated. The Shapiro–Wilk test indicated normal distribution of the data and therefore one-way ANOVA was performed to detect a difference in the values of APACHE II and eGFR between the three levels of epileptic activity (positive, negative, possible). It was found that the APACHE II score was significantly different between the three groups (p = 0.034). A statistically significant difference was also found for the CKD-EPI eGFR (p = 0.031). For the APACHE II score, the absence of epileptic activity (i.e., the "negative" group) differed from the "possible" (p = 0.033) and the "positive" epileptic activity (p = 0.027). The patients with the lowest APACHE II score belonged to the group without epileptic seizures.

Regarding the CKD-EPI eGFR values, the three epilepsy groups showed significant differences ($p = 0.031$). Similarly, the group of patients without seizures was statistically different from the "possible" ($p = 0.034$) and the "positive" group ($p = 0.022$). Statistical comparisons were made for all other characteristics, but no significant differences were found.

It should be clarified that the abovementioned analysis was not the primary purpose of the study and for this reason no power assessment was made focusing on this. In addition, this assessment only attempted to link disease severity with renal function and not to explore their relationship with levetiracetam pharmacokinetics. In any case, it should be underlined that due to the limited sample size and therefore the number of patients within each group, these findings should be considered as exploratory and further adequately powered studies are necessary to confirm the results.

3.2. Population Pharmacokinetic Model

Population pharmacokinetic analysis was initially performed separately for each patient group. However, due to the limited number of ICU patients in group B, no model could be derived. Therefore, in order not to waste the information of group B patients, all individuals were pooled into one group to increase the ability to develop a robust pharmacokinetic model, but each patient's individual characteristics (e.g., renal function) were assessed and all available data were used. Furthermore, this pooling was possible because there was no statistically significant difference (p-value > 0.05) in levetiracetam concentration values between the two groups.

A one-compartment model with first order elimination was finally selected as the best model for the description of the levetiracetam concentration—time profiles. The estimates of the pharmacokinetic parameters of this best model are given in Table 2. It is of note that the volume of distribution of the central compartment (V) was 47.19 L and the clearance (CL) was 1.45 L/h. Of all the covariates examined for their influence on the kinetics of levetiracetam, only CKD-EPI eGFR was found to be significant for the clearance of levetiracetam. The coefficient of CKD-EPI eGFR with respect to clearance was 0.25, implying that better renal function (in terms of eGFR) leads to higher clearance of levetiracetam. The residual error model that led to the best results was the proportional error model with b = 0.099. The percentage relative standard errors for all estimated parameters had relatively low values. The significant covariate in the final model was the CKD-EPI eGFR (Wald test $p = 0.00164$), which is the Chronic Kidney Disease Epidemiology Collaboration equation, developed in an effort to provide a more precise formula for estimating glomerular filtration rate using serum creatinine, age, sex, and race data.

Table 2. Population parameters of the final pharmacokinetic model of levetiracetam.

Parameters (Units)	Estimate	Standard Error	Relative Standard Error (%)	p-Value
		Fixed effects		
V (L)	47.19	5.096	10.8	-
Cl (L/h)	1.45	0.221	15.2	-
beta_eGFR	0.25	0.034	13.6	0.00164
		Random effects		
ω_V	0.184	0.037	20.1	-
ω_Cl	0.251	0.0454	18.1	-
		Error model parameters		
b	0.099	0.0124	12.5	-

Key: V, volume of distribution; Cl, clearance; ω_V, between-subject variability value for V; ω_Cl; between-subject variability value for Cl; beta_eGFR, allometric scaling factor for CKD-EPI eGFR (centered around median) on Cl; b, proportional component of the error model.

Figure 1 shows the graphical evaluation of the final model. The visual representation of the predictive power shows that the prediction interval of the developed model includes the experimental concentration data in all cases. Similarly, the good predictive ability of the model is also illustrated in Figure 2, where the population predicted versus observed concentration values are almost linearly correlated (Figure 2a) and the NPDE versus time (Figure 2b) and concentration (Figure 2c) plots show an adequate performance. Additional goodness-of-fit plots are shown in the supplementary material (Figure S1a–c).

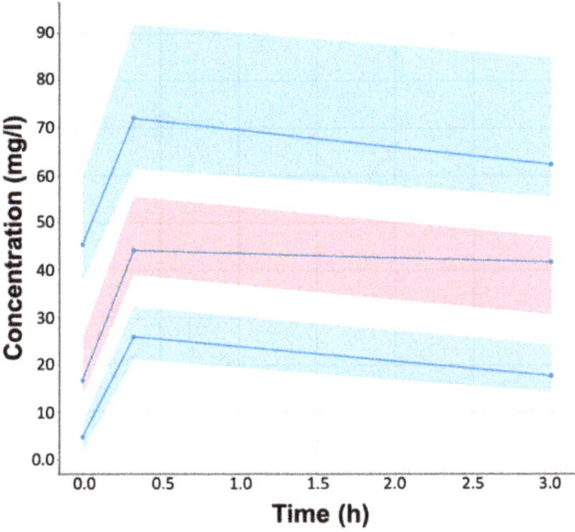

Figure 1. Visual predictive check plot for the final best levetiracetam pharmacokinetic model. The blue lines refer to the 10th, 50th, 90th percentile of empirical data and the shaded areas refer to the predicted 90% confidence intervals around each zone (10th, 50th, 90th percentiles). A number of 1000 Monte Carlo simulations were used.

The shrinkage of the volume and clearance estimates (parameter values were randomly sampled from the conditional distribution) was adequate and equal to 18.3% and −3.23%, respectively. The relationship between CKD-EPI eGFR and levetiracetam clearance is shown in Figure 3. An almost linear relationship (*levetiracetam clearance = 0.01534·(CKD-EPI eGFR) − 0.21031*) is observed with a correlation coefficient of 0.76. This finding is further evidence that an increase in CKD-EPI eGFR is associated with a subsequent increase in levetiracetam clearance.

3.3. Simulations

To determine the effects of renal function (expressed by the significant covariate CKD-EPI eGFR) on levetiracetam levels, simulations were performed based on the developed pharmacokinetic model. The utilized model parameters were those listed in Table 2, while two physiological/pathological situations were further simulated: Subjects with normal renal function [23] and subjects with 50% clearance restriction. It can be seen from Figure 4 that increased levetiracetam concentrations are observed with increasing renal function impairment. In healthy subjects, the mean maximum concentration was 31.1 mg/L (Figure 4a), which increased to 37.67 mg/L and 44.12 mg/L in patients with a 50% reduction in eGFR (Figure 4b) and patients with identical physiological characteristics to those in the study (Figure 4c), respectively. This means that the dosing regimens administered resulted in concentrations above the upper value of the reference range of 12–46 mg/L in almost half of the study patients [14].

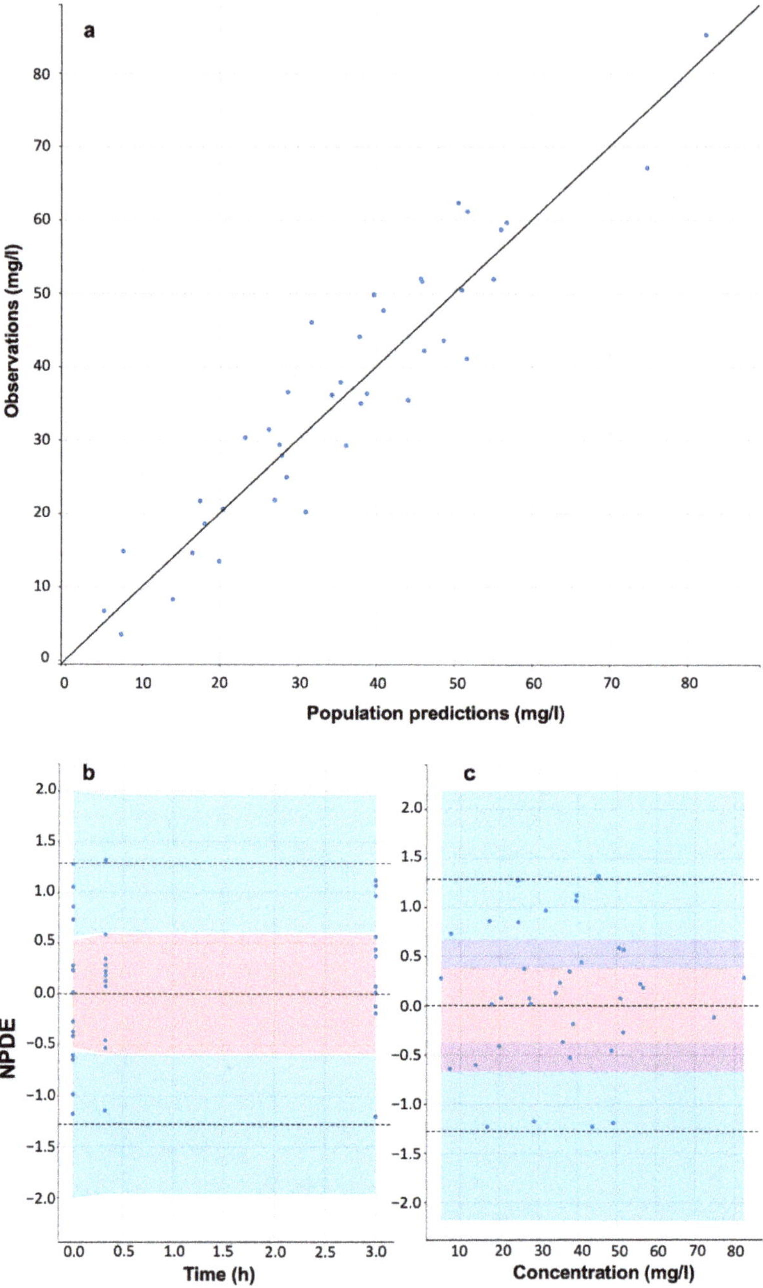

Figure 2. Goodness-of-fit plots for the final best model. (**a**) Observed vs. population predicted by the model concentrations of levetiracetam. The closed circles refer to the (predicted, observed) pairs and the solid line expresses the ideal situation of unity (i.e., y = x). (**b**) Normalized prediction distribution errors (NPDE) vs. time, and (**c**) NPDE vs. concentration. The dotted lines refer to the predicted median (at y = 0) and the 90% predicted percentiles, while the band indicates the 90% prediction interval.

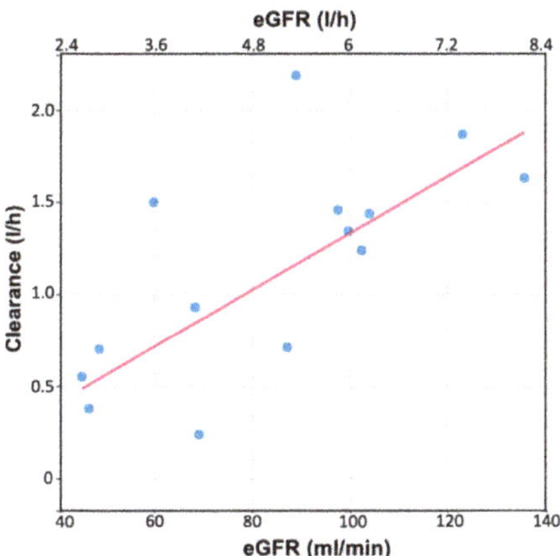

Figure 3. Relationship between levetiracetam clearance and the estimated glomerular filtration rate (eGFR) according to the Chronic Kidney Disease Epidemiology collaboration mathematical formula (CKD-EPI eGFR). The eGFR axis is expressed in dual units: mL/min and L/h. The correlation coefficient between levetiracetam clearance and eGFR was equal to 0.76.

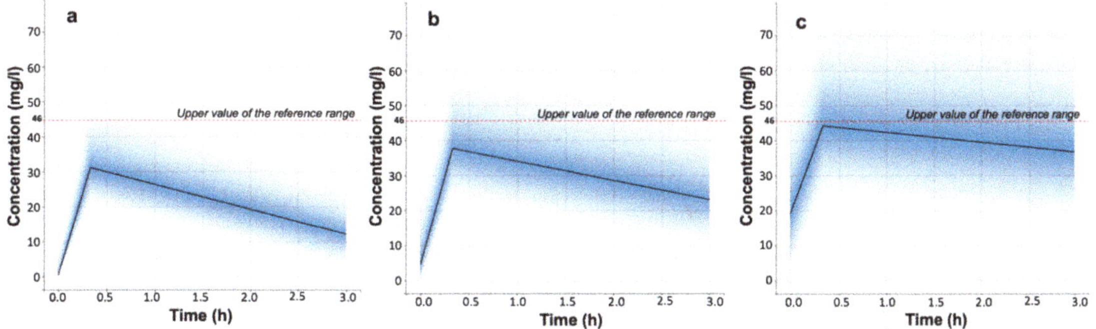

Figure 4. Simulated concentration vs. time profiles of levetiracetam in normal subjects (**a**), renal impairment 50% (**b**), and for the patients of the study (**c**). In each group, 500 individuals were generated. In the case of the normal renal function subjects (**a**) the utilized model parameters were obtained from study [23]. For the 50% renal impairment the model parameters were similar to (**a**), except for clearance, which was set at 50% of the reported value. For the study patients, the parameter values listed in Table 2 were used. The dosing regimen for each subject was identical to that actually used in the ICU (Table 1).

4. Discussion

The aim of this study was to investigate the pharmacokinetics of levetiracetam in critically ill intensive care adult patients and to identify possible pathophysiological factors affecting the kinetics of levetiracetam and, consequently, clinical outcome. A clinical study was conducted on 14 patients from an intensive care unit of a tertiary care hospital in whom levetiracetam was administered for either prophylactic or therapeutic reasons. Eleven patients had stable renal function according to the physicians' assessment, while the remaining three had acute renal injury. Having stable renal function does not mean that their renal function was normal, i.e., a glomerular filtration rate greater than 90 mL/min. In these 11 patients, the median eGFR was 97.7 mL/min and three patients (3 of 11) had an eGFR of less than 70 mL/min. In the patient group with acute kidney injury, the median eGFR was 45.3 mL/min. Overall, only five patients had an eGFR greater than 90 mL/min (i.e., normal renal function), and the remaining nine patients ranged from mild to severe loss of renal function.

Using measurements of serum concentrations in these patients, a population pharmacokinetic model for levetiracetam was developed. The final model for levetiracetam thus derived, referred to a linear one-compartment model with intravenous infusion administration and first-order elimination. This one-compartment model was consistent with other published levetiracetam studies conducted in children, adults, and elderly patients (Table S1) [23–27,41–43]. It should be mentioned that loading doses of levetiracetam were used. However, in the model development and the simulations (Figure 4) performed in this study, loading doses were not considered because they do not alter steady-state plasma levels, but only shorten the time required to reach these plateau levels.

Among the covariates tested, only measures of renal function showed a significant effect on clearance. Among the different eGFR measures explored, CKD-EPI eGFR was selected because it resulted in the lowest p-value ($p = 0.00164$, Table 2) and is one of the most commonly used eGFR measures. No significant relationships were found for other scores describing disease severity (such as APACHE II or SOFA), likely due to the small number of patients participating in the study. Levetiracetam clearance in this study (1.45 L/h, Table 2) was lower than the values reported in the literature (from 2.17 L/h to 6.87 L/h, Table S1). This finding may be attributed to the fact that our study included patients with impaired renal function. Moreover, the clearance estimates would also increase with increasing eGFR of our patients, as beta_eGFR had a positive sign (equal to 0.25, Table 2) [44]. In the study by Sime et al., which also recruited critically ill adult patients, the estimated clearance was 2.51 L/h [7]. Higher levetiracetam clearance estimates (e.g., 5.9 L/h) are reported in studies of epilepsy patients who are not critically ill, such as the large study by Ito et al. [41]. In that study the median eGFR estimate was 97.6 (mL/min)/1.73m^2, while in our study the median eGFR estimate of all participants was much lower 81.62 (mL/min)/1.73 m^2. The range of eGFR values in our study was from 18.8 mL/min to 123.3 mL/min, whereas for the study of Ito et al. the reported eGFR was as high as 189 mL/min [41]. The issue that our patients had poorer renal function compared with other studies reporting higher levetiracetam clearance studies, may explain the low clearance values found in our study.

The individual fitting plots for the 14 patients are shown in Figure S2. These fitting results demonstrate the adequacy of the descriptiveness of the developed model. Using only three blood samples per patient and data from only 14 subjects, it was possible to develop a model that describes the pharmacokinetics of levetiracetam. For comparison purposes, simulations were performed for the other relevant models in the literature. The model estimates (structural model, mean model parameter, between-subject variabilities, error model) were those reported in the literature (summarized in Table S1) and were kept fixed, while the covariates were related to our study patients. Then, it was attempted to examine the adequacy of fitting of these literature models to the concentration–time data of the 14 patients in this study. VPCs and normalized prediction distribution errors (NPDEs) plots were constructed (Figures S3 and S4 in the Supplementary Material). Figures S3 and S4 reveal that some literature models adequately fit the C-t data of our study, The best overall performance was observed for the models of Karatza et al., Pigeolet et al., Ito et al., and Chhun et al. [23,26,41,43]. However, for all these models, the performance was not better than for our model.

The positive correlation (correlation coefficient equal to 0.76) between eGFR and levetiracetam clearance, shown in Figure 3, indicates a lower elimination capacity in patients with renal dysfunction. This result is consistent with the prescribing information for levetiracetam, which recommends adjusting the levetiracetam dose based on renal function [24,41,42]. At this point, it should be mentioned that several other covariates related to patient disease severity have been tested for their potential role in influencing levetiracetam pharmacokinetics. However, none of these covariates were found to have a significant effect on any of the model parameters of levetiracetam (i.e., $p > 0.05$).

In a further step, the developed population pharmacokinetic model was used to evaluate the effects of impaired renal function on the kinetics of levetiracetam (Figure 4). In the simulated "healthy" subjects, the dosing regimens resulted in levetiracetam levels that were within the therapeutic range (Figure 4a), which is consistent with the literature [40]. The same levetiracetam doses resulted in significantly increased concentration levels in patients with impaired renal function, such as a 50% decrease in clearance and in the study patients (Figure 4b,c), as well as a higher risk of drug toxicity. It is worth noting that this observation is consistent with recommendations in the literature regarding treatment with levetiracetam. It is of note that if seizures cannot be controlled even with high concentrations of levetiracetam, increasing the dose above 2000 mg/day will not benefit the patient; instead, clinicians should consider switching to an antiepileptic drug with a different mechanism of action [23,42]. These findings highlight the importance of dose adjustment in patients depending on their eGFR [45,46]. In general, monitoring levetiracetam serum concentrations as part of therapeutic drug monitoring, ensures more effective and safer dosing regimens.

The relationship between patients' epileptic activity (group 1: no epileptic activity, group 2: possible epileptic activity, group 3: confirmed epileptic activity), APACHE II score, and eGFR levels has also been investigated. Despite the fact that only a few subjects were included in the study, statistically significant ($p = 0.034$) differences in the APACHE II score between the three groups of patients (no epileptic activity, possible, and confirmed epileptic activity) were found. The same groups also differed statistically significantly in terms of Cockcroft–Gault eGFR ($p = 0.031$).

The study had some limitations, one of which was the small number of patients who participated. Due to the small sample size, correlations between variables and their influence on pharmacokinetics may not have been detected (e.g., body weight) [45,46]. Other studies have been published with large numbers of subjects of epileptic adult patients receiving levetiracetam [25,26,40,41]. However, in this study special emphasis was based on exploring the most appropriate index for renal function in the case of critically ill patients. For this reason, this study explored seven indices for glomerular filtration rate (urea, creatinine, renal impairment stage, Cockroft–Gault_eGFR, MDRD_eGFR, CKD-EPI_eGFR, and Jeliffe_eGFR). Even using the data from this small sample, one covariate (the eGFR values collected from CKD-EPI) was found to perform best compared with all other indices and to play a significant role in the pharmacokinetics of levetiracetam. To increase the convergence of the model, five Markov chains were run in parallel. All available data were replicated, resulting in a number of 50 patients and improving the precision of the estimates.

5. Conclusions

A clinical trial was conducted on fourteen critically ill patients in an intensive care unit. Levetiracetam was administered intravenously either for prophylaxis or treatment of seizures. Blood samples were collected and plasma levels of levetiracetam were quantified. A population pharmacokinetic analysis was performed that included a number of patient-related somatometric characteristics and pathophysiological variables. A one-compartment model with intravenous infusion and first-order elimination kinetics was found to best describe the concentration–time data. Several renal function measures were estimated and evaluated in the modeling. Among them, CKD-EPI eGFR showed a statistically significant effect on levetiracetam clearance. With impaired renal function (especially at lower CKD-EPI eGFR values), levetiracetam clearance decreased. Simulations of dosing regimens revealed that even typically administered doses of levetiracetam may result in significantly increased concentrations and a higher risk of drug toxicity in patients with impaired renal function. Therefore, renal function should be considered when developing levetiracetam dosing regimens. Other measures describing disease severity, such as SOFA and APACHE II, had no effect on levetiracetam kinetics. Finally, when analyzing the association between

APACHE II score and epileptic activity, it was found that patients with the lowest APACHE II score had a lower risk of experiencing epileptic seizures.

Supplementary Materials: The following are available online at https://www.mdpi.com/article/10.3390/app12031208/s1, Table S1: Previously published population pharmacokinetic modeling results for levetiracetam, Figure S1: Goodness-of-fit plots for the final best model (a. Observed vs. predicted by the model individual concentrations of levetiracetam, b. Individual Weighted Residuals (IWRES) vs. time, and c. Individual Weighted Residuals vs. concentration), Figure S2: Levetiracetam concentration vs. time plots for the 14 patients of the study, Figure S3: Visual predictive check plots for the literature models, Figure S4: Normalized prediction distribution errors (NPDE) vs. time and concentration for the literature models.

Author Contributions: Conceptualization, S.-L.M., N.M., A.K., D.K., H.S. and V.K.; methodology, S.-L.M., N.M. and V.K.; data curation, R.K. and V.K.; investigation, N.M., A.K., D.K., H.S.; writing original draft, S.-L.M., R.K. and V.K.; supervision, S.-L.M., N.M. and V.K. All authors have read and agreed to the published version of the manuscript.

Funding: This research received no external funding.

Institutional Review Board Statement: The study was conducted according to the guidelines of the Declaration of Helsinki and approved by the Ethics Committee of Thriassio General Hospital of Elefsina (27 July 2020).

Informed Consent Statement: Written informed consent was obtained from the patients to publish this paper.

Data Availability Statement: The data presented in this study are available on request from the corresponding author.

Conflicts of Interest: The authors declare no conflict of interest.

References

1. UpToDate®. Levetiracetam: Drug Information. Available online: https://www.uptodate.com/contents/levetiracetam-druginformation?topicRef=128153&source=see_link (accessed on 26 October 2021).
2. Rossetti, A.O.; Bromfield, E.B. Levetiracetam in the treatment of status epilepticus in adults: A study of 13 episodes. *Eur. Neurol.* **2005**, *54*, 34–38. [CrossRef] [PubMed]
3. Patel, N.C.; Landan, I.R.; Levin, J.; Szaflarski, J.; Wilner, A.N. The use of levetiracetam in refractory status epilepticus. *Seizure* **2006**, *15*, 137–141. [CrossRef] [PubMed]
4. DJohn, J.; Ibrahim, R.; Patel, P.; DeHoff, K.; Kolbe, N. Administration of Levetiracetam in Traumatic Brain Injury: Is it Warranted? *Cureus* **2020**, *12*, e9117. [CrossRef] [PubMed]
5. Hazama, A.; Ziechmann, R.; Arul, M.; Krishnamurthy, S.; Galgano, M.; Chin, L.S. The Effect of Keppra Prophylaxis on the Incidence of Early Onset, Post-traumatic Brain Injury Seizures. *Cureus* **2018**, *10*, e2674. [CrossRef] [PubMed]
6. Zangbar, B.; Khalil, M.; Gruessner, A.; Joseph, B.; Friese, R.; Kulvatunyou, N.; Wynne, J.; Latifi, R.; Rhee, P.; O'Keeffe, T. Levetiracetam Prophylaxis for Post-traumatic Brain Injury Seizures is Ineffective: A Propensity Score Analysis. *World J. Surg.* **2016**, *40*, 2667–2672. [CrossRef] [PubMed]
7. Sime, F.B.; Roberts, J.A.; Jeffree, R.L.; Pandey, S.; Adiraju, S.; Livermore, A.; Butler, J.; Parker, S.L.; Wallis, S.C.; Lipman, J.; et al. Population Pharmacokinetics of Levetiracetam in Patients with Traumatic Brain Injury and Subarachnoid Hemorrhage Exhibiting Augmented Renal Clearance. *Clin. Pharmacokinet.* **2021**, *60*, 655–664. [CrossRef]
8. Wright, C.; Downing, J.; Mungall, D.; Khan, O.; Williams, A.; Fonkem, E.; Garrett, D.; Aceves, J.; Kirmani, B. Clinical pharmacology and pharmacokinetics of levetiracetam. *Front. Neurol.* **2013**, *4*, 192. [CrossRef]
9. Ramael, S.; Daoust, A.; Otoul, C.; Toublanc, N.; Troenaru, M.; Lu, Z.S.; Stockis, A. Levetiracetam intravenous infusion: A randomized, placebo-controlled safety and pharmacokinetic study. *Epilepsia* **2006**, *47*, 1128–1135. [CrossRef]
10. Patsalos, P.N. Clinical Pharmacokinetics of Levetiracetam. *Clin. Pharmacokinet.* **2004**, *43*, 707–724. [CrossRef] [PubMed]
11. Toublanc, N.; Okagaki, T.; Boyce, M.; Chan, R.; Mugitani, A.; Watanabe, S.; Yamamoto, K.; Yoshida, K.; Andreas, J.O. Pharmacokinetics of the antiepileptic drug levetiracetam in healthy Japanese and Caucasian volunteers following intravenous administration. *Eur. J. Drug Metab. Pharmacokinet.* **2015**, *40*, 461–469. [CrossRef]
12. Sirsi, D.; Safdieh, J.E. The safety of levetiracetam. *Expert Opin. Drug Saf.* **2007**, *6*, 241–250. [CrossRef] [PubMed]
13. Sourbron, J.; Chan, H.; Wammes-van der Heijden, E.A.; Klarenbeek, P.; Wijnen, B.; de Haan, G.J.; van der Kuy, H.; Evers, S.; Majoie, M. Review on the relevance of therapeutic drug monitoring of levetiracetam. *Seizure* **2018**, *62*, 131–135. [CrossRef] [PubMed]

14. Patsalos, P.N.; Berry, D.J.; Bourgeois, B.F.; Cloyd, J.C.; Glauser, T.A.; Johannessen, S.I.; Leppik, I.E.; Tomson, T.; Perucca, E. Antiepileptic drugs–best practice guidelines for therapeutic drug monitoring: A position paper by the subcommission on therapeutic drug monitoring, ILAE Commission on Therapeutic Strategies. *Epilepsia* **2008**, *49*, 1239–1276. [CrossRef] [PubMed]
15. Jacob, S.; Nair, A.B. An Updated Overview on Therapeutic Drug Monitoring of Recent Antiepileptic Drugs. *Drugs R & D* **2016**, *16*, 303–316. [CrossRef]
16. Contin, M.; Mohamed, S.; Albani, F.; Riva, R.; Baruzzi, A. Levetiracetam clinical pharmacokinetics in elderly and very elderly patients with epilepsy. *Epilepsy Res.* **2012**, *98*, 130–134. [CrossRef] [PubMed]
17. Glauser, T.A.; Mitchell, W.G.; Weinstock, A.; Bebin, M.; Chen, D.; Coupez, R.; Stockis, A.; Lu, Z.S. Pharmacokinetics of levetiracetam in infants and young children with epilepsy. *Epilepsia* **2007**, *48*, 1117–1122. [CrossRef] [PubMed]
18. Fountain, N.B.; Conry, J.A.; Rodríguez-Leyva, I.; Gutierrez-Moctezuma, J.; Salas, E.; Coupez, R.; Stockis, A.; Lu, Z.S. Prospective assessment of levetiracetam pharmacokinetics during dose escalation in 4- to 12-year-old children with partial-onset seizures on concomitant carbamazepine or valproate. *Epilepsy Res.* **2007**, *74*, 60–69. [CrossRef]
19. Nei, S.D.; Wittwer, E.D.; Kashani, K.B.; Frazee, E.N. Levetiracetam Pharmacokinetics in a Patient Receiving Continuous Venovenous Hemofiltration and Venoarterial Extracorporeal Membrane Oxygenation. *Pharmacotherapy* **2015**, *35*, e127–e130. [CrossRef]
20. New, A.M.; Nei, S.D.; Kashani, K.B.; Rabinstein, A.A.; Frazee, E.N. Levetiracetam Pharmacokinetics During Continuous Venovenous Hemofiltration and Acute Liver Dysfunction. *Neurocrit. Care* **2016**, *25*, 141–144. [CrossRef] [PubMed]
21. Cook, A.M.; Arora, S.; Davis, J.; Pittman, T. Augmented renal clearance of vancomycin and levetiracetam in a traumatic brain injury patient. *Neurocrit. Care* **2013**, *19*, 210–214. [CrossRef]
22. Hiemke, C.; Bergemann, N.; Clement, H.W.; Conca, A.; Deckert, J.; Domschke, K.; Eckermann, G.; Egberts, K.; Gerlach, M.; Greiner, C.; et al. Consensus Guidelines for Therapeutic Drug Monitoring in Neuropsychopharmacology: Update 2017. *Pharmacopsychiatry* **2018**, *51*, e1. [CrossRef] [PubMed]
23. Karatza, E.; Markantonis, S.L.; Savvidou, A.; Verentzioti, A.; Siatouni, A.; Alexoudi, A.; Gatzonis, S.; Mavrokefalou, E.; Karalis, V. Pharmacokinetic and pharmacodynamic modeling of levetiracetam: Investigation of factors affecting the clinical outcome. *Xenobiotica* **2020**, *50*, 1090–1100. [CrossRef] [PubMed]
24. Keppra SmPC, *Keppra UCB Pharma Limited, Summary of Product Characteristics*; Keppra SmPC, UCB Pharma S.A.: Brussels, Belgium, 2016; Available online: https://www.medicines.org.uk/emc/product/2293/smpc#gref (accessed on 26 October 2021).
25. Hernández-Mitre, M.P.; Medellín-Garibay, S.E.; Rodríguez-Leyva, I.; Rodríguez-Pinal, C.J.; Zarazúa, S.; Jung-Cook, H.H.; Roberts, J.A.; Romano-Moreno, S.; Milán-Segovia, R. Population Pharmacokinetics and Dosing Recommendations of Levetiracetam in Adult and Elderly Patients with Epilepsy. *J. Pharm. Sci.* **2020**, *109*, 2070–2078. [CrossRef] [PubMed]
26. Pigeolet, E.; Jacqmin, P.; Sargentini-Maier, M.L.; Stockis, A. Population pharmacokinetics of levetiracetam in Japanese and Western adults. *Clin. Pharmacokinet.* **2007**, *46*, 503–512. [CrossRef] [PubMed]
27. Toublanc, N.; Sargentini-Maier, M.L.; Lacroix, B.; Jacqmin, P.; Stockis, A. Retrospective population pharmacokinetic analysis of levetiracetam in children and adolescents with epilepsy: Dosing recommendations. *Clin. Pharmacokinet.* **2008**, *47*, 333–341. [CrossRef] [PubMed]
28. Spencer, D.D.; Jacobi, J.; Juenke, J.M.; Fleck, J.D.; Kays, M.B. Steady-state pharmacokinetics of intravenous levetiracetam in neurocritical care patients. *Pharmacotherapy* **2011**, *31*, 934–941. [CrossRef] [PubMed]
29. Uges, J.W.; van Huizen, M.D.; Engelsman, J.; Wilms, E.B.; Touw, D.J.; Peeters, E.; Vecht, C.J. Safety and pharmacokinetics of intravenous levetiracetam infusion as add-on in status epilepticus. *Epilepsia* **2009**, *50*, 415–421. [CrossRef]
30. Klein, P.; Herr, D.; Pearl, P.L.; Natale, J.; Levine, Z.; Nogay, C.; Sandoval, F.; Trzcinski, S.; Atabaki, S.M.; Tsuchida, T.; et al. Results of phase 2 safety and feasibility study of treatment with levetiracetam for prevention of posttraumatic epilepsy. *Arch. Neurol.* **2012**, *69*, 10. [CrossRef]
31. Jarvie, D.; Mahmoud, S.H. Therapeutic Drug Monitoring of Levetiracetam in Select Populations. *J. Pharm. Pharm. Sci.* **2018**, *21*, 149s–176s. [CrossRef]
32. ICH Harmonized Tripartite Guideline for Good Clinical Practice, E6(R2). Available online: https://www.ema.europa.eu/en/documents/scientific-guideline/ich-e-6-r2-guideline-good-clinical-practice-step-5_en.pdf (accessed on 17 January 2022).
33. ICH Topic E9 Statistical Principles for Clinical Trials. Note for Guidance on Statistical Principles for Clinical Trials (CPMP/ICH/363/96). Available online: https://www.ema.europa.eu/en/documents/scientific-guideline/ich-e-9-statistical-principles-clinical-trials-step-5_en.pdf (accessed on 17 January 2022).
34. Regulation (EU) 2016/679 of the European Parliament and of the Council of 27 April 2016 on the Protection of Natural Persons with Regard to the Processing of Personal Data and on the Free Movement of Such Data, and Repealing Directive 95/46/EC (General Data Protection Regulation. Available online: https://eur-lex.europa.eu/legal-content/EN/TXT/PDF/?uri=CELEX:32016R0679 (accessed on 17 January 2022).
35. Directive (EU) 2016/680 of the European Parliament and of the Council of 27 April 2016 on the Protection of Natural Persons with Regard to the Processing of Personal Data by Competent Authorities for the Purposes of the Prevention, Investigation, Detection or Prosecution of Criminal Offences or the Execution of Criminal Penalties, and on the Free Movement of Such Data, and Repealing Council Framework Decision 2008/977/JHA. Available online: https://eur-lex.europa.eu/legal-content/EN/TXT/PDF/?uri=CELEX:32016R0679 (accessed on 17 January 2022).
36. Jeong, S. Scoring Systems for the Patients of Intensive Care Unit. *Acute Crit. Care* **2018**, *33*, 102–104. [CrossRef]

37. Fleisher, G.; Ludwig, S.; Bachur, R.; Gorelick, M.; Shaw, K. *Textbook of Pediatric Emergency Medicine (Textbook of Pediatric Medicine (Fleisher))*, 6th ed.; Lippincott Williams & Wilkins: Philadelphia, PA, USA, 2010; p. 1882.
38. Liu, X.; Lv, L.; Wang, C.; Shi, C.; Cheng, C.; Tang, H.; Chen, Z.; Ye, Z.; Lou, T. Comparison of prediction equations to estimate glomerular filtration rate in Chinese patients with chronic kidney disease. *Intern Med. J.* **2012**, *42*, e59–e67. [CrossRef] [PubMed]
39. Levey, A.S.; Stevens, L.A.; Schmid, C.H.; Zhang, Y.L.; Castro, A.F.; Feldman, H.I.; Kusek, J.W.; Eggers, P.; Van Lente, F.; Greene, T.; et al. CKD-EPI (Chronic Kidney Disease Epidemiology Collaboration), A new equation to estimate glomerular filtration rate. *Ann. Intern Med.* **2009**, *150*, 604–612. [CrossRef] [PubMed]
40. Levey, A.S.; Inker, L.A.; Coresh, J. GFR estimation: From physiology to public health. *Am. J. Kidney Dis.* **2014**, *63*, 820–834. [CrossRef] [PubMed]
41. Ito, S.; Yano, I.; Hashi, S.; Tsuda, M.; Sugimoto, M.; Yonezawa, A.; Ikeda, A.; Matsubara, K. Population Pharmacokinetic Modeling of Levetiracetam in Pediatric and Adult Patients With Epilepsy by Using Routinely Monitored Data. *Ther. Drug Monit.* **2016**, *38*, 371–378. [CrossRef] [PubMed]
42. Rhee, S.J.; Shin, J.W.; Lee, S.; Moon, J.; Kim, T.J.; Jung, K.Y.; Park, K.I.; Lee, S.T.; Jung, K.H.; Yu, K.S.; et al. Population pharmacokinetics and dose-response relationship of levetiracetam in adult patients with epilepsy. *Epilepsy Res.* **2017**, *132*, 8–14. [CrossRef] [PubMed]
43. Chhun, S.; Jullien, V.; Rey, E.; Dulac, O.; Chiron, C.; Pons, G. Population pharmacokinetics of levetiracetam and dosing recommendation in children with epilepsy. *Epilepsia* **2009**, *50*, 1150–1157. [CrossRef] [PubMed]
44. Himmelfarb, J.; Alp Ikizler, T. *Chronic Kidney Disease, Dialysis, and Transplantation: A Companion to Brenner and Rector's the Kidney*, 4th ed.; Elsevier: Amsterdam, The Netherlands, 2019.
45. Li, Z.R.; Wang, C.Y.; Zhu, X.; Jiao, Z. Population Pharmacokinetics of Levetiracetam: A Systematic Review. *Clin. Pharmacokinet.* **2021**, *60*, 305–318. [CrossRef]
46. Methaneethorn, J.; Leelakanok, N. Population Pharmacokinetics of Levetiracetam: A Systematic Review. *Curr. Clin. Pharmacol.* **2021**, ahead of print. [CrossRef]

MDPI

Article

Combining Yoga Exercise with Rehabilitation Improves Balance and Depression in Patients with Chronic Stroke: A Controlled Trial

Yen-Ting Lai [1,2,3], Chien-Hung Lin [4,5,6], City C. Hsieh [7], Jung-Cheng Yang [2,3], Han-Hsing Tsou [8,9,10], Chih-Ching Lin [4,11], Szu-Yuan Li [4,11], Hsiang-Lin Chan [12,†] and Wen-Sheng Liu [4,6,8,9,13,14,*,†]

1 Department of Physical Medicine and Rehabilitation, National Taiwan University College of Medicine, Taipei 100, Taiwan; csmclaiyt@gmail.com
2 Department of Physical Medicine and Rehabilitation, National Taiwan University Hospital Hsin-Chu Branch, Hsinchu 300, Taiwan; w3yjcw@gmail.com
3 Department of Physical Medicine and Rehabilitation, National Taiwan University Hsin-Chu Hospital, Hsinchu 300, Taiwan
4 Faculty of Medicine, School of Medicine, National Yang Ming Chiao Tung University, Hsinchu 300, Taiwan; chlin5@vghtpe.gov.tw (C.-H.L.); lincc2@vghtpe.gov.tw (C.-C.L.); syli@vghtpe.gov.tw (S.-Y.L.)
5 Department of Pediatrics, Taipei Veterans General Hospital, Taipei 112, Taiwan
6 College of Science and Engineering, Fu Jen Catholic University, New Taipei City 242, Taiwan
7 Department of Kinesiology, National Tsing Hua University, Hsinchu 300, Taiwan; chsieh@mail.nd.nthu.edu.tw
8 Institute of Food Safety and Health Risk Assessment, National Yang Ming Chiao Tung University, Taipei 112, Taiwan; hhtsou@nycu.edu.tw
9 Institute of Food Safety and Health Risk Assessment, National Yang-Ming University, Taipei 112, Taiwan
10 Kim Forest Enterprise Co., Ltd., New Taipei City 221, Taiwan
11 Division of Nephrology, Department of Medicine, Taipei Veterans General Hospital, Taipei 112, Taiwan
12 Department of Child Psychiatry, Chang Gung Memorial Hospital and University, Taoyuan 333, Taiwan; ivyhlin70@kimo.com
13 Division of Nephrology, Department of Medicine, Taipei City Hospital, Zhongxing Branch, Taipei 103, Taiwan
14 Departmentof Special Education, University of Taipei, Taipei 100, Taiwan
* Correspondence: dat23@tpech.gov.tw; Tel.: +886-2-979305704
† These authors contributed equally to this manuscript.

Citation: Lai, Y.-T.; Lin, C.-H.; Hsieh, C.C.; Yang, J.-C.; Tsou, H.-H.; Lin, C.-C.; Li, S.-Y.; Chan, H.-L.; Liu, W.-S. Combining Yoga Exercise with Rehabilitation Improves Balance and Depression in Patients with Chronic Stroke: A Controlled Trial. *Appl. Sci.* **2022**, *12*, 922. https://doi.org/10.3390/app12020922

Academic Editors: Francesco Cappello and Magdalena Gorska-Ponikowska

Received: 1 December 2021
Accepted: 10 January 2022
Published: 17 January 2022

Abstract: Background: We combined yoga with standard stroke rehabilitation and compared it to the rehabilitation alone for depression and balance in patients. Methods: Forty patients aged from 30 to 80 who had suffered a stroke 90 or more days previously were divided evenly with age stratification and patients' will (hence not randomized). In the intervention group 16 completed 8-week stroke rehabilitation combined with 1 h of yoga twice weekly. Another 19 patients completed the standard rehabilitation as the control group. Results: The yoga group showed significant improvement in depression (Taiwanese Depression Questionnaire, $p = 0.002$) and balance (Berg Balance Scale, $p < 0.001$). However, the control group showed improvement only in balance ($p = 0.001$) but not in depression ($p = 0.181$). Further analysis showed both sexes benefitted in depression, but men had a greater improvement in balance than women. Depression in left-brain lesion patients improved more significantly than in those with right-brain lesion, whereas balance improved equally despite lesion site. For patients under or above the age of 60, depression and balance both significantly improved after rehabilitation. Older age is significantly related to poor balance but not depression. Conclusions: Combining yoga with rehabilitation has the potential to improve depression and balance. Factors related to sex, brain lesion site and age may influence the differences.

Keywords: stroke; yoga; balance disability; depression

1. Introduction

Stroke is highly prevalent in high-income countries (1015–1184 cases per 100,000 people in 2013) [1] Imbalance and post-stroke depression are common among stroke patients and may severely impair their activities of daily living [2,3]. The majority of stroke patients (around 83%) had a balance disability [4]. In a meta-analysis, the percentage of post-stroke depression was about 31% [5], and depression is associated with falls and negatively correlated with functional status [6]. The risk of falls for stroke patients is 1.77 times compared to healthy adults so post-stroke rehabilitation is essential [7].

Exercise increases serotonin which works against depression [8]. Carek et al. reported both aerobic and anaerobic exercises are effective [9]. Increased exercise frequency and group therapy seem be beneficial [10]. The practice of exercises promotes improvements in the levels of depression in people who suffered an ischemic stroke [11]. Franklin et al. reported yoga practices might be also protective [12]. Yoga is widely integrated into physical therapies [13]. However, few reports have focused on the impact of yoga in stroke patients [14–16]. Yoga can lower blood pressure, serum cholesterol and increase in internal reasons for exercise [17,18]. Furthermore, yoga consists of dynamic/static postures to stretch muscles and joints, and improve balance and coordination [19]. Bastille and Gill-Body [14] reported an 8-week intervention (1.5-h yoga sessions two times per week) for patients with chronic post-stroke hemiparesis improved balance. Chronic stroke is defined as the condition 90 days after stroke [20]. About 95% of patients achieved their best neurological recovery within 11 weeks, while minor stroke related to faster recovery [21].

Previous studies enrolled patients who began yoga "after" completing the post stroke rehabilitation program. However, no study focused combining yoga into a standard rehabilitation. We compared depression and balance among chronic stroke patients who underwent combining yoga and standard rehabilitation and those with standard rehabilitation alone.

2. Materials and Methods

2.1. Ethic Statement

Institutional Review Board (IRB)/Ethics Committee approval was obtained before the trial began, and the study was conducted in full compliance with the Declaration of Helsinki. The study was approved by the Ethics Committee of National Taiwan University Hospital Hsin-Chu Branch, 201210006IRB. All written consents were obtained from study participants. Clinical Trial Registration Information: Yoga Exercise for Improving Balance in Patients with Subacute and Chronic Stroke, NCT01806922. URL: https://clinicaltrials.gov/ct2/show/NCT01806922 (accessed on 11 February 2014).

2.2. Inclusion and Exclusion Criteria

The inclusion criteria: (1) Patients with stroke more then 90 days, (2) can stand independently for 1 min, (3) aged from 30 to 80 years. Exclusion criteria: (1) Those under other therapy (such as acupuncture), (2) diagnosis of other psychiatric disease or use of a psychotropic agent, (3) difficult to follow instructions, and (4) other contraindications like cardiopulmonary disease.

2.3. Protocol

Between onset and enrollment, all patients received the same regular rehabilitation at our hospital. The participants were first stratified into several age groups (5 years apart), then divided each age group by half into the following two groups (first by their will and then by assignment). The intervention group was trained with 60 min of yoga exercise, two times per week, combined with standard stroke rehabilitation for 8 weeks. The control group only underwent standard rehabilitation (Figure 1).

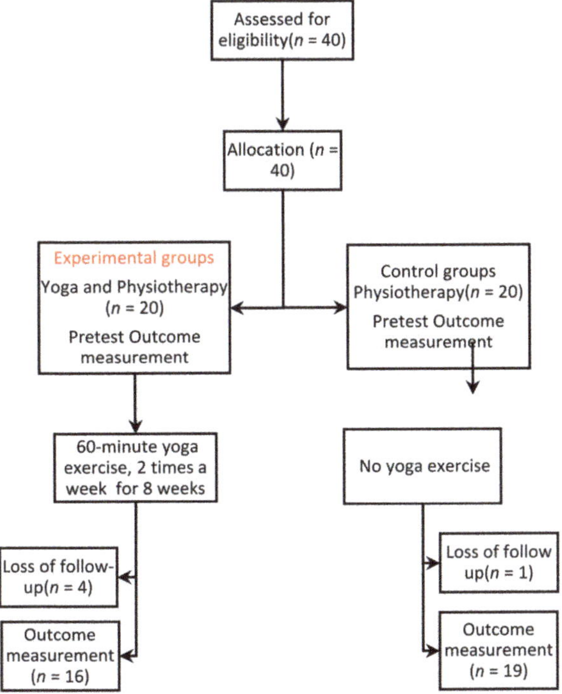

Figure 1. Flow diagram of the study.

Yoga Exercise Combined Standardized Rehabilitation Program (Intervention Group): In the intervention group, yoga was taught by a registered yoga therapist. A standard protocol was developed with 10 min of warm-up, 40 min of main activity, and 10 min cool-down exercises. All sessions focused on breathing, deep relaxation, mediation, posture, and trunk/limb stretching based on a study showed yoga improved balance in chronic stroke patients [15] (see Supplementary Materials Table S1–S3).

Standardized Rehabilitation Program (Control Group):All participants participated in the program based on the Taiwan Guideline for Stroke Rehabilitation [22] focused on dynamic balance exercises and weight-shift training given by the same therapists.

2.4. Assessment of Outcome

The outcome (depression and balance) were measured with the Taiwanese Depression Questionnaire (TDQ) [23] and Berg Balance Scale (BBS) [24] by the same therapist. (For TDQ: the lower the score, the better the mood. As for BBS: the higher the score, the better the balance.)

The TDQ is an 18-item questionnaire, and each item applies a 4-point scale for the response. It is a culturally specific self-rating instrument for screening of depression. Subjects are guided to rate each item on a scale from 0 to 3 based on "how often you felt during the past week". TDQ scores range from 0 to 54. Previous studies indicate good psychometric properties in the TDQ [25,26]. The cutoff scores for the screening of depression was 15 [27].

Balance was assessed with the Berg Balance Scale (BBS), which is the gold standard for functional balance testing [28]. It is a 14-item scale and can measure static and dynamic balance after stroke with good reliability and validity [28]. The total score is 56 points. The risk of falls in older adults was high when the score was less than 36 [29], and the risk increased by 6% to 8% with every one-point reduction within the range of 46 to 54 points [30].

2.5. Statistical Analysis

We performed statistical analyses using SPSS for Windows (Version 19.0, IBM Corp, Armonk, NY, USA). Continuous data are expressed as mean ± standard deviation. The independent *t*-test was used to compare the differences between the two groups (indicated with dotted arrows in the figure). We analyzed the measured values before and after the treatment by paired *t*-tests (indicated with bold arrows in the figure). A *p*-value of less than 0.05 is considered to be statistical significant (indicated with an asterisk in the figure). The sample size was determined based on an effective size to detect differences in different groups. If we permitted a 5% chance of a type I error ($\alpha = 0.05$), with a power of 80%, assuming the difference among the two groups was at least equal to the standard derivation, then approximately 32 patients would be required, while 35 patients completed the study. Subgroup analysis was performed on sex, brain lesion site and age.

3. Results

3.1. Primary Analysis of the Effect for Adding Yoga

Forty people enrolled initially with 20 to the intervention (yoga) group and 20 to the control group (Figure 1). In the yoga group, 16 completed the study. (One failed due to feeling exhausted, and three due to loss of follow-up thus being excluded). In the control group, 19 completed the study, and one loss of follow-up due to recurrent stroke. The completion rate of did not differ significantly. Age, sex, and brain lesion site (Table 1) were also similar. The TDQ score did not differ at baseline, whereas BBS scores were significantly better in the yoga group (Table 1). Overall, the TDQ of all patients decreased (improved) (15.23 ± 9.91 to 11.71 ± 8.66; $p = 0.001$) and the BBS increased (improved) from 44.71 ± 7.64 to 46.60 ± 7.46 ($p < 0.001$) after intervention.

Table 1. Comparison of baseline clinical and 8-week post-assessment of depression (Taiwanese Depression Questionnaire, TDQ) and balance (Berg Balance Scale, BBS), between control and experimental groups.

	All (n = 35)	With Yoga (n = 16)	Without Yoga (n = 19)	p
Study completion rate, −/+	35/40	16/20	19/20	0.151
Categorical variables, Chi-square test				
Sex, female:male	15:20	6:10	9:10	0.734
Stroke side, left:right †	17:17	11:5	6:12	0.084
Age < 60:age ≥ 60, years	19:16	11:5	8:11	0.176
Continuous variables, *t*-test				
Age	59.0± 10.10	56.8 ± 9.11	60.9 ± 10.743	0.241
T onset (days)	691.9± 20.17	560.6 ± 538.57	802.4 ± 1153.00	0.447
Taiwanese Depression Questionnaire (TDQ) score				
Baseline TDQ	15.23± 9.91	14.06 ± 8.85	16.21 ± 10.86	0.531
Post intervention TDQ	11.71 ± 8.66	8.25 ± 5.19	14.63 ± 9.99	0.028 *
The change of TDQ	−3.51 ± 5.84	−5.81 ± 6.13	−1.57 ± 4.94	0.806
Paired t (p)	0.001 *	0.002 *	0.181	
Berg Balance Scale (BBS) score				
Baseline BBS	44.71 ± 7.64	48.56 ± 4.42	41.47 ± 8.01	<0.001 *
Post intervention BBS	46.60 ± 7.46	51.13 ± 4.25	42.79 ± 7.85	<0.001 *
The change of BBS	1.87 ± 1.79	2.56 ± 1.96	1.31 ± 1.45	0.029 *
Paired *t*-test (p)	<0.001 *	<0.001 *	0.001 *	

Data are given as means ± SD unless otherwise indicated. † One patient was eliminated in this analysis due to this patient suffered from bilateral stroke. * $p < 0.05$, T onset: days after stroke. D0: for baseline TDQ; D1 for post-assessment TDQ; delta D: the change from D0 to D1. B0: for baseline BBS; B1 for post-assessment BBS; delta B: the change from D0 to D1.

The TDQ scores improved significantly only in the yoga group (14.06 ± 8.85 to 8.25 ± 5.19, $p = 0.002$). In contrast, patients in the control group had little improvement ($p = 0.181$) (Figure 2A) (bold black arrow). The baseline TDQ did not differ ($p = 0.531$). But the post-study TDQ was significantly lower (8.25 ± 5.19 vs. 14.63 ± 9.99, $p = 0.028$) in the yoga group (Figure 2A) (Dotted arrow).

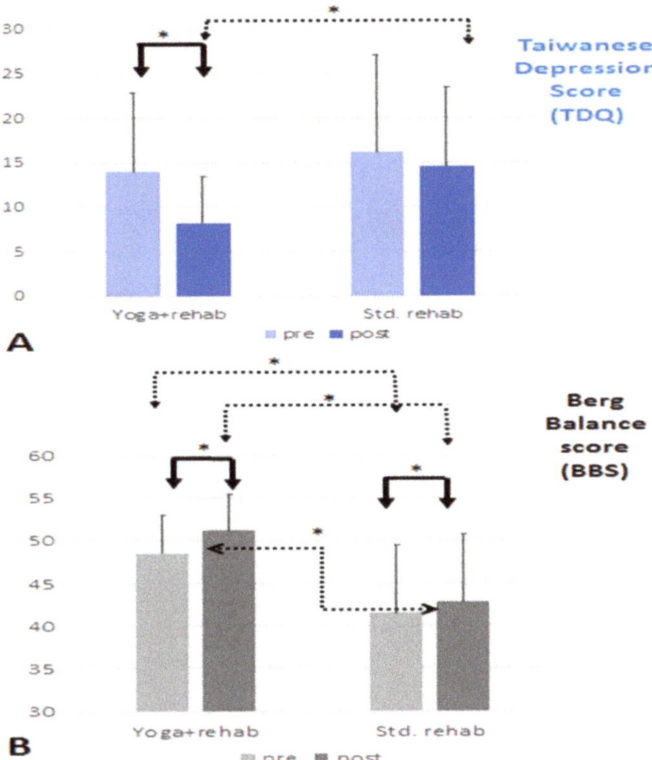

Figure 2. (**A**): Bold black arrows show TDQ score significant dropped (improved) in the yoga group after the intervention compared with their own baseline ($p = 0.002$). Dotted arrow showed the post intervention TDQ score is significantly lower in the yoga group compared to the control standard rehabilitation group. ($p = 0.028$). (**B**): Bold black arrows show BBS score significant increased (improved) in yoga ($p < 0.001$) and control group ($p = 0.001$) after the intervention compared with their own baseline BBS score. The dotted arrow shows the pre- and post-study values of the BBS in the yoga group are higher than that of the control group (both $p < 0.001$) Screwed dotted arrow shows the post intervention increase of BBS score is significantly larger in the yoga group compared to the control standard rehabilitation group ($p = 0.029$). (*: $p < 0.05$).

Both groups showed improvement in balance. The BBS score increased from 48.56 ± 4.42 to 51.13 ± 4.25 ($p < 0.001$) in the yoga group, while from 41.47 ± 8.01 to 42.79 ± 7.85 ($p = 0.001$) (Figure 2B) for the control group. (Bold black arrows) Although the pre- and post-study BBS scores in the yoga group were higher (dotted arrows), the improvement in the yoga group is also significantly larger than that of the control group (2.56 ± 1.96 vs. 1.31 ± 1.45, $p = 0.029$) (Table 1) (Figure 2B) (screwed dotted arrow).

3.2. Subgroup Analysis

The experiment group and control group were combined and divided by gender (female vs. male), brain lesion side (left vs. right) and age (age <60 vs. 60 or more) sequentially to see how these factors affected the result of rehabilitation (Table 2).

Table 2. Subgroup analysis for sex, lesion side, and age before and after the rehabilitation in comparison Taiwanese Depression Questionnaire (TDQ) scores and Berg Balance Scale (BBS) scores.

Subgroup Variable	Women $n = 15$	Men $n = 20$	p	Left $n = 17$	Right $n = 17$	p	Age < 60 $n = 19$	Age ≥ 60 $n = 16$	p
Age (years)	60.88 ± 12.42	57.64 ± 7.99	0.386	55.79 ± 7.81	62.06 ± 11.57	0.733	51.48 ± 5.91	68.00 ± 5.56	<0.001
T onset (days)	426.67 ± 376.12	376.12 ± 1146.20	0.103	1002.76 ± 1212.40	404.65 ± 358.96	0.60	734.00 ± 951.33	641.81 ± 910.08	0.773
Taiwanese Depression Questionnaire (TDQ) score									
D0	16.73 ± 9.18	14.10 ± 10.51	0.445	13.47 ± 9.28	16.47 ± 10.61	0.387	15.74 ± 8.93	14.63 ± 11.23	0.746
D1	12.93± 8.10	8.10 ± 9.16	0.479	8.94 ± 5.68	13.76 ± 10.26	0.102	12.32 ± 7.97	11.00 ± 9.64	0.661
Delta D	−3.80 ± 5.62	−3.30 ± 6.13	0.806	−4.52 ± 5.95	−2.70 ± 5.87	0.376	−3.42 ± 6.41	−3.62 ± 5.28	0.920
Paired t (p)	0.020 *	0.027 *		0.006 *	0.076		0.032 *	0.015 *	
Berg Balance Scale (BBS) score (BBS) score									
B0	44.60 ± 8.16	44.80 ± 7.08	0.939	46.41 ± 6.21	44.12 ± 7.34	0.333	48.58 ± 4.77	40.13 ± 7.55	0.001 *
B1	45.73 ± 8.53	47.25 ± 7.05	0.569	48.82 ± 6.24	45.35 ± 7.80	0.162	50.37 ± 4.65	42.13 ± 8.18	0.002 *
Delta B	1.13 ± 1.45	2.45 ± 1.84	0.029 *	2.41 ± 1.90	1.23 ± 1.48	0.053	1.78 ± 1.96	2.00 ± 1.63	0.735
Paired t (p)	0.009 *	<0.001 *		<0.001 *	0.003 *		0.001 *	<0.001 *	

* $p < 0.05$, T onset: days after stroke. D0: for baseline TDQ; D1 for post-assessment TDQ; delta D: the change from D0 to D1. B0: for baseline BBS; B1 for post-assessment BBS; delta B: the change from D0 to D1.

3.2.1. Gender

The mean age and the duration since stroke did not statistically differ by sex (Table 1). The TDQ scores improved after the study in both men and women. The TDQ scores dropped (improved) from 16.73 ± 9.18 to 12.93 ± 8.10 ($p = 0.020$) for women and from 14.10 ± 10.51 to 8.10 ± 9.16 ($p = 0.027$) for men (Figure 3A) (bold black arrow). For balance, the BBS score among all women improved from 44.60 ± 8.16 to 45.73 ± 8.53 ($p = 0.009$) and from 44.80 ± 7.08 to 47.25 ± 7.05 among men ($p < 0.001$) (Figure 3B). Thus, rehabilitation was effective despite sex. However, significantly greater improvement in BBS was observed in men ($p = 0.029$) (Figure 3B) (screwed dotted arrow). Balance among men seems to benefit more than that of women.

3.2.2. Brain Lesion Side

Further analysis of the stroke site in the brain is shown in Table 2 and illustrated in Figure 3B. For depression, patients with left-sided brain lesions showed significant improvement (TDQ scores decreased from 13.47 ± 9.28 to 8.94 ± 5.68, $p = 0.006$) (Figure 3C) (bold black arrow). In contrast, patients with right-sided brain lesions showed no significant improvement in TDQ ($p = 0.076$) (Figure 3C).

For balance, unlike depression, patients showed significant improvement despite the lesion site. Overall, BBS scores of patients with left-brain and right-brain lesions increased from 46.41 ± 6.21 to 48.82 ± 6.24 ($p < 0.001$) and from 44.12 ± 7.34 to 45.35 ± 7.80 ($p = 0.003$), respectively (Figure 3D) (bold black arrow). In summary, TDQ scores for depression improved only in patients with left brain lesions, but balance improved in all subgroups despite the lesion site.

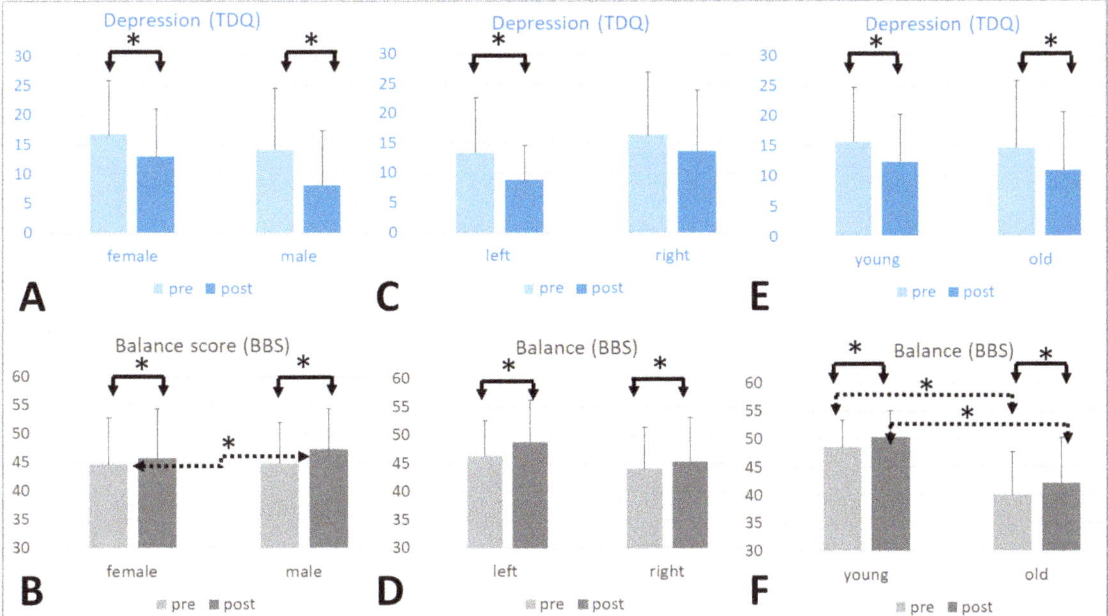

Figure 3. (**A**): Bold black arrow shows TDQ score significant dropped (improved) in the male and female groups after the intervention compared with their baseline TDQ score. (**B**): The improvement of TDQ is mainly contributed by the yoga groups in both genders. (**C**): Bold black arrow shows TDQ score significant dropped (improved) in the left brain lesion group after the intervention compared with their baseline. ($p = 0.006$) (**D**): The improvement of TDQ is mainly contributed by patients with left brain lesion in the yoga groups. ($p = 0.011$) (**E**): Bold black arrow shows TDQ score significant dropped (improved) in the young group ($p = 0.032$) and old group ($p = 0.015$) after the intervention compared with their baseline. (**F**): The improvement of TDQ is mainly contributed by young patients in the yoga groups ($p = 0.010$). (*: $p < 0.05$).

3.2.3. Age

Table 2 also shows the age effect. We divided the patients by the age of 60 because the mean age is about 60 years (55.79 ± 7.81). Overall, depression improved in both the young and older groups. The TDQ score dropped from 15.74 ± 8.93 to 12.32 ± 7.97 ($p = 0.032$) in the young and from 14.63 ± 11.23 to 11.00 ± 9.64 ($p = 0.015$) in the older group (Figure 3E) (bold black arrow).

As for balance, both young and older patients benefitted from rehabilitation. Generally, the BBS score improved from 48.58 ± 4.77 to 50.3 ± 4.65 ($p = 0.001$) in young patients and from 40.13 ± 7.55 to 42.13 ± 8.18 ($p < 0.001$) among older patients (Bold black arrows). Additionally, the BBS scores of older patients were much lower than those of the younger ones. (Figure 3F) (dotted arrows).

3.3. Factors Correlated with Balance

Further analysis showed age is a big contributor to poor balance (linear regression for age vs. baseline BBS score, $p = 0.005$, $R^2 = 0.464$; age vs. post-intervention BBS score, $p = 0.006$, $R^2 = 0.456$) (Figure 4).

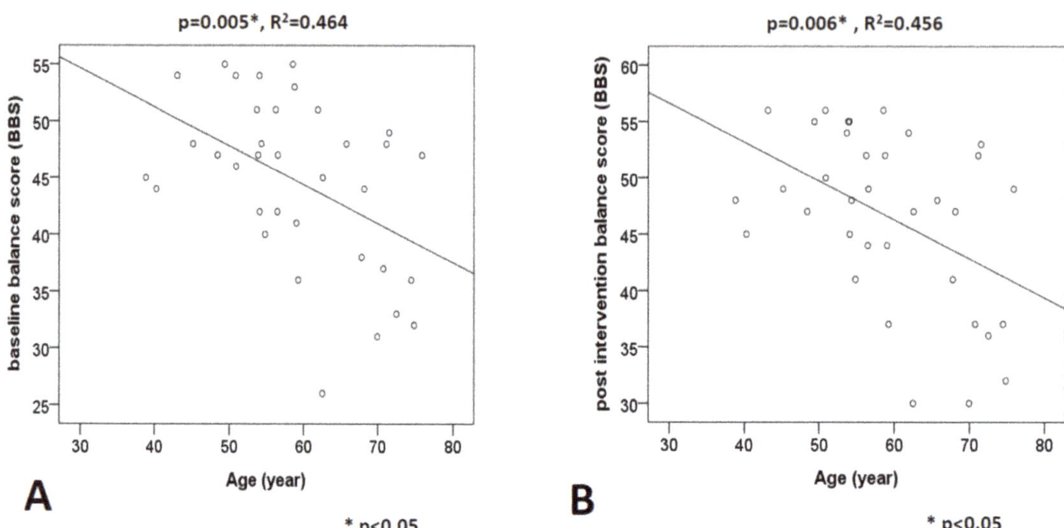

Figure 4. (**A**) Age is associated with lower baseline balance score (BBS) ($p = 0.005$, $R^2 = 0.464$), (**B**) Age is associated with lower post intervention balance score (BBS) ($p = 0.006$, $R^2 = 0.456$) (*: $p < 0.05$)).

Although the baseline TDQ and BBS scores are not correlated ($p = 0.25$), the post-intervention TDQ score is significantly correlated to the BBS score ($p = 0.016$) (Figure 5).

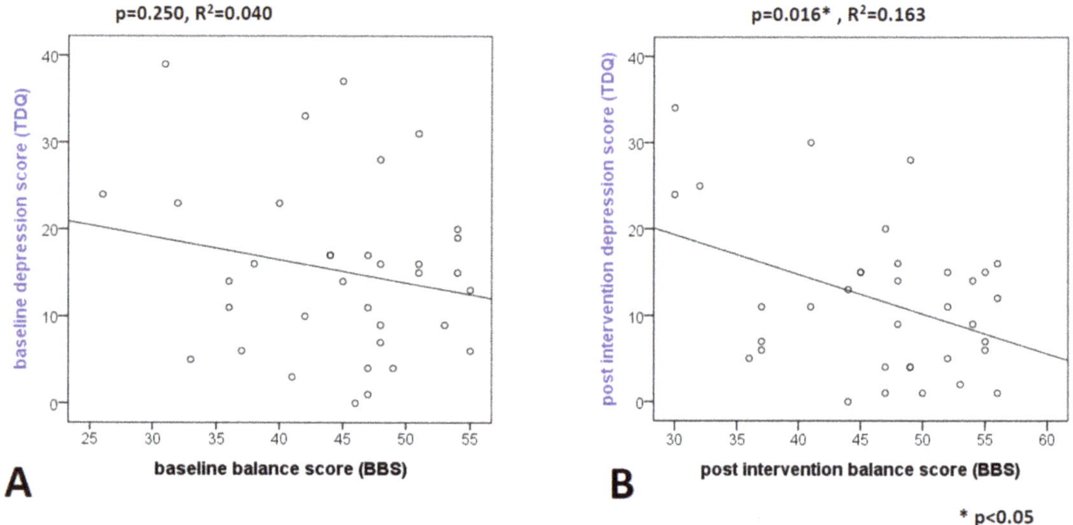

Figure 5. (**A**) Baseline balance score is not associated with baseline depression score (TDQ) ($p = 0.250$, $R^2 = 0.040$), (**B**) post intervention balance is associated with post intervention depression score (TDQ) ($p = 0.016$, $R^2 = 0.163$). (*: $p < 0.05$).

4. Discussion

All patients had significant improvement in depression and balance after rehabilitation (Table 1). Overall, a decrease in TDQ score ($p = 0.001$) and an increase in BBS score ($p < 0.001$) were noted. We found the TDQ scores of patients in the yoga group were much lower than those in the control group after the intervention while the baseline TDQ scores were similar (Figure 2A). Only patients in the yoga group showed significant improvement in depression after the intervention which indicates yoga provided anti-depression effect.

Because this was not a randomized trial, the baseline BBS score is much higher in the yoga group (with a better physical condition) compared to the control group even before intervention. BBS scores remained much higher after the intervention (Figure 2B). The reason may be that healthier people have higher intention to be volunteers for epidemiological research or a challenging task and have better survival rates than the general population, thereby producing bias referred to as the "healthy volunteer effect" [31]. Although both groups showed improvements in BBS scores (Figure 2B), the change of improvement in BBS was also significantly larger in the yoga group ($p = 0.029$). This finding indicates yoga still provided extra benefits compared to standard rehabilitation.

We compared the differences in the effects of rehabilitation on both sexes (Table 2 and Figure 3). Overall, we found that depression was significantly improved equally (Figure 3A). Unlike depression, balance showed a significantly greater improvement in men generally than in women ($p = 0.029$) (Figure 3B). One possible explanation is that men have poorer flexibility than women due to a lack of hormones such as estrogen or relaxin. Relaxin is known for relaxing the tendons and joints in pregnancy and delivery in women [32]. Yoga has been proven to improve flexibility [33], and flexibility is related to balance [34]. It may be that women have better flexibility in the first place, so the improvement of flexibility and balance after yoga may not be as good as man. In other words, the mind factor (anti-depression) of yoga is similar for both sexes but the body factor (balance) is quite different with better improvements in men.

The effect of brain lesion site is shown in Table 2 and Figure 3B. Only patients with left-sided brain lesions showed a significant improvement in depression (Figure 3C). This is probably due to left-sided brain lesion being associated with aphasia because the linguistic center is located the left brain while aphasia is highly associated with depression [35]. Yoga may lead to an improvement in aphasia [36]. Unlike depression, balance was uniformly improved in all subgroups of patients, despite of brain lesion side. These results are compatible with the brain structure whereby motor cortex is symmetrical, but the language center is located in the left brain [35].

Age is also an influencing factor. For patients younger than 60, depression ($p = 0.032$) and balance ($p = 0.001$) significantly improved. (Figure 3E). For older patients aged 60 or more, depression showed significant improvement ($p = 0.015$), and balance improved as well ($p < 0.001$) (Figure 3F). Therefore, rehabilitation is important for both young and older patients.

Further analysis of factors associated with balance, we found the BBS score is much lower in the older group, as age is significantly inverse related to the balance (Figure 4). This could be because young patients have better neuroplasticity and muscle power [37]. Aging is a big contributor to poor balance (linear regression for age vs. baseline BBS score, $p = 0.005$, $R^2 = 0.464$; age vs. post-intervention BBS score, $p = 0.006$, $R^2 = 0.456$) (Figure 4A,B). Nonetheless, age is not associated with depression (Age vs. baseline TDQ score, $p = 0.551$; age vs. post-intervention BBS score, $p = 0.439$). Therefore, old age may not guarantee depression.

On the other hand, although the baseline depression score (TDQ) and balance (BBS) scores are not correlated ($p = 0.25$), the post-intervention TDQ score is significantly correlated to the BBS score ($p = 0.016$). This may indicates that better post-intervention physical condition may contribute to a better quality of life and mood after completing the rehabilitation (Figure 5A,B). However, there are more factors other than age (such as gender, brain lesion side) affecting the outcome of rehabilitation.

In summary, for the subgroup analysis of overall rehabilitation effect, we found both sexes benefited in depression and men benefited more in balance than women. Patients with left-brain lesions improved more in depression than those with right lesions after rehabilitation, while balance equally improved regardless of the lesion site. For patients under or above the age of 60, depression and balance both significantly improved after rehabilitation. Older age was significantly related to poor balance but not depression. Post-intervention higher balance (BBS) score was significantly correlated to lower TDQ score and thus a better mood.

Our first limitation was that this was a pilot study with small recruitment number, but we still obtained many significant results. Second, the inclusion criteria limit each subject must be able to stand independently for 1 min, which indicated participants were all relatively mildly functionally impaired. Whether the results can be applied to more disabled people is unknown. Third, in order to respect participants' will and their capability, our study was not randomized. (However, if done in a randomized manner, the completion rate could be much lower in the intervention group.) Also, the long-term effect after 8 weeks was not confirmed. A larger randomized controlled study with longer follow up period is necessary in the future. Finally, we should add one more group with standard rehabilitation of 300 min (5 h) each week to see which group performs better. However, patients always get bored and become impatient during repetitive exercise in the real clinical setting. That is why we add recreational sport (such as yoga) as an adjuvant exercise because yoga is relaxing and interesting.

Because of the COVID-19 pandemic, group exercise may increase the infection risk. Whether remote video-assisted yoga training provides the same benefit as the current group exercise requires further research.

5. Conclusions

Generally, improvements were noted for balance in patients with standard rehabilitation but improvement in depression was only observed among patients when yoga was combined. Yoga provided more benefit for men than women in balance, for patients with left side brain lesion in depression, and for patients aged under or above 60 in both balance and depression. Older age was significantly related to poor balance but not depression. Post-intervention higher balance (BBS) score was significantly correlated to lower TDQ score and thus a better mood. Our study showed that combining yoga in the current standard rehabilitation is safe and has the potential to improve mood and balance, which provides useful information for stroke patients in tailor-made rehabilitation combing yoga.

Supplementary Materials: The following supporting information can be downloaded at: https://www.mdpi.com/article/10.3390/app12020922/s1. Table S1: Standard stroke rehabilitation; Table S2: Yoga class schedule (16 yoga sessions) in each week; Table S3: Yoga pose.

Author Contributions: Y.-T.L. made substantial contributions to the conception and design; acquisition, analysis and interpretation of data; drafting of the manuscript. C.-H.L., H.-H.T., S.-Y.L. and C.-C.L. revised the manuscript critically for important intellectual content. C.C.H. and J.-C.Y. provided clinical care for the patient and critically revised the manuscript for important intellectual content. W.-S.L. and H.-L.C. served as primary consultants in the management of the patient and made substantial contributions to the conception and design; analysis and interpretation of data; drafting of the manuscript; and revising the manuscript critically for important intellectual content; and gave final approval of the version to be published. All authors have read and agreed to the published version of the manuscript.

Funding: This trial was supported by the Department of Physical Medicine and Rehabilitation, National Taiwan University Hospital Hsin-Chu Branch (Protocol No.: HCH-102-22), the Ministry of Science and Technology (MOST 103-2410-H-134-022-MY), Taiwan and Taipei City Hospital (TPCH-108-27), and Department of Health of Taipei City Government (10901-62-023).

Institutional Review Board Statement: Institutional Review Board (IRB)/Ethics Committee approval was obtained before the trial began, and the study was conducted in full compliance with the Declaration of Helsinki. The study was approved by the Ethics Committee of National Taiwan University Hospital Hsin-Chu Branch, 201210006IRB. Clinical Trial Registration Information: Yoga Exercise for Improving Balance in Patients with Subacute & Chronic Stroke, NCT01806922. URL: https://clinicaltrials.gov/ct2/show/NCT01806922 (accessed on 11 February 2014).

Informed Consent Statement: Informed consent was obtained from all subjects involved in the study. Written informed consent has been obtained from the patient(s) to publish this paper.

Conflicts of Interest: The authors declare no conflict of interest. The funders had no role in the design of the study; in the collection, analyses, or interpretation of data; in the writing of the manuscript, or in the decision to publish the results.

References

1. Benjamin, E.J.; Blaha, M.J.; Chiuve, S.E.; Cushman, M.; Das, S.R.; Deo, R.; de Ferranti, S.D.; Floyd, J.; Fornage, M.; Gillespie, C.; et al. Heart Disease and Stroke Statistics—2017. *Circulation* **2017**, *135*, e146–e603. [CrossRef]
2. Pahlman, U.; Gutierrez-Perez, C.; Savborg, M.; Knopp, E.; Tarkowski, E. Cognitive function and improvement of balance after stroke in elderly people: The Gothenburg cognitive stroke study in the elderly. *Disabil. Rehabil.* **2011**, *33*, 1952–1962. [CrossRef] [PubMed]
3. Robinson, R.G.; Jorge, R.E. Post-Stroke Depression: A Review. *Am. J. Psychiatry* **2016**, *173*, 221–231. [CrossRef] [PubMed]
4. Tyson, S.F.; Hanley, M.; Chillala, J.; Selley, A.; Tallis, R.C. Balance disability after stroke. *Phys. Ther.* **2006**, *86*, 30–38. [CrossRef]
5. Hackett, M.L.; Pickles, K. Part I: Frequency of depression after stroke: An updated systematic review and meta-analysis of observational studies. *Int. J. Stroke* **2014**, *9*, 1017–1025. [CrossRef] [PubMed]
6. Hsieh, L.P.; Kao, H.J. Depressive symptoms following ischemic stroke: A study of 207 patients. *Acta Neurol. Taiwan.* **2005**, *14*, 187–190.
7. Simpson, L.A.; Miller, W.C.; Eng, J.J. Effect of stroke on fall rate, location and predictors: A prospective comparison of older adults with and without stroke. *PLoS ONE* **2011**, *6*, e19431. [CrossRef] [PubMed]
8. Young, S.N. How to increase serotonin in the human brain without drugs. *J. Psychiatry Neurosci.* **2007**, *32*, 394–399.
9. Carek, P.J.; Laibstain, S.E.; Carek, S.M. Exercise for the treatment of depression and anxiety. *Int. J. Psychiatry Med.* **2011**, *41*, 15–28. [CrossRef]
10. Troeung, L.; Egan, S.J.; Gasson, N. A waitlist-controlled trial of group cognitive behavioural therapy for depression and anxiety in Parkinson's disease. *BMC Psychiatry* **2014**, *14*, 19. [CrossRef]
11. Aidar, F.J.; Jacó de Oliveira, R.; Gama de Matos, D.; Chilibeck, P.D.; de Souza, R.F.; Carneiro, A.L.; Machado Reis, V. A randomized trial of the effects of an aquatic exercise program on depression, anxiety levels, and functional capacity of people who suffered an ischemic stroke. *J. Sports Med. Phys. Fitness* **2018**, *58*, 1171–1177. [CrossRef] [PubMed]
12. Franklin, R.A.; Butler, M.P.; Bentley, J.A. The physical postures of yoga practices may protect against depressive symptoms, even as life stressors increase: A moderation analysis. *Psychol. Health Med.* **2018**, 1–10. [CrossRef] [PubMed]
13. Ebnezar, J.; Nagarathna, R.; Yogitha, B.; Nagendra, H.R. Effects of an integrated approach of hatha yoga therapy on functional disability, pain, and flexibility in osteoarthritis of the knee joint: A randomized controlled study. *J. Altern. Complementary Med.* **2012**, *18*, 463–472. [CrossRef] [PubMed]
14. Bastille, J.V.; Gill-Body, K.M. A yoga-based exercise program for people with chronic poststroke hemiparesis. *Phys. Ther.* **2004**, *84*, 33–48. [CrossRef]
15. Schmid, A.A.; Van Puymbroeck, M.; Altenburger, P.A.; Schalk, N.L.; Dierks, T.A.; Miller, K.K.; Damush, T.M.; Bravata, D.M.; Williams, L.S. Poststroke balance improves with yoga: A pilot study. *Stroke* **2012**, *43*, 2402–2407. [CrossRef] [PubMed]
16. Immink, M.A.; Hillier, S.; Petkov, J. Randomized controlled trial of yoga for chronic poststroke hemiparesis: Motor function, mental health, and quality of life outcomes. *Top. Stroke Rehabil.* **2014**, *21*, 256–271. [CrossRef]
17. Hartley, L.; Dyakova, M.; Holmes, J.; Clarke, A.; Lee, M.S.; Ernst, E.; Rees, K. Yoga for the primary prevention of cardiovascular disease. *Cochrane Database Syst. Rev.* **2014**, Cd010072. [CrossRef]
18. Anne, E.; Cox, S.U.-F.; Amy, N.; D'Hondt-Taylor, M. The role of state mindfulness during yoga in predicting self-objectification and reasons for exercise. *Psychol. Sport Exerc.* **2016**, *22*, 321–327. [CrossRef]
19. Tsai, H.-G.; Chang, T.-C.; Chen, H.-Y.; Chen, Y.-C.; Liu, C.-Y. The Application of YOGA Activity in Psychiatric Occupational Therapy. *J. Taiwan Occup. Therapy Res. Pract.* **2008**, *4*, 69–77. [CrossRef]
20. Maraka, S.; Jiang, Q.; Jafari-Khouzani, K.; Li, L.; Malik, S.; Hamidian, H.; Zhang, T.; Lu, M.; Soltanian-Zadeh, H.; Chopp, M.; et al. Degree of corticospinal tract damage correlates with motor function after stroke. *Ann. Clin. Transl. Neurol.* **2014**, *1*, 891–899. [CrossRef] [PubMed]
21. Jorgensen, H.S.; Nakayama, H.; Raaschou, H.O.; Vive-Larsen, J.; Stoier, M.; Olsen, T.S. Outcome and time course of recovery in stroke. Part II: Time course of recovery. The Copenhagen Stroke Study. *Arch. Phys. Med. Rehabil.* **1995**, *76*, 406–412. [CrossRef]
22. Der-Sheng, H.; Chia-Wei, L.; Lu, L.; Ming-Yen, H.; Chueh-Hung, W.; Huey-Wen, L.; Ke-Vin, C.; Shin-Liang, P.; Tyng-Guey, W.; Chein-Wei, C. Taiwan Guideline for Stroke Rehabilitation. *Taiwan J. Phys. Med. Rehabil.* **2016**, *44*, 1–9. [CrossRef]
23. Lee, Y.; Yang, M.J.; Lai, T.J.; Chiu, N.M.; Chau, T.T. Development of the Taiwanese Depression Questionnaire. *Change Gung Med. J.* **2000**, *23*, 688–694.

24. Berg, K.; Wood-Dauphinee, S.; Williams, J.I. The Balance Scale: Reliability assessment with elderly residents and patients with an acute stroke. *Scand. J. Rehabil. Med.* **1995**, *27*, 27–36. [PubMed]
25. Huang, Y.C.; Hsu, S.T.; Hung, C.F.; Wang, L.J.; Chong, M.Y. Mental health of caregivers of individuals with disabilities: Relation to Suicidal Ideation. *Compr. Psychiatry* **2018**, *81*, 22–27. [CrossRef]
26. Su, S.F.; Chang, M.Y.; He, C.P. Social Support, Unstable Angina, and Stroke as Predictors of Depression in Patients With Coronary Heart Disease. *J. Cardiovasc. Nurs.* **2018**, *33*, 179–186. [CrossRef]
27. Lee, Y.; Wu, Y.-S.; Chien, C.-Y.; Fang, F.-M.; Hung, C.-F. Use of the Hospital Anxiety and Depression Scale and the Taiwanese Depression Questionnaire for screening depression in head and neck cancer patients in Taiwan. *Neuropsychiatr. Dis. Treat.* **2016**, *12*, 2649–2657. [CrossRef]
28. Langley, F.A.; Mackintosh, S.F.H. Functional balance assessment of older community dwelling adults: A systematic review of the literature. *Int. J. All. Health Sci. Pract.* **2007**, *5*, 1–11. [CrossRef]
29. Berg, K.O.; Wood-Dauphinee, S.L.; Williams, J.I.; Maki, B. Measuring balance in the elderly: Validation of an instrument. *Can. J. Public Health* **1992**, *83* (Suppl. S2), S7–S11.
30. Maeda, N.; Kato, J.; Shimada, T. Predicting the probability for fall incidence in stroke patients using the Berg Balance Scale. *J. Int. Med. Res.* **2009**, *37*, 697–704. [CrossRef]
31. Froom, P.; Melamed, S.; Kristal-Boneh, E.; Benbassat, J.; Ribak, J. Healthy volunteer effect in industrial workers. *J. Clin. Epidemiol.* **1999**, *52*, 731–735. [CrossRef]
32. Hansen, M.; Kjaer, M. Sex Hormones and Tendon. *Adv. Exp. Med. Biol.* **2016**, *920*, 139–149. [CrossRef]
33. Bucht, H.; Donath, L. Sauna Yoga Superiorly Improves Flexibility, Strength, and Balance: A Two-Armed Randomized Controlled Trial in Healthy Older Adults. *Int. J. Environ. Res. Public Health* **2019**, *16*, 3721. [CrossRef] [PubMed]
34. Overmoyer, G.V.; Reiser, R.F., 2nd. Relationships between lower-extremity flexibility, asymmetries, and the Y balance test. *J. Strength Cond. Res.* **2015**, *29*, 1240–1247. [CrossRef]
35. Ashaie, S.A.; Hurwitz, R.; Cherney, L.R. Depression and Subthreshold Depression in Stroke-Related Aphasia. *Arch. Phys. Med. Rehabil.* **2019**, *100*, 1294–1299. [CrossRef]
36. Lynton, H.; Kligler, B.; Shiflett, S. Yoga in stroke rehabilitation: A systematic review and results of a pilot study. *Top. Stroke Rehabil.* **2007**, *14*, 1–8. [CrossRef] [PubMed]
37. Tolahunase, M.R.; Sagar, R.; Faiq, M.; Dada, R. Yoga- and meditation-based lifestyle intervention increases neuroplasticity and reduces severity of major depressive disorder: A randomized controlled trial. *Restor. Neurol. Neurosci.* **2018**, *36*, 423–442. [CrossRef] [PubMed]

 applied sciences

Article

Development of a Simple Spheroid Production Method Using Fluoropolymers with Reduced Chemical and Physical Damage

Hidetaka Togo [1], Kento Yoshikawa-Terada [2], Yudai Hirose [1], Hideo Nakagawa [1], Hiroki Takeuchi [2,*] and Masanobu Kusunoki [1,*]

[1] Graduate School of Biology-Oriented Science and Technology, Kindai University, 930 Nishimitani, Kinokawa 649-6493, Wakayama, Japan; 1944730001m@waka.kindai.ac.jp (H.T.); 2133730002c@waka.kindai.ac.jp (Y.H.); nakagawa@waka.kindai.ac.jp (H.N.)

[2] Department of Obstetrics and Gynecology, Mie University Graduate School of Medicine, 2-174 Edobashi, Tsu 514-8507, Mie, Japan; yoshikawa-k@med.mie-u.ac.jp

* Correspondence: h-takeuchi@med.mie-u.ac.jp (H.T.); kusunoki@waka.kindai.ac.jp (M.K.); Tel.: +81-59-232-1111 (H.T.); +81-736-77-3888 (M.K.)

Abstract: Establishing an in vitro–based cell culture system that can realistically simulate in vivo cell dynamics is desirable. It is thus necessary to develop a method for producing a large amount of cell aggregates (i.e., spheroids) that are uniform in size and quality. Various methods have been proposed for the preparation of spheroids; however, none of them satisfy all requirements, such as cost, size uniformity, and throughput. Herein, we successfully developed a new cell culture method by combining fluoropolymers and dot patterned extracellular matrix substrates to achieve size-controlled spheroids. First, the spheroids were spontaneously formed by culturing them two-dimensionally, after which the cells were detached with a weak liquid flow and cultured in suspension without enzyme treatment. Stable quality spheroids were easily produced, and it is expected that the introduction and running costs of the technique will be low; therefore, this method shows potential for application in the field of regenerative medicine.

Keywords: spheroid; organoid; van der Waals force; fluoropolymers; adhesion; microstamp

Citation: Togo, H.; Yoshikawa-Terada, K.; Hirose, Y.; Nakagawa, H.; Takeuchi, H.; Kusunoki, M. Development of a Simple Spheroid Production Method Using Fluoropolymers with Reduced Chemical and Physical Damage. *Appl. Sci.* **2021**, *11*, 10495. https://doi.org/10.3390/app112110495

Academic Editor: Francisco Arrebola

Received: 19 September 2021
Accepted: 4 November 2021
Published: 8 November 2021

1. Introduction

Cell culture, an important technique in basic and preclinical research, is a cost-effective way of reducing the number of experimental animals during the drug discovery process [1]. Although two-dimensional (2D) cell culture is a universal and valuable method, it has its limitations [2]; for example, the cells proliferate and grow in a single layer and thus lack the cell–cell and extracellular matrix (ECM) interactions present in native tissues and cells. In addition, the cells are stretched and undergo a cytoskeletal rearrangement, acquiring artificial polarity and expressing abnormal genes and proteins [3]. Unlike tissues in an in vivo environment, 2D systems cannot provide a complex and dynamic microenvironment for cells, leading to potentially invalid findings [4,5]. For these reasons, the majority of compounds discovered in 2D cell culture during drug discovery fail in clinical trials [6]. Therefore, it is important to establish an in vitro cell culture system that can more realistically simulate the dynamics of cells in vivo.

Three-dimensional (3D) culture systems are an excellent in vitro model that can mimic in vivo processes, such as embryogenesis, morphogenesis, and organogenesis [7]. Spheroids and organoids are cellular aggregates with complex cell–cell adhesion that generate gradients of nutrients, gases, growth factors, and signaling factors. These structures can replicate the cellular microenvironment observed in real tissues and simulate in vivo cell dynamics more realistically. Human induced pluripotent stem cells (iPSCs), the discovery of which has helped stem cell research make great strides, can be used to generate spheroids and organoids, which were employed in the elucidation of pathological conditions [8–12] and drug discovery [8,13–15]. Recently, organoids derived from iPSCs generated from patients with intractable diseases were used to elucidate the disease mechanisms [16,17]. In the field of drug discovery, the validation of drug efficacy using organoids derived from human iPSCs in the early stage of in vitro testing is expected to reduce the possibility of candidate compound failure before initiating clinical trials. Indeed, drug testing using iPSC-derived organoids has shown comparable results to those of drugs already used for treatment [18]. Because the use of spheroids and organoids is expected to increase in the future, developing methods for their efficient production is important.

Various 3D culture methods have been tried and implemented. Conventional 3D culture methods can be broadly classified into four types: micropatterned forced flotation, hanging drop, swirling culture or spinner flask, and force-driven methods [19]. The micropatterned forced flotation method involves well plates in which cells deposited inside the microwells adhere to each other in three dimensions [20–22], as well as low-adhesion plates in which areas with different cell adhesion properties are patterned to prevent cell aggregation [23]. Microfluidic systems are a type of micropatterned forced flotation that combines both the target cells and matrix components to create spheroids on microfluidic chips with microgel beads [24,25]. The hanging drop method utilizes the surface tension of the culture medium to form droplets, while the interface with the air is used as a microwell [26–28]. The spinner flask or swirling culture (bioreactor) is a method of spheroidization via collision and contact between cells through continuous swirling and agitation of the culture medium in which cells are dispersed [29,30]; this method also includes microfluidic types on a smaller scale [31–33]. Force-driven methods include magnetic methods where nanoparticles, such as iron oxide, gold, and poly-L-lysine, are attached to the cell surface. Spheroids are formed via the following: magnetic forces [34]; electric methods in which an external force, such as an electric field, is applied [35]; and acoustic tweezers [36]. Other methods have also been developed in recent years; for example, 3D printing can artificially create 3D shapes [37–39], using a computer-controlled arrangement of heterogeneous cells with a gel-like culture medium as the filler. Although these methods exhibit excellent features, such as dimensional control of the spheroid, reduction in damage and irritation to the cells, simplicity of the procedure, high success rate, mass production, high throughput, and low cost, none of them satisfy all these requirements simultaneously. A new type of classification is microfluidic devices. Microfluidic devices are devices that have microscopic grooves or tubes within. These are used for various applications, such as forming small bioreactors [32,40] or circulatory systems [31] to recapitulate in vivo, using flow paths. Some of these microfluidic methods are included in the micropatterned forced flotation method.

Fluoropolymers have a low surface free energy and are generally regarded as materials that do not allow cell adhesion. They have a high biocompatibility and inertness, due to these properties. Therefore, fluoropolymers are used in artificial blood vessels and other components that are implanted in the body. In general environments, these polymers are very stable, due to the high binding energy of C–F bonds, and they are, thus, a safe and non-toxic material for use in the living body. However, their high inertness prevents cell adhesion, which is a problem for biomaterials. As a result, various surface modifications have been investigated to take advantage of their high biocompatibility and inert properties [41–46]. To improve adhesion, plasma treatment [47–51], UV irradiation [52–55], γ-radiation [56,57], ion introduction [58–61], and polydopamine treatment [62–66] have been employed for surface modification; however, most of these methods require high energy to break the strong C–F bonds and add functional groups to improve adhesion.

We are working on the control of cell adhesion by manipulating the surface properties of fluoropolymers for 3D cell patterning and have found that extracellular matrix (ECM) substrates deposited on fluoropolymers show weak cell adhesion in this study. In this study, we took advantage of this property and developed a new spheroid fabrication method that simultaneously satisfies all of the above requirements for spheroid production by forming an ECM dot pattern on the fluoropolymer, which allows cells to adhere on it, retain it during cell growth, and detach from the fluoropolymer without employing damaging enzymatic treatments.

2. Materials and Methods

2.1. Spheroid Culture Scaffold Fabrication

A fluoropolymer surface was formed by coating a glass substrate with CYTOP™ (CTL-107MK; AGC Chemicals, Tokyo, Japan). CYTOP™ is a fluoropolymer product with properties similar to polytetrafluoroethylene (PTFE) [67,68]. This resin has fluoropolymer properties, such as water repellency, oil repellency, and chemical resistance, as well as high transparency, which are difficult to obtain with crystalline resins, such as PTFE. It can also be used for cell culture observation using an inverted microscope because it has no fluorescence properties. The material was immersed in a dip solution consisting of CYTOP™ and a 16-fold dilution of a thinner solution (CT-Solv.100; AGC Chemicals) to coat the glass substrate (Table 1). The pull-out and curing conditions were as recommended by the manufacturer. The material was pulled out at a speed of approximately 1 mm/s and cured in an oven at 100 °C for at least 90 min.

2.2. Preparation of a Stamp for Patterning Formation

The dot patterns of the ECM were fabricated using a microstamp made out of polydimethylsiloxane (PDMS; SYLGARD™184; Dow Toray, Tokyo, Japan). PDMS has high biological safety [69–73] and is also used in engineering to transfer fine patterns [74]. To fabricate the microstamp, PDMS mixed with a hardener was poured into a mold made of PTFE, after which the mold was defoamed. After defoaming, the stamps were heat-cured at 90 °C for 90 min. The dimensions of the stamps were 100 μm in spheroid diameter (the size at which spheroid necrosis does not occur [75–77]), 800 μm in dot size diameter, and a 7 × 7 matrix arrangement with a pitch of 1500 μm. The dot diameter was defined as the area where the volume of a cell sheet cultured in a monolayer on the dot pattern is equivalent to a spheroid 100 μm in diameter, assuming that one cell is a sphere with a diameter of 10 μm. The dimensions of the molds and microstamps were measured using an OLS-5000 (Olympus, Tokyo, Japan).

2.3. Patterning of the Extracellular Matrix

ECM substrates, including Matrigel Growth Factor Reduced (354230; Corning, NY, USA), iMatrix-511 silk (892021; Nippi, Tokyo, Japan), fibronectin (33016015; Thermo Fisher Scientific, Waltham, MA, USA), vitronectin (A31804; Thermo Fisher Scientific, Waltham, MA, USA), and collagen IV (ASC-4-104-01; Nippi), were diluted in RPMI 1640 (30264-56; Nacalai Tesque, Kyoto, Japan) 2-, 4-, 8-, and 16-fold. PDMS stamps dipped in ECM diluent were stamped onto the substrates, after which the ECM was transferred to the fluoropolymer surface with the formed pattern and incubated at 37 °C and 5% CO_2 for 1 h. After incubation, the substrates were washed with RPMI 1640.

2.4. Cell Culture

HepG2 (TKG02058) and MCF7 (TKG0479) cells were provided by the Cell Resource Center for Biomedical Research, Institute of Development, Aging and Cancer, Tohoku University, Japan. The cells were cultured in Dulbecco's modified Eagle's medium (DMEM; 5919, Nissi, Tokyo, Japan.) supplemented with 10% fetal bovine serum (FBS; BioWest, Nuaillé, France), 2 mM L-Alanyl-L-glutamine solution (01102-82, Nacalai Tesque) and Penicillin-Streptomycin Mixed Solution (26252-94, Nacalai Tesque) (10%FBS/DMEM.) for 37 °C at 5% CO_2 and 95% air. The cells were passaged every 3 days at a ratio of 1:3, using TE dissociation buffer containing 0.25% trypsin (35547-64; Nacalai Tesque) and 0.04% EDTA (15105-35; Nacalai Tesque).

2.5. Spheroid Preparation

The ECM was transferred to the fluoropolymer-coated substrate, using a microstamp (Figure 1a). Cells were then evenly seeded at 1.0×10^4 cells/cm^2 on the substrate and allowed to evenly adhere (Figure 1b). After 7 days, the cells proliferated and formed colonies only on the ECM-coated area (Figure 1c). These cell colonies were detached by gently pipetting, followed by culturing in suspension for 7 days until spheroid formation (Figure 1d). All cultures were performed on fluoropolymer-coated glass substrate in a 5-well petri dish (NM-CD-5F, Naka Medical, Tachikawa, Japan).

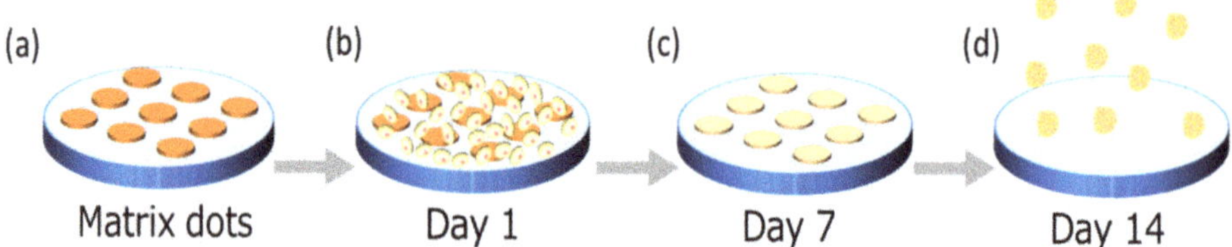

Figure 1. Schematic diagram of the spheroid production process. (**a**) Matrix transferred by stamping. (**b**) Cell adhesion occurring one day after seeding. (**c**) Cell colony formation 7 days after seeding. (**d**) After detaching the cells, spheroids formed in suspension by day 14.

Spheroids were produced by the hanging drop method. The average number of cells in the dot pattern colonies was estimated, and they were seeded to form a spheroid. Briefly, the cells were grown until confluent. Single cells were then dissociated using TE buffer. The number of cells was counted, and spheroids were prepared by seeding 8×10^3 cells/drop using the hanging drop method.

2.6. Evaluation of Cell Patterning

All ECM and cell patterning were observed using a phase-contrast microscope and then photographed and dimensioned using FLOYD-4K (Wraymer, Osaka, Japan). The diameter of the ECM was measured, but as cell colonies and spheroids do not form a perfect circle, the equivalent diameter of a circle was calculated from the area. Photographs were acquired of the coated ECM dimensions, colony size after cell seeding (day 1), cell sheet dimensions left as a pattern on the ECM (day 7), and spheroid size (days 10–14).

2.7. Measurement of Spheroid Viability Using Trypan Blue Staining

The cells of the 2D culture and spheroids were dissociated using TE buffer, and a single cell suspension was obtained. The cells in the suspension were then stained with trypan blue (35535-02, Nacalai Tesque) to determine the number of viable cells. Student *t*-test was performed to compare survival rates between the 2D culture and spheroids.

3. Results

3.1. Water Repellency of Fluoropolymers

The contact angle measurements were conducted to measure the surface free energy, which is mainly responsible for fluoropolymer inertness in cells. The contact angles for a 2-, 4-, 8-, and 16-fold dilution were 113.36°, 115.00°, 114.31°, and 112.58°, respectively, showing good water repellency, even at a 16-fold dilution. In consideration of economic efficiency, we used a 16-fold dilution of CYTOP™ for coating (Table 1).

Table 1. Contact angle measurement results for the different CYTOP™ dilutions.

Dilution Rate	2x	4x	8x	16x
Contact angle (degree)	113.36	115.00	114.31	112.58
Left angle (degree)	114.09	115.18	114.53	112.51
Right angle (degree)	112.62	114.81	114.09	112.65
Height from top to base (mm)	3.97	3.98	4.00	4.03
Base line length (mm)	5.48	5.30	5.48	5.61
Base area (mm^2)	23.60	22.03	23.60	24.67
Drop volume (μL)	60.06	55.21	60.74	63.64
Wetting energy (mN/m)	−28.86	−30.76	−29.97	−27.95
Spreading coefficient (mN/m)	101.66	103.56	102.77	100.75
Work of adhesion (mN/m)	43.94	42.04	42.83	44.85
Photo				

3.2. Dimensional Reproducibility of the Microstamps

After fabricating the microstamp with PDMS, we measured the dimensions of the top surface of the dots and found an average of 825.24 μm, σ of 4.96 μm, and the shape diameter that was close to the target of 800 μm with small variations (Figure 2).

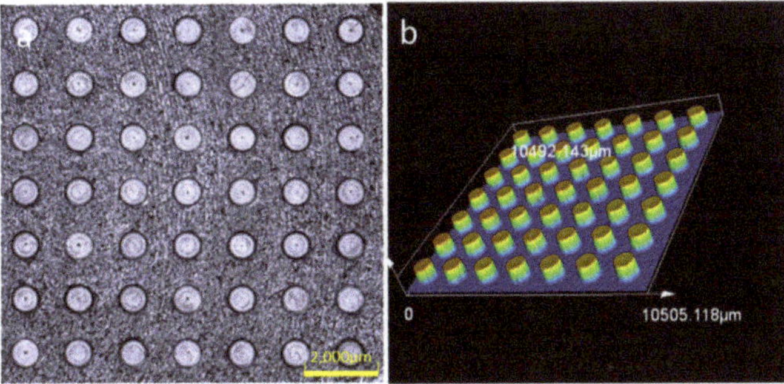

Figure 2. Observation of the top surface of a micro stamp using a laser microscope. (**a**) Optical imaging of a microstamp; (**b**) 3D image by a laser.

3.3. Culturing HepG2 Cells on the Matrigel Pattern

Matrigel was transferred to the fluoropolymer-coated substrate using a microstamp with an average pattern diameter of 898.74 ± 140.69 μm (Figure 3a,e: matrix transferred using a microstamp). HepG2 cells were seeded on this substrate and found to be uniformly adhered to the entire surface by day 1 (Figure 3b). The medium was gently changed every 2–3 days to avoid detaching the cells. By day 7, the cells had formed circular colonies based on the stamp pattern with an average colony diameter of 930.11 ± 178.47 μm

(Figure 3c,e). After detaching the colonies from the substrate by weak pipetting and culturing in suspension, they spontaneously formed spheroids by day 8 and completely spheroidized by day 14. The average spheroid size was 466.23 ± 75.25 μm (Figure 3d,e, and Supplementary Figure S1).

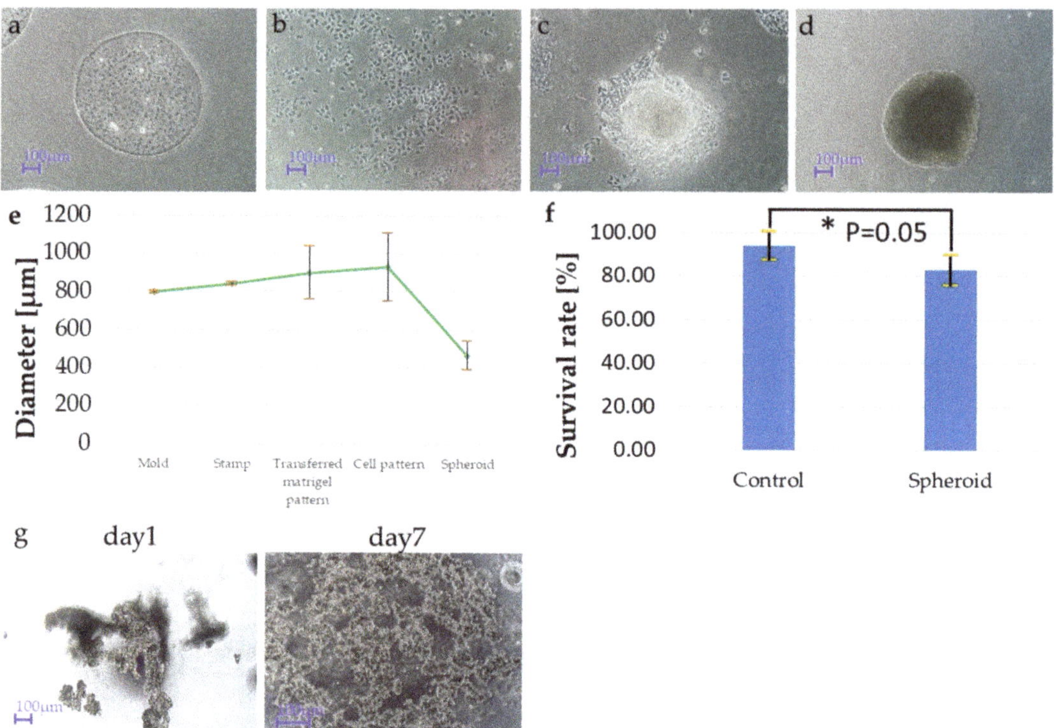

Figure 3. Bright-field image of HepG2 cultured on Matrigel dots and chart of size on every stage. (**a**) Stamped ECM (Matrigel) on the fluoropolymer surface. (**b**) HepG2 cells 1 day after seeding. (**c**) HepG2 cells grown to a colony of the same size as that of the Matrigel pattern on day 7. (**d**) Day 14 spheroids derived from detached HepG2 cells grown in suspension after culturing in 2D for 7 days on Matrigel. (**e**) The average diameter of the dot patterns during each step as shown in (**a**–**d**). Error bars indicate the standard deviation. (**f**) Viability confirmation by trypan blue staining. Student *t*-test was performed to compare survival rates between the 2D and spheroid ($p = 0.05$). p value < 0.05 was considered statistically significant. Data are expressed as means ± standard deviations. The asterisks (*) indicate a statistically significant difference. (**g**) HepG2 on days 1 and 7 following the hanging-drop method.

Trypan blue staining was performed to determine the viability of the prepared spheroids. The survival rate of the HepG2 cells as estimated by trypan blue staining was 93.95 ± 6.56% in 2D culture and 82.84 ± 6.99% in the spheroids derived by our method (Figure 3f). Next, spheroid formation was monitored for 7 days using the hanging-drop method; however, no HepG2-derived spheroids were observed (Figure 3g).

3.4. Culturing HepG2 Cells on Other Extracellular Matrix Patterns

We used other ECM substrates and the MCF7 cell line to determine whether spheroids formed using other combinations besides Matrigel and HepG2 cells. Using the other ECM substrates, including iMatrix (fragmented laminin), vitronectin, fibronectin, and collagen IV dot patterns also formed. The seeded HepG2 cells successfully adhered to these ECM substrates, after which colonies formed by day 7 and spheroids by day 14 (Figure 4(a1–a5)). Similar to HeG2 cells, seeded MCF7 cells formed colonies 7 days after seeding and spheroids by day 14 after the suspension culture (Figure 4(b1–b5)). When the cells were continuously cultured in 2D without detaching by day 7, the cells continued to proliferate and became over confluent on the ECM dots by day 10. Subsequently, the cells no longer proliferated in the monolayer, due to the influence of non-adhesive areas outside the colony pattern but instead proliferated toward the upper part of the colony and partially formed a spheroid. This phenomenon was also observed with HepG2 cells, indicating that it is difficult to control spheroid diameter within the area of the pattern when using cell lines with high proliferation ability.

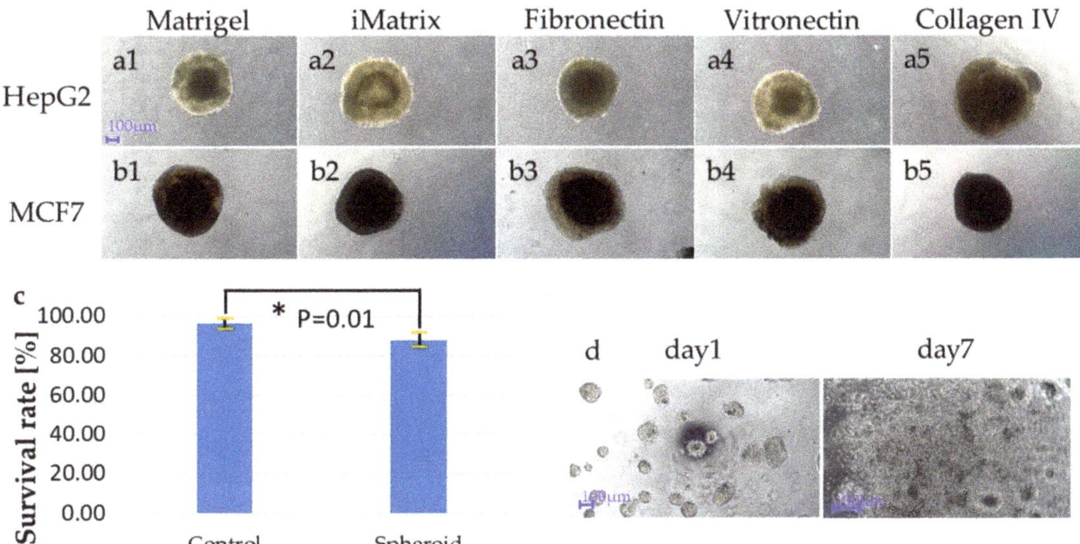

Figure 4. Spheroids derived from HepG2 or MCF7 cells after culturing in 2D for 7 days on different ECM substrates. HepG2 cells seeded on Matrigel (**a1**), Matrix (**a2**), fibronectin (**a3**), vitronectin (**a4**), and collagen IV (**a5**). MCF7 cells seeded on Matrigel (**b1**), Matrix (**b2**), fibronectin (**b3**), vitronectin (**b4**), and collagen IV (**b5**). (**c**) Viability confirmation by trypan blue staining. Student *t*-test was performed to compare the survival rates between the 2D and spheroid ($p = 0.01$). p value < 0.05 was considered statistically significant. Data are expressed as means ± standard deviations. The asterisks (*) indicate a statistically significant difference. (**d**) MCF7 on days 1 and 7 following the hanging-drop method.

In MCF7 and matrigel, trypan blue staining was performed to determine the viability of spheroids from colonies; the survival rate of MCF7 in trypan blue was 96.09 ± 2.82% in 2D culture and 87.89 ± 3.62% in the spheroids derived by our method (Figure 4c). Next, spheroid formation was monitored for 7 days, using the hanging drop method. MCF7-derived spheroids with a diameter of approximately 100–200 um were observed on day 1; however, no larger-sized spheroids were observed (Figure 4d).

4. Discussion

In the present study, all tested ECM substrates were successfully scaffolded on a fluoropolymer, and both HepG2 and MCF7 cell lines formed spheroids after culturing, without negatively affecting the cell population. The key to achieving this was the combination of the fluoropolymer and ECM dot pattern. Fluoropolymers are generally known to be difficult regarding adhesion; thus, many studies have attempted to improve adhesion by surface modification [78]. However, we showed that cells can adhere to the substrate via an ECM that can be detached easily by pipetting.

The survival rate of the spheroids derived using our method was significantly lower than that of the cells in the 2D culture. The reason for this was speculated to be the lack of space for cell growth inside the spheroid and the lack of nutrient circulation in the culture medium. For cancer spheroids, however, a slight decrease in survival are permitted; there are, accordingly, no disadvantages of our method. In addition, the hanging drop method could not produce large-sized spheroids in 7 days, but our method was effective in producing such large spheroids. It was also found to be more efficient because increasing the number of seeded cells could reduce the number of days required for spheroid formation (data not shown). These results suggest that our method can also be applied to spheroids derived from cancer cells.

We assume that our method would be useful in cases where cells of interest are induced from pluripotent stem cells, such as iPSCs, to produce spheroids. This method involves the process of culturing in 2D for a period of time to proliferate and induce, and the process of creating spheroids to confirm the functionality of the cells of interest. Therefore, the first half of the experiment (about 7 days) was assumed to be the cell proliferation and differentiation stage and the second half was assumed to involve the production of spheroids. We assume that the greatest benefit of producing spheroids in this way is the elimination of chemical and physical stimuli. The cell line was passaged using an enzyme but was not used to produce spheroids. Since the cell line used here is passaged every 2–4 days, we minimized the effect of enzyme treatment on cells by assigning 7 days for colony formation. Thus, when increasing the seeding amount and attempting spheroidization in a short period of time, it is necessary to consider the effect of enzyme treatment.

Fluoropolymers have a high biocompatibility and inertness due to their non-polarity and strong C–F bonds. They are, thus, a safe and non-toxic material for use in the living body. Their high inertness entails advantages, such as a low adhesion in vivo, while their hampered in vivo settling is a problem for biomaterials. Their symmetrical molecular structure makes them chemically resistant and stable against most chemicals, including acids, alkalis, and organic chemicals. However, they can react slightly with molten alkali metals and their solutions as well as fluorine and chlorine trifluoride at high temperatures. Therefore, the ECM used in this study was not expected to form strong chemical bonds. In addition, the non-polar molecular structure does not allow for electrical bonds, such as hydrogen bonds. Nevertheless, the cells were able to adhere to the ECM that was transferred onto the fluoropolymer (Figure 3a,c). Hydrophobic interactions [79] and the surface charge of the fluoropolymer [78] were reported as possible mechanisms underlying this adhesion. However, we believe that van der Waals forces are instead responsible because of the weak adhesive force that allowed cells to peel off over time regardless of the ECM type. During formation of the ECM dots, the ECM floating in the culture medium settled down due to gravity and gathered on the substrate surface after the gel-like protein molecules (matrix components) were stamped. As a result, the distance between the protein molecules and the fluoropolymer surface approached zero, and subsequently, the protein molecules adhered to the fluoropolymer surface via van der Waals forces. Van der Waals forces are extremely weak compared with chemical bonds and hydrogen bonds. Therefore, when the cells proliferated and the tension between them increased, the cells began to detach from the edges of the cell colonies and were easily peeled off from the ECM-coated fluoropolymer surface upon pipetting with a weak liquid flow.

If the ECM indeed adheres via van der Waals forces, the question is why the fluoropolymer does not allow adhesion of other substances and is considered an inert material. This may be because the objects to which the fluoropolymer adheres are usually considered solids. The surfaces of solids possess microscopic irregularities, which are shown in Figure 5a. There are many gaps between the solid and the surface of the fluoropolymer. Because van der Waals forces are only generated when the distance is very short, the area ratio of the contact point to the entire interface is extremely small with solids; if the contact area/external force ratio is large, the adhesive force increases. Therefore, in many cases, it is impossible to obtain an attractive force sufficient to hold the entire solid against external forces. In contrast, when the object to be adhered is a liquid (Figure 5b), there is no gap, and the molecules are in close contact with each other. However, the molecules in the liquid move over long distances, and the van der Waals forces are not strong enough to overcome their kinetic energy and maintain the molecules at the interface. If the attractive forces are strong, the contact angle should be small, but this is not the case. Furthermore, because of the flow phenomenon, i.e., the shape of the entire liquid can freely change, it is not possible for the liquid to stay in place. In contrast, water (H_2O) can remain on a glass surface because of the presence of a larger number of hydrogen bonds than in silicone oil. Moreover, when highly viscous silicone oil with extremely low polarity is dropped onto a fluoropolymer, the surface becomes wet easily, and the oil remains on the surface, even if it is physically wiped off. The ECM used in the present study [79–84] is a gel with properties between those of a solid and liquid (Figure 5c); it can be deformed to the extent that no gaps are created between it and the fluoropolymer surface. We thus concluded that the protein molecules adhered to the fluoropolymer surface via van der Waals forces.

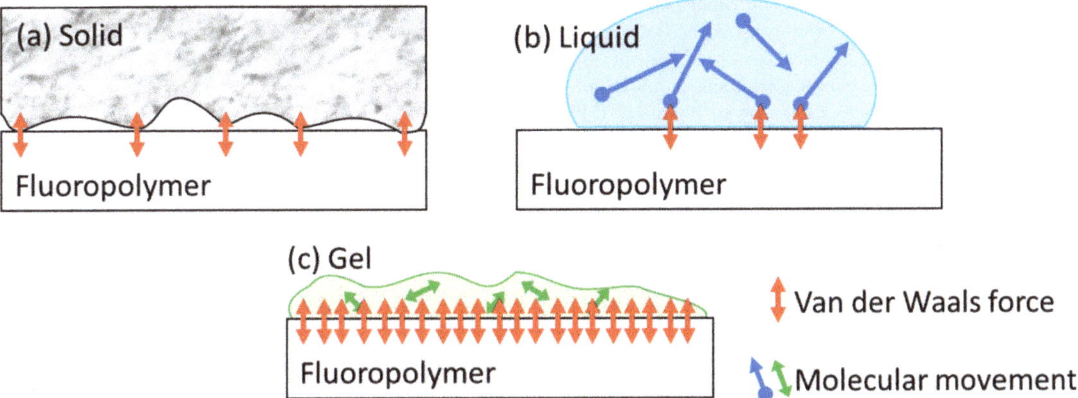

Figure 5. Application of force to the fluoropolymer depending on the state of the material placed on its surface. (**a**) Only point contact is possible due to the fine surface irregularities of solids, and thus, the area contributing to van der Waals forces is small. (**b**) Molecular motion in a spherically curled liquid as the surface tension exceeds the attractive van der Waals forces. (**c**) A gel with low surface tension and low fluidity is more densely packed, and the van der Waals forces act over a large area.

In this study, the dimensions of the spheroids were too large compared to the dimensions of the 2D matrix dots (Figure 3e). The reason is considered to be that cell lines were used here. Seven days after seeding, it was observed that the number of cell in the colony exceeded that in the 2D confluent state and proliferate in the 3D layer. If the colonies were detached by pipetting before that, the spheroid size could be controlled. In the future, the proposed method will be employed to spheroids using stem cells, and in that case, such a problem could be automatically avoided. At that time, we expect that it will also be possible to induce all cells in a planar manner at the 2D culture stage, and then suspend them while maintaining the generated cell polarity to create 3D cells using our method. In a conventional method, when a differentiation factor is added to stem cells after spheroidization, the differentiation factor may act only on the surface of the spheroid, and the induction of target cells may be insufficient in the central part. In a conventional method, when a differentiation factor is added to stem cells after spheroidization, the differentiation factors may act only on the surface of the spheroid, which can lead to insufficient induction of the target cells in the central part. In another conventional method, when a differentiation factor is added to stem cells before spheroidization, the cell substrate will be degraded during detachment via enzymatic treatment, and the cell polarity formed by 2D culture will be destroyed. Furthermore, enzymatic treatment destroys cell–cell adhesion factors, which may adversely affect the efficiency of spheroid production. Our proposed method can solve those problems and is intended to be used for drug discovery and regenerative medicine research in the future [75,85]. We believe that this method will enable to obtain spheroids of stable quality in a simple and low-cost manner, which will contribute to the development of medicine.

5. Conclusions

We have developed a technique for generating spheroids that consists of 2D cell culture on ECM dot patterns deposited on a fluoropolymer scaffold, followed by cell colony detachment from the substrate and suspension culture. The ECM substrate was weakly bound onto the fluoropolymer surface, and thus, the adherent cell colonies were easily detached by gentle pipetting; we theorized that van der Waals forces were responsible for this weak adhesion. Spheroid formation was successful with all tested combinations of five ECM substrates (Matrigel, iMatrix, fibronectin, vitronectin, and collagen IV) and two cell lines (HepG2 and MCF7). When forming spheroids during the differentiation process of pluripotent stem cells, in contrast to conventional methods, our technique enabled spheroidization without chemical (via enzymatic treatment) or physical stimulation, and thus, a more stable spheroid production could be obtained. For spheroids such as cancer cells and cells established from tissues, however, the chemical and physical stimuli are the same as those for the conventional method. Moreover, the technique is relatively uncomplicated and is expected to be low in cost. In future studies, we will test the effect of the ECM pattern size on the spheroid size and assess differentiation induction using this scaffold to establish size controllability and a method for iPSC spheroid/organoid production. We anticipate that such spheroid/organoids can be used for drug discovery and regenerative medicine.

Supplementary Materials: The following are available online at https://www.mdpi.com/article/10.3390/app112110495/s1, Figure S1: Detailed measurement at each stage.

Author Contributions: Conceptualization, H.T. (Hidetaka Togo), H.T. (Hiroki Takeuchi) and M.K.; methodology, H.T. (Hidetaka Togo), H.T. (Hiroki Takeuchi) and M.K.; validation, H.T. (Hidetaka Togo), K.Y.-T. and Y.H.; formal analysis, H.T. (Hidetaka Togo); investigation, H.T. (Hidetaka Togo), K.Y.-T. and Y.H.; resources, H.T. (Hiroki Takeuchi) and M.K.; data curation, H.T. (Hidetaka Togo) and K.Y.-T.; writing—original draft preparation, H.T. (Hidetaka Togo); writing—review and editing, H.T. (Hiroki Takeuchi) and M.K.; visualization, H.T. (Hidetaka Togo); supervision, H.N.; project administration, M.K.; funding acquisition, H.T. (Hiroki Takeuchi) and M.K. All authors have read and agreed to the published version of the manuscript.

Funding: This research was supported by JST A-STE, grant number JPMJTR20UC (to M.K.) and JSPS KAKENHI grant numbers JP18K16768, 20K18188 (to H.T. (Hiroki Takeuchi)).

Institutional Review Board Statement: Not applicable.

Informed Consent Statement: Not applicable.

Conflicts of Interest: The authors declare no conflict of interest.

References

1. Bhadriraju, K.; Chen, C.S. Engineering cellular microenvironments to improve cell-based drug testing. *Drug Discov. Today* **2002**, *7*, 612–620. [CrossRef]
2. Breslin, S.; O'Driscoll, L. Three-dimensional cell culture: The missing link in drug discovery. *Drug Discov. Today* **2013**, *18*, 240–249. [CrossRef]
3. Nath, S.; Devi, G.R. Three-dimensional culture systems in cancer research: Focus on tumor spheroid model. *Pharmacol. Ther.* **2016**, *163*, 94–108. [CrossRef]
4. Riedl, A.; Schlederer, M.; Pudelko, K.; Stadler, M.; Walter, S.; Unterleuthner, D.; Unger, C.; Kramer, N.; Hengstschläger, M.; Kenner, L.; et al. Comparison of cancer cells in 2D vs 3D culture reveals differences in AKT-mTOR-S6K signaling and drug responses. *J. Cell Sci.* **2017**, *130*, 203–218. [CrossRef] [PubMed]
5. Birgersdotter, A.; Sandberg, R.; Ernberg, I. Gene expression perturbation in vitro–a growing case for three-dimensional (3D) culture systems. *Semin. Cancer Biol.* **2005**, *15*, 405–412. [CrossRef]

6. Goetz, M.P.; Rae, J.M.; Suman, V.J.; Safgren, S.L.; Ames, M.M.; Visscher, D.W.; Reynolds, C.; Couch, F.J.; Lingle, W.L.; Flockhart, D.A.; et al. Pharmacogenetics of tamoxifen biotransformation is associated with clinical outcomes of efficacy and hot flashes. *J. Clin. Oncol.* **2005**, *23*, 9312–9318. [CrossRef] [PubMed]
7. Achilli, T.M.; Meyer, J.; Morgan, J.R. Advances in the formation, use and understanding of multi-cellular spheroids. *Expert Opin. Biol.* **2012**, *12*, 1347–1360. [CrossRef]
8. Egawa, N.; Kitaoka, S.; Tsukita, K.; Naitoh, M.; Takahashi, K.; Yamamoto, T.; Adachi, F.; Kondo, T.; Okita, K.; Asaka, I.; et al. Drug screening for ALS using patient-specific induced pluripotent stem cells. *Sci. Transl. Med.* **2012**, *4*, 145ra104. [CrossRef]
9. Kondo, T.; Asai, M.; Tsukita, K.; Kutoku, Y.; Ohsawa, Y.; Sunada, Y.; Imamura, K.; Egawa, N.; Yahata, N.; Okita, K.; et al. Modeling Alzheimer's disease with iPSCs reveals stress phenotypes associated with intracellular Aβ and differential drug responsiveness. *Cell Stem Cell* **2013**, *12*, 487–496. [CrossRef]
10. Hino, K.; Ikeya, M.; Horigome, K.; Matsumoto, Y.; Ebise, H.; Nishio, M.; Sekiguchi, K.; Shibata, M.; Nagata, S.; Matsuda, S.; et al. Neofunction of ACVR1 in fibrodysplasia ossificans progressiva. *Proc. Natl. Acad. Sci. USA* **2015**, *112*, 15438–15443. [CrossRef] [PubMed]
11. Ishii, T.; Ishikawa, M.; Fujimori, K.; Maeda, T.; Kushima, I.; Arioka, Y.; Mori, D.; Nakatake, Y.; Yamagata, B.; Nio, S.; et al. In Vitro Modeling of the Bipolar Disorder and Schizophrenia Using Patient-Derived Induced Pluripotent Stem Cells with Copy Number Variations of PCDH15 and RELN. *eNeuro* **2019**, *6*. [CrossRef]
12. Shiba, M.; Higo, S.; Kondo, T.; Li, J.; Liu, L.; Ikeda, Y.; Kohama, Y.; Kameda, S.; Tabata, T.; Inoue, H.; et al. Phenotypic recapitulation and correction of desmoglein-2-deficient cardiomyopathy using human-induced pluripotent stem cell-derived cardiomyocytes. *Hum. Mol. Genet.* **2021**, *30*, 1384–1397. [CrossRef] [PubMed]
13. Kondo, T.; Imamura, K.; Funayama, M.; Tsukita, K.; Miyake, M.; Ohta, A.; Woltjen, K.; Nakagawa, M.; Asada, T.; Arai, T.; et al. iPSC-Based Compound Screening and In Vitro Trials Identify a Synergistic Anti-amyloid β Combination for Alzheimer's Disease. *Cell Rep.* **2017**, *21*, 2304–2312. [CrossRef] [PubMed]
14. Hino, K.; Zhao, C.; Horigome, K.; Nishio, M.; Okanishi, Y.; Nagata, S.; Komura, S.; Yamada, Y.; Toguchida, J.; Ohta, A.; et al. An mTOR Signaling Modulator Suppressed Heterotopic Ossification of Fibrodysplasia Ossificans Progressiva. *Stem Cell Rep.* **2018**, *11*, 1106–1119. [CrossRef] [PubMed]
15. Yamaguchi, A.; Ishikawa, K.I.; Inoshita, T.; Shiba-Fukushima, K.; Saiki, S.; Hatano, T.; Mori, A.; Oji, Y.; Okuzumi, A.; Li, Y.; et al. Identifying Therapeutic Agents for Amelioration of Mitochondrial Clearance Disorder in Neurons of Familial Parkinson Disease. *Stem Cell Rep.* **2020**, *14*, 1060–1075. [CrossRef] [PubMed]
16. Shimizu, T.; Mae, S.I.; Araoka, T.; Okita, K.; Hotta, A.; Yamagata, K.; Osafune, K. A novel ADPKD model using kidney organoids derived from disease-specific human iPSCs. *Biochem. Biophys. Res. Commun.* **2020**, *529*, 1186–1194. [CrossRef]
17. Shiihara, M.; Ishikawa, T.; Saiki, Y.; Omori, Y.; Hirose, K.; Fukushige, S.; Ikari, N.; Higuchi, R.; Yamamoto, M.; Morikawa, T.; et al. Development of a system combining comprehensive genotyping and organoid cultures for identifying and testing genotype-oriented personalised medicine for pancreatobiliary cancers. *Eur. J. Cancer* **2021**, *148*, 239–250. [CrossRef]
18. Crespo, M.; Vilar, E.; Tsai, S.Y.; Chang, K.; Amin, S.; Srinivasan, T.; Zhang, T.; Pipalia, N.H.; Chen, H.J.; Witherspoon, M.; et al. Colonic organoids derived from human induced pluripotent stem cells for modeling colorectal cancer and drug testing. *Nat. Med.* **2017**, *23*, 878–884. [CrossRef]
19. Cui, X.; Hartanto, Y.; Zhang, H. Advances in multicellular spheroids formation. *J. R. Soc. Interface* **2017**, *14*, 20160877. [CrossRef]
20. Ivascu, A.; Kubbies, M. Rapid generation of single-tumor spheroids for high-throughput cell function and toxicity analysis. *J. Biomol. Screen.* **2006**, *11*, 922–932. [CrossRef]
21. Masuda, T.; Takei, N.; Nakano, T.; Anada, T.; Suzuki, O.; Arai, F. A microfabricated platform to form three-dimensional toroidal multicellular aggregate. *Biomed. Microdevices* **2012**, *14*, 1085–1093. [CrossRef] [PubMed]
22. Thomsen, A.R.; Aldrian, C.; Bronsert, P.; Thomann, Y.; Nanko, N.; Melin, N.; Rücker, G.; Follo, M.; Grosu, A.L.; Niedermann, G.; et al. A deep conical agarose microwell array for adhesion independent three-dimensional cell culture and dynamic volume measurement. *Lab Chip* **2017**, *18*, 179–189. [CrossRef] [PubMed]
23. Uchida, H.; Machida, M.; Miura, T.; Kawasaki, T.; Okazaki, T.; Sasaki, K.; Sakamoto, S.; Ohuchi, N.; Kasahara, M.; Umezawa, A.; et al. A xenogeneic-free system generating functional human gut organoids from pluripotent stem cells. *JCI Insight* **2017**, *2*, e86492. [CrossRef] [PubMed]
24. Chan, H.F.; Zhang, Y.; Ho, Y.P.; Chiu, Y.L.; Jung, Y.; Leong, K.W. Rapid formation of multicellular spheroids in double-emulsion droplets with controllable microenvironment. *Sci. Rep.* **2013**, *3*, 3462. [CrossRef]
25. Jang, M.; Koh, I.; Lee, S.J.; Cheong, J.H.; Kim, P. Droplet-based microtumor model to assess cell-ECM interactions and drug resistance of gastric cancer cells. *Sci. Rep.* **2017**, *7*, 41541. [CrossRef]
26. Yoshii, Y.; Waki, A.; Yoshida, K.; Kakezuka, A.; Kobayashi, M.; Namiki, H.; Kuroda, Y.; Kiyono, Y.; Yoshii, H.; Furukawa, T.; et al. The use of nanoimprinted scaffolds as 3D culture models to facilitate spontaneous tumor cell migration and well-regulated spheroid formation. *Biomaterials* **2011**, *32*, 6052–6058. [CrossRef]
27. Yamazaki, H.; Gotou, S.; Ito, K.; Kohashi, S.; Goto, Y.; Yoshiura, Y.; Sakai, Y.; Yabu, H.; Shimomura, M.; Nakazawa, K. Micropatterned culture of HepG2 spheroids using microwell chip with honeycomb-patterned polymer film. *J. Biosci. Bioeng.* **2014**, *118*, 455–460. [CrossRef]
28. Snyman, C.; Elliott, E. An optimized protocol for handling and processing fragile acini cultured with the hanging drop technique. *Anal. Biochem.* **2011**, *419*, 348–350. [CrossRef]

29. Qian, X.; Nguyen, H.N.; Song, M.M.; Hadiono, C.; Ogden, S.C.; Hammack, C.; Yao, B.; Hamersky, G.R.; Jacob, F.; Zhong, C.; et al. Brain-Region-Specific Organoids Using Mini-bioreactors for Modeling ZIKV Exposure. *Cell* **2016**, *165*, 1238–1254. [CrossRef]
30. Yabe, S.G.; Fukuda, S.; Nishida, J.; Takeda, F.; Nashiro, K.; Okochi, H. Induction of functional islet-like cells from human iPS cells by suspension culture. *Regen* **2019**, *10*, 69–76. [CrossRef]
31. Hsiao, A.Y.; Torisawa, Y.S.; Tung, Y.C.; Sud, S.; Taichman, R.S.; Pienta, K.J.; Takayama, S. Microfluidic system for formation of PC-3 prostate cancer co-culture spheroids. *Biomaterials* **2009**, *30*, 3020–3027. [CrossRef]
32. Baudoin, R.; Griscom, L.; Prot, J.M.; Legallais, C.; Leclerc, E. Behavior of HepG2/C3A cell cultures in a microfluidic bioreactor. *Biochem. Eng. J.* **2011**, *53*, 172–181. [CrossRef]
33. Borys, B.S.; Le, A.; Roberts, E.L.; Dang, T.; Rohani, L.; Hsu, C.Y.; Wyma, A.A.; Rancourt, D.E.; Gates, I.D.; Kallos, M.S. Using computational fluid dynamics (CFD) modeling to understand murine embryonic stem cell aggregate size and pluripotency distributions in stirred suspension bioreactors. *J. Biotechnol.* **2019**, *304*, 16–27. [CrossRef]
34. Souza, G.R.; Molina, J.R.; Raphael, R.M.; Ozawa, M.G.; Stark, D.J.; Levin, C.S.; Bronk, L.F.; Ananta, J.S.; Mandelin, J.; Georgescu, M.M.; et al. Three-dimensional tissue culture based on magnetic cell levitation. *Nat. Nanotechnol.* **2010**, *5*, 291–296. [CrossRef]
35. Sebastian, A.; Buckle, A.M.; Markx, G.H. Tissue engineering with electric fields: Immobilization of mammalian cells in multilayer aggregates using dielectrophoresis. *Biotechnol. Bioeng.* **2007**, *98*, 694–700. [CrossRef] [PubMed]
36. Chen, K.; Wu, M.; Guo, F.; Li, P.; Chan, C.Y.; Mao, Z.; Li, S.; Ren, L.; Zhang, R.; Huang, T.J. Rapid formation of size-controllable multicellular spheroids via 3D acoustic tweezers. *Lab Chip* **2016**, *16*, 2636–2643. [CrossRef]
37. Heid, S.; Boccaccini, A.R. Advancing bioinks for 3D bioprinting using reactive fillers: A review. *Acta Biomater.* **2020**, *113*, 1–22. [CrossRef] [PubMed]
38. Aguilar, I.N.; Olivos, D.J., 3rd; Brinker, A.; Alvarez, M.B.; Smith, L.J.; Chu, T.G.; Kacena, M.A.; Wagner, D.R. Scaffold-free bioprinting of mesenchymal stem cells using the Regenova printer: Spheroid characterization and osteogenic differentiation. *Bioprinting* **2019**, *15*, e00050. [CrossRef]
39. Daly, A.C.; Kelly, D.J. Biofabrication of spatially organised tissues by directing the growth of cellular spheroids within 3D printed polymeric microchambers. *Biomaterials* **2019**, *197*, 194–206. [CrossRef]
40. Leclerc, E.; Sakai, Y.; Fujii, T. Perfusion culture of fetal human hepatocytes in microfluidic environments. *Biochem. Eng. J.* **2004**, *20*, 143–148. [CrossRef]
41. Awaja, F.; Gilbert, M.; Kelly, G.; Fox, B.; Pigram, P.J. Adhesion of polymers. *Prog. Polym. Sci.* **2009**, *34*, 948–968. [CrossRef]
42. Roina, Y.; Auber, F.; Hocquet, D.; Herlem, G. ePTFE functionalization for medical applications. *Mater. Today Chem.* **2021**, *20*, 100412–100443. [CrossRef]
43. Bacakova, L.; Filova, E.; Parizek, M.; Ruml, T.; Svorcik, V. Modulation of cell adhesion, proliferation and differentiation on materials designed for body implants. *Biotechnol. Adv.* **2011**, *29*, 739–767. [CrossRef] [PubMed]
44. Wang, H.; Kwok, D.T.; Wang, W.; Wu, Z.; Tong, L.; Zhang, Y.; Chu, P.K. Osteoblast behavior on polytetrafluoroethylene modified by long pulse, high frequency oxygen plasma immersion ion implantation. *Biomaterials* **2010**, *31*, 413–419. [CrossRef] [PubMed]
45. Lih, E.; Oh, S.H.; Joung, Y.K.; Lee, J.H.; Han, D.K. Polymers for cell/tissue anti-adhesion. *Prog. Polym. Sci.* **2015**, *44*, 28–61. [CrossRef]
46. Feng, S.; Zhong, Z.; Wang, Y.; Xing, W.; Drioli, E. Progress and perspectives in PTFE membrane: Preparation, modification, and applications. *J. Membr. Sci.* **2018**, *549*, 332–349. [CrossRef]
47. Bax, D.V.; McKenzie, D.R.; Weiss, A.S.; Bilek, M.M. The linker-free covalent attachment of collagen to plasma immersion ion implantation treated polytetrafluoroethylene and subsequent cell-binding activity. *Biomaterials* **2010**, *31*, 2526–2534. [CrossRef]
48. Bilek, M.M.M.; Vandrovcová, M.; Shelemin, A.; Kuzminova, A.; Kylián, O.; Biederman, H.; Bačáková, L.; Weiss, A.S. Plasma treatment in air at atmospheric pressure that enables reagent-free covalent immobilization of biomolecules on polytetrafluoroethylene (PTFE). *Appl. Surf. Sci.* **2020**, *518*, 146128. [CrossRef]
49. Cho, Y.K.; Park, D.; Kim, H.; Lee, H.; Park, H.; Kim, H.J.; Jung, D. Bioactive surface modifications on inner walls of poly-tetra-fluoro-ethylene tubes using dielectric barrier discharge. *Appl. Surf. Sci.* **2014**, *296*, 79–85. [CrossRef]
50. Hajian, H.; Wise, S.G.; Bax, D.V.; Kondyurin, A.; Waterhouse, A.; Dunn, L.L.; Kielty, C.M.; Yu, Y.; Weiss, A.S.; Bilek, M.M.; et al. Immobilisation of a fibrillin-1 fragment enhances the biocompatibility of PTFE. *Colloids Surf. B Biointerfaces* **2014**, *116*, 544–552. [CrossRef] [PubMed]
51. Pu, F.R.; Williams, R.L.; Markkula, T.K.; Hunt, J.A. Expression of leukocyte–endothelial cell adhesion molecules on monocyte adhesion to human endothelial cells on plasma treated PET and PTFE in vitro. *Biomaterials* **2002**, *23*, 4705–4718. [CrossRef]
52. Gumpenberger, T.; Heitz, J.; Bäuerle, D.; Kahr, H.; Graz, I.; Romanin, C.; Svorcik, V.; Leisch, F. Adhesion and proliferation of human endothelial cells on photochemically modified polytetrafluoroethylene. *Biomaterials* **2003**, *24*, 5139–5144. [CrossRef]
53. Mikulikova, R.; Moritz, S.; Gumpenberger, T.; Olbrich, M.; Romanin, C.; Bacakova, L.; Svorcik, V.; Heitz, J. Cell microarrays on photochemically modified polytetrafluoroethylene. *Biomaterials* **2005**, *26*, 5572–5580. [CrossRef]
54. Svorcík, V. Cell proliferation on UV-excimer lamp modified and grafted polytetrafluoroethylene. *Nucl. Instrum. Methods Phys. Res. Sect. B Beam Interact. Mater. At.* **2004**, *217*, 307–313. [CrossRef]
55. Ahad, I.U.; Butruk, B.; Ayele, M.; Budner, B.; Bartnik, A.; Fiedorowicz, H.; Ciach, T.; Brabazon, D. Extreme ultraviolet (EUV) surface modification of polytetrafluoroethylene (PTFE) for control of biocompatibility. *Nucl. Instrum. Methods Phys. Res. Sect. B Beam Interact. Mater. At.* **2015**, *364*, 98–107. [CrossRef]

56. Pérez-Calixto, M.; Diaz-Rodriguez, P.; Concheiro, A.; Alvarez-Lorenzo, C.; Burillo, G. Amino-functionalized polymers by gamma radiation and their influence on macrophage polarization. *React. Funct. Polym.* **2020**, *151*, 104568. [CrossRef]
57. Rosado, D.; Meléndez-Ortiz, H.I.; Ortega, A.; Gallardo-Vega, C.; Burillo, G. Modification of poly(tetrafluoroethylene) with polyallylamine by gamma radiation. *Radiat. Phys. Chem.* **2020**, *172*, 108766. [CrossRef]
58. Colwell, J.M.; Wentrup-Byrne, E.; Bell, J.M.; Wielunski, L.S. A study of the chemical and physical effects of ion implantation of micro-porous and nonporous PTFE. *Surf. Coat. Technol.* **2003**, *168*, 216–222. [CrossRef]
59. Gao, A.; Hang, R.; Li, W.; Zhang, W.; Li, P.; Wang, G.; Bai, L.; Yu, X.F.; Wang, H.; Tong, L.; et al. Linker-free covalent immobilization of heparin, SDF-1alpha, and CD47 on PTFE surface for antithrombogenicity, endothelialization and anti-inflammation. *Biomaterials* **2017**, *140*, 201–211. [CrossRef]
60. Kondyurina, I.; Shardakov, I.; Nechitailo, G.; Terpugov, V.; Kondyurin, A. Cell growing on ion implanted polytetrafluorethylene. *Appl. Surf. Sci.* **2014**, *314*, 670–678. [CrossRef]
61. Sommani, P.; Tsuji, H.; Kojima, H.; Sato, H.; Gotoh, Y.; Ishikawa, J.; Takaoka, G.H. Irradiation effect of carbon negative-ion implantation on polytetrafluoroethylene for controlling cell-adhesion property. *Nucl. Instrum. Methods Phys. Res. Sect. B Beam Interact. Mater. At.* **2010**, *268*, 3231–3234. [CrossRef]
62. Hou, Y.; Deng, X.; Xie, C. Biomaterial surface modification for underwater adhesion. *Smart Mater. Med.* **2020**, *1*, 77–91. [CrossRef]
63. Song, H.; Yu, H.; Zhu, L.; Xue, L.; Wu, D.; Chen, H. Durable hydrophilic surface modification for PTFE hollow fiber membranes. *React. Funct. Polym.* **2017**, *114*, 110–117. [CrossRef]
64. Talon, I.; Schneider, A.; Ball, V.; Hemmerle, J. Functionalization of PTFE Materials Using a Combination of Polydopamine and Platelet-Rich Fibrin. *J. Surg Res.* **2020**, *251*, 254–261. [CrossRef] [PubMed]
65. Ku, S.H.; Ryu, J.; Hong, S.K.; Lee, H.; Park, C.B. General functionalization route for cell adhesion on non-wetting surfaces. *Biomaterials* **2010**, *31*, 2535–2541. [CrossRef] [PubMed]
66. Yu, C.; Yang, H.; Wang, L.; Thomson, J.A.; Turng, L.-S.; Guan, G. Surface modification of polytetrafluoroethylene (PTFE) with a heparin-immobilized extracellular matrix (ECM) coating for small-diameter vascular grafts applications. *Mater. Sci. Eng. C* **2021**, *128*, 112301. [CrossRef]
67. Lee, S.-H.; Lee, C.-S.; Shin, D.-S.; Kim, B.-G.; Lee, Y.-S.; Kim, Y.-K. Micro protein patterning using a lift-off process with fluorocarbon thin film. *Sens. Actuators B Chem.* **2004**, *99*, 623–632. [CrossRef]
68. Ueno, H.; Inoue, M.; Okonogi, A.; Kotera, H.; Suzuki, T. Correlation between Cells-on-Chips materials and cell adhesion/proliferation focused on material's surface free energy. *Colloids Surf. A Physicochem. Eng. Asp.* **2019**, *565*, 188–194. [CrossRef]
69. Celik, N.; Sahin, F.; Ruzi, M.; Yay, M.; Unal, E.; Onses, M.S. Blood repellent superhydrophobic surfaces constructed from nanoparticle-free and biocompatible materials. *Colloids Surf. B Biointerfaces* **2021**, *205*, 111864. [CrossRef] [PubMed]
70. Yadhuraj, S.R.; Babu Gandla, S.; Omprakash, S.S.; Sudarshan, B.G.; Prasanna Kumar, S.C. Design and Development of Micro-channel using PDMS for Biomedical Applications. *Mater. Today Proc.* **2018**, *5*, 21392–21397. [CrossRef]
71. Yadhuraj, S.R.; Babu Gandla, S.; Sudarshan, B.G.; Prasanna Kumar, S.C. Preparation and Study of PDMS Material. *Mater. Today Proc.* **2018**, *5*, 21406–21412. [CrossRef]
72. Liu, X.; Zhang, X.; Chen, Q.; Pan, Y.; Liu, C.; Shen, C. A simple superhydrophobic/superhydrophilic Janus-paper with enhanced biocompatibility by PDMS and candle soot coating for actuator. *Chem. Eng. J.* **2021**, *406*, 126532. [CrossRef]
73. Li, Z.; Liu, H.; Xu, X.; Ma, L.; Shang, S.; Song, Z. Surface modification of silicone elastomer with rosin acid-based quaternary ammonium salt for antimicrobial and biocompatible properties. *Mater. Des.* **2020**, *189*, 108493. [CrossRef]
74. Rogers, J.A.; Bao, Z.; Meier, M.; Dodabalapur, A.; Schueller, O.J.A.; Whitesides, G.M. Printing, molding, and near-field photolithographic methods for patterning organic lasers, smart pixels and simple circuits. *Synth. Met.* **2000**, *115*, 5–11. [CrossRef]
75. Han, M.A.; Jeon, J.H.; Shin, J.Y.; Kim, H.J.; Lee, J.S.; Seo, C.W.; Yun, Y.J.; Yoon, M.Y.; Kim, J.T.; Yang, Y.I.; et al. Intramyocardial delivery of human cardiac stem cell spheroids with enhanced cell engraftment ability and cardiomyogenic potential for myocardial infarct repair. *J. Control Release* **2021**, *336*, 499–509. [CrossRef] [PubMed]
76. Anada, T.; Fukuda, J.; Sai, Y.; Suzuki, O. An oxygen-permeable spheroid culture system for the prevention of central hypoxia and necrosis of spheroids. *Biomaterials* **2012**, *33*, 8430–8441. [CrossRef]
77. Nishimura, Y.; Wang, P.C. Possibility of culturing the early developing kidney cells by utilizing simulated microgravity environment. *Biochem. Biophys. Res. Commun.* **2021**, *573*, 9–12. [CrossRef] [PubMed]
78. Makohliso, S.A.; Giovangrandi, L.; Leonard, D.; Mathieu, H.J.; Ilegems, M.; Aebischer, P. Application of Teflon-AF® thin films for bio-patterning of neural cell adhesion. *Biosens. Bioelectron.* **1998**, *13*, 1227–1235. [CrossRef]
79. Hausmann, A.; Sanciolo, P.; Vasiljevic, T.; Weeks, M.; Schroën, K.; Gray, S.; Duke, M. Fouling of dairy components on hydrophobic polytetrafluoroethylene (PTFE) membranes for membrane distillation. *J. Membr. Sci.* **2013**, *442*, 149–159. [CrossRef]
80. Andreatta, F.; Lanzutti, A.; Aneggi, E.; Gagliardi, A.; Rondinella, A.; Simonato, M.; Fedrizzi, L. Degradation of PTFE non-stick coatings for application in the food service industry. *Eng. Fail. Anal.* **2020**, *115*, 104652. [CrossRef]
81. Zhao, Q. Effect of surface free energy of graded NI–P–PTFE coatings on bacterial adhesion. *Surf. Coat. Technol.* **2004**, *185*, 199–204. [CrossRef]
82. Zhao, Q.; Liu, Y.; Wang, C.; Wang, S.; Müller-Steinhagen, H. Effect of surface free energy on the adhesion of biofouling and crystalline fouling. *Chem. Eng. Sci.* **2005**, *60*, 4858–4865. [CrossRef]

83. Krsmanovic, M.; Biswas, D.; Ali, H.; Kumar, A.; Ghosh, R.; Dickerson, A.K. Hydrodynamics and surface properties influence biofilm proliferation. *Adv. Colloid Interface Sci.* **2021**, *288*, 102336. [CrossRef] [PubMed]
84. Liu, Y.; Zhao, Q. Influence of surface energy of modified surfaces on bacterial adhesion. *Biophys. Chem.* **2005**, *117*, 39–45. [CrossRef] [PubMed]
85. Marrazzo, P.; Pizzuti, V.; Zia, S.; Sargenti, A.; Gazzola, D.; Roda, B.; Bonsi, L.; Alviano, F. Microfluidic Tools for Enhanced Characterization of Therapeutic Stem Cells and Prediction of Their Potential Antimicrobial Secretome. *Antibiotics* **2021**, *10*, 750. [CrossRef] [PubMed]

 applied sciences

MDPI

Article

Application of Thermodynamics and Protein–Protein Interaction Network Topology for Discovery of Potential New Treatments for Temporal Lobe Epilepsy

Chang Yu [1], Edward A. Rietman [2], Hava T. Siegelmann [2], Marco Cavaglia [1] and Jack A. Tuszynski [3,4,*]

1 Biology Department, University of Massachusetts, Amherst, MA 01003, USA; chang_yu@brown.edu (C.Y.); marco.cavaglia@polito.it (M.C.)
2 The Biologically Inspired Neural & Dynamical Systems (BINDS) Laboratory, Department of Computer Science, University of Massachusetts, Amherst, MA 01003, USA; erietman@gmail.com (E.A.R.); hava@cs.umass.edu (H.T.S.)
3 Dipartimento di Ingegneria Meccanica e Aerospaziale (DIMEAS), Politecnico di Torino, I-10129 Turin, Italy
4 Department of Physics, University of Alberta, Edmonton, AB T6G 2E9, Canada
* Correspondence: jackt@ualberta.ca

Featured Application: This work describes the use of new methodology based on Gibbs homology analysis for the identification of potential protein targets as well as their inhibitors for the development of therapeutic options for various diseases. In the past, similar approaches have been proposed and partially validated for various types of cancer. Here, we apply the method that combines thermodynamic measures with protein–protein interaction network topology to temporal lobe epilepsy. Our results identify a number of potential therapeutic targets.

Abstract: In this paper, we propose a bioinformatics-based method, which introduces thermodynamic measures and topological characteristics aimed to identify potential drug targets for pharmacoresistant epileptic patients. We apply the Gibbs homology analysis to the protein–protein interaction network characteristic of temporal lobe epilepsy. With the identification of key proteins involved in the disease, particularly a number of ribosomal proteins, an assessment of their inhibitors is the next logical step. The results of our work offer a direction for future development of prospective therapeutic solutions for epilepsy patients, especially those who are not responding to the current standard of care.

Keywords: epilepsy; systems biology; protein–protein interactions; CNS; Gibbs homology; drug targets; anti-epileptic drugs

Citation: Yu, C.; Rietman, E.A.; Siegelmann, H.T.; Cavaglia, M.; Tuszynski, J.A. Application of Thermodynamics and Protein–Protein Interaction Network Topology for Discovery of Potential New Treatments for Temporal Lobe Epilepsy. *Appl. Sci.* **2021**, *11*, 8059. https://doi.org/10.3390/app11178059

Academic Editors: Francesco Cappello and Magdalena Gorska-Ponikowska

Received: 2 July 2021
Accepted: 27 August 2021
Published: 31 August 2021

Publisher's Note: MDPI stays neutral with regard to jurisdictional claims in published maps and institutional affiliations.

1. Introduction

It has been estimated that about 50 million people worldwide suffer from epilepsy [1]. In 2015, about 3.4 million people had active epilepsy in the U.S. alone [2]. Epilepsy is one of the most common and most disabling neurological disorders, characterized by recurrent unprovoked excessive brain activity [3]. The current understanding of the neurophysiological mechanisms of epilepsy is largely based on extensive investigations of neuronal cells. However, glial cells have also been demonstrated to play a fundamental role in triggering seizures. Accurate diagnosis of epilepsy is challenging and therapeutic strategies span a range from single, to multiple drug courses of administration, to respective neurosurgical procedures and dietary therapy. In spite of the fact that several dozen antiepileptic drugs (AEDs) are available, there are still approximately 20–30% of patients who do not respond satisfactorily to these AEDs [3]. To develop new, effective treatments, studies focus on the development of the central nervous system (CNS) and neuronal activities in vivo to understand the causes and mechanisms of the disease initiation and progression [4]. However, to find effective treatments for patients who are refractory to current AEDs, it is necessary to study each case more precisely and individually. Hippocampal biopsy tissue of pharmaco-resistant Temporal Lobe Epileptic (TLE) patients is an extraordinarily useful substrate to study molecular mechanisms related to structural and cellular abnormalities in epilepsy. Several genes and signaling cascade alterations have already been reported in the literature based on the TLE hippocampus analysis [5]. The availability of gene expression profiling provides a detailed insight into the disease for each individual patient. Methods, such as microarray and RNA sequencing, developed in the recent decade are used to obtain genome-wide mRNA expression data. One of the advantages of using such methods is that this type of investigation is biomarker driven. Especially with whole genome-wide data, the comprehensive information on the status of each functioning unit, its interacting complex and interaction pathways is suitable for analysis, which can reveal potentially important molecular mechanisms and novel therapeutic targets. By investigating genome-wide expression data, we can study the integrated results of all possible causative changes for each patient. Thus, the results of such an analysis are considered to be precise and biomarker-driven [6].

Epilepsy is a complex neuro-pathology that arises due to different etiologies, having various localizations, often occurring in conjunction with other diseases. Irrespective of what triggers the paroxysm, the electrical discharge during seizure is the common clinical manifestation for all forms of epilepsy, suggesting an underlying common molecular mechanism [7]. Although revealing the precise origin of epilepsy is still part of the ongoing investigations, it was shown that the cause may vary from de novo genetic mutations [8] to traumatic brain injury [9,10]. While gene mutations may naturally lead to altered downstream pathway behavior, brain injury was also shown to cause chronically altered gene expression signatures of genes that were linked to epilepsy [9]. Recent studies revealed some genetic causes of epilepsy, such as gene SCN1A mutations, which affect sodium channels [8] and tuberous sclerosis complex (TSC) mutations, which lead to the dysregulation of the mechanistic target of rapamycin (mTOR) pathway [11], both of which lead to epileptic conditions. Mechanisms of epileptogenesis consist of genetic and epigenetic alterations occurring in both neuronal and astroglia cells [12]. Abnormal activity of astrocytes during epileptic events has been extensively reported to play a major role due to their K+ buffering role in the extracellular milieu [13].

Clinical diagnosis of epilepsy starts with an identification of the type of seizure and proceeds with EEG and neuroimaging studies. Accurate diagnosis is pivotal for the adoption of an appropriate therapy. Anti-epileptic drugs (AEDs) represent first-line treatment for epilepsy and despite the availability of more than 20 such drugs, approximately 30% of patients do not respond to this type of therapy [12]. Among all epilepsy types, temporal lobe epilepsy (TLE) is the most common drug-resistant form of epilepsy in adults. Current AEDs mainly act by directing transmembrane ion channel function or by promoting γ-aminobutyric acid (GABA)-mediated inhibition to decrease the electrical activity of the brain [3,14]. For instance, phenytoin and lacosamide inhibit sodium channel activation; retigabine opens potassium channels; ethosuximide and lamotrigine block calcium channels; tiagabine inhibits GABA reuptake and phenobarbital and benzodiazepines enhance GABA receptors [3]. The AEDs generally focus on stabilizing and elevating the threshold of the CNS against hyperexcitability. On the other hand, the role of GABA neurotransmitters and GABA receptors in inhibiting activity in the central nervous system [15], and the role of glutamate neurotransmitters in aberrant hyperexcitability [16] have been shown to represent possible mechanisms which can be explored in order to develop new treatments. With extensive research focused on each of the above approaches, great progress has been made in deciphering epilepsy's pathophysiology at a molecular level. The better the understanding of the complex molecular mechanisms involved in the disease initiation and progression, the more unrealistic it appears that a drug affecting a single neurotransmitter receptor or an ion channel will be found to effectively treat epilepsy. Instead, combinations of pharmacological agents designed for an individual patient with a known expression profile should lead to better personalized clinical outcomes. In order to optimize such drug combinations, sophisticated quantitative analyses of protein–protein interaction networks should be implemented involving the key biomarker proteins of interest and the results of these analysis validated experimentally. In this paper we develop such a computational modelling effort for epilepsy that is based on thermodynamic measures of protein–protein network characterization, having previously provided a general approach [17] and also applied it specifically to other diseases such as various types and stages of cancer [18–26].

2. Materials and Methods

The theoretical underpinnings for our understanding of the thermodynamics and bioenergetics of brain development started by investigating the molecular biology of human diseases from a systems and network biology perspective. These studies were developed over a several-year period [17–26]. Here, we only provide a brief summary of these approaches. The transcriptome and other -omic (e.g., proteomic, genomic, metabolomic) measures can represent the collective energetic state of a cell. By the use of the word "energetic", we mean it from a thermodynamics perspective where one uses thermodynamic functions of state such as entropy, Gibbs free energy, enthalpy, and internal energy. In particular, for open thermodynamic systems such as the human body Gibbs free energy is a suitable thermodynamic function of state, which can be computed from the so-called chemical potential for the statistical system such as a network of proteins expressed by a living cell. There, a chemical potential can be found for all interacting molecules in a cell, in particular, a chemical potential of all the proteins that interact with each other. This can be imagined to represent a rugged landscape, not dissimilar to Waddington's epigenetic landscape [27,28]. We provide mathematical expressions for the Gibbs free energy of a cellular protein–protein interaction network below.

To perform these calculations, we need input data and a method of calculating the Gibbs free energy. The method we propose uses mRNA transcriptome data or RNA-seq data as a surrogate for actual measurements of protein concentration values. This assumption is largely valid since Kim et al. [29] and Wihelm et al. [30] have shown an 83% correlation between mass spectrometry proteomic information and transcriptomic information for multiple tissue types. Further, Guo et al. [31] found a Spearman correlation of 0.8 in comparing RNAseq and mRNA transcriptome from TCGA human cancer data [32]. Therefore, we have decided to use this highly correlated proxy for protein expression data in our calculations.

Given a set of transcriptome data as representative of protein concentration values, we overlay that on the graph of the human protein–protein interaction network from BioGrid [33]. This means we assign to each protein representing a node of the network, the scaled (between 0 and 1), transcriptome value (or RNAseq value). The edges in this network correspond to protein–protein interactions and they define a unique topology for a given protein–protein interaction network. As shown in our previous work [18–25], each disease studied so far is characterized by a unique network topology. From the data extracted for a given protein–protein interaction network, we compute the Gibbs free energy of each protein–protein interaction using the relation:

$$G_i = c_i \ln \frac{c_i}{\sum_j c_j}, \tag{1}$$

where c_i is the "concentration" of the protein i, normalized, or rescaled, to be between 0 and 1. The sum in the denominator is taken over all protein neighbors (i.e., those that interact with it) of i, and including i. Therefore, the denominator can be considered related to a degree-entropy, although its functional form is much simplified since it does not include logarithmic terms. Carrying out this mathematical operation essentially transforms the "concentration" value assigned to each protein to a corresponding contribution to the Gibbs free energy. Thus, we replace the scalar value of transcriptome to a scalar function the Gibbs free energy. Thus, the equation represents the relationship between concentrations of proteins and the corresponding Gibbs energy.

The Gibbs free energy is a negative number, thus associated with each protein on the network that is a negative potential energy well. When plotted in 3D space where the vertical axis corresponds to the Gibbs free energy and the points in the horizontal plane represent protein coordinates, this results in a rugged energy landscape shown schematically in Figure 1. If we use what is called a topological filtration on this landscape, we essentially move a filtration plane up from the deepest energy well. As the filtration plane is moved up, larger-and-larger energetic subnetworks are captured. For convenience, we stop the filtration at energy threshold 32 meaning 32 nodes in the energetic subnetwork are retained. We call these subnetworks Gibbs-homology networks. This is not a magic number. The threshold of 32 was selected for convenience in showing networks visually. Incidentally, if we attempted to build a "network" by ranking the gene expression values, we would find disconnected nodes and not a connected network.

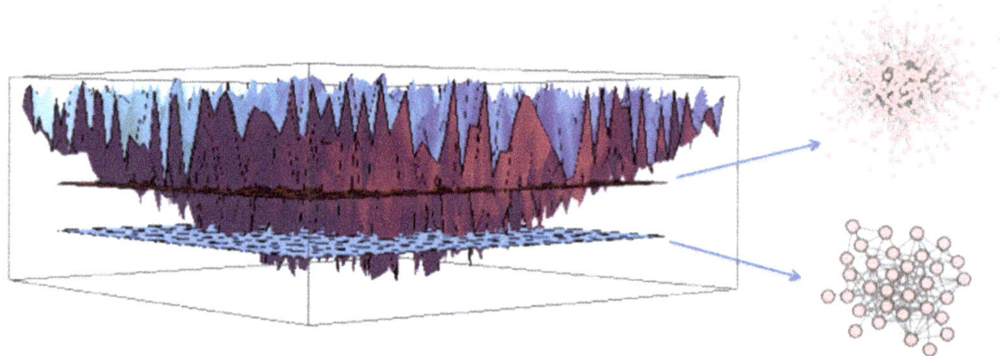

Figure 1. As the "filtration plane" moves up from the bottom, more-and-more nodes are captured in larger-and-larger energetic subnetworks shown on the right-hand side of the figure.

We now compute the Betti centrality, a topological measure, on the 32-node energetic networks as described in detail by Benzekry et al. [20]. The concept is easily explained as follows. In networks, there are holes, or rings, of various sizes. In these energetic pathways, protein–protein interaction networks, the proteins form interaction rings. In densely connected, but not fully connected, networks the rings, or holes, may consist of triangles and larger rings of interaction. To find the Betti centrality we ask ourselves the following question. Which protein when removed from the network will change the overall total number of rings the most? The total number of rings is called the Betti number. Given a network G consisting of edges, e, and vertices, v, the Betti centrality is given by

$$B(v_i) = B(G) - B(G - \{v_i\}). \tag{2}$$

Hence, the difference between the total Betti number $B(G)$ and the Betti number of the network after removing node i, gives the Betti centrality for node labeled i. We compute this for all nodes in the threshold-32 energetic network. Often there will be two or more proteins in the network that have equivalent Betti centrality. We discuss this equivalence and the Betti centrality with respect to the brain region data below in this manuscript.

3. Data Sources

Patient information for TLE is available in ref. [34] using Dataset GSE63808. Note that the data from GSE63808 only include epilepsy patients and did not include healthy controls. Obviously, it would be unethical to collect temporal lobe tissue form healthy people.

Computing the Betti centrality for energy threshold 32, we find eleven proteins as the most energetically significant overall. These are ranked in the Pareto chart shown in Figure 2. (Note that Pareto ranking is a common statistical method for displaying differences in data. A Pareto diagram is a simple bar chart that ranks related measures in decreasing order of occurrence, Pareto ranking is based on the principle of non-dominated sorting also called Pareto dominance, which can clearly be seen in the present case). Since there can be one or two (sometimes three) Betti centrality nodes with equivalent energies, the number of centrality nodes in the Pareto chart adds up to greater than 131 (which is the total number of patients whose data have been accessed in this study).

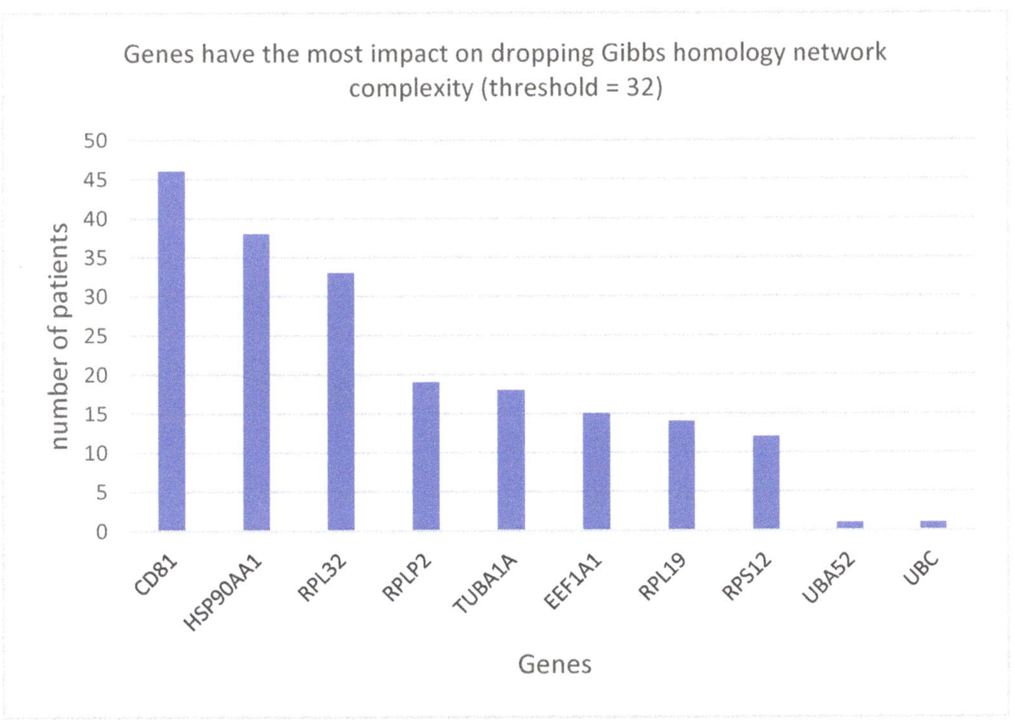

Figure 2. Genes having the most impact on dropping Gibbs homology network complexity (threshold = 32), with counts corresponding to the number of patients having the particular gene as target.

4. Results

By using the Pareto ranking, the most important node in the network is found to be CD81. (See Figure S1 for Pareto ranking of control patients.) A Gibbs Homology network at threshold 32 in which CD81 has the highest Betti centrality is shown in Figure 3 (highlighted in yellow) and the nearest-neighbor nodes with smaller, though important Gibbs energies, are shown highlighted in green.

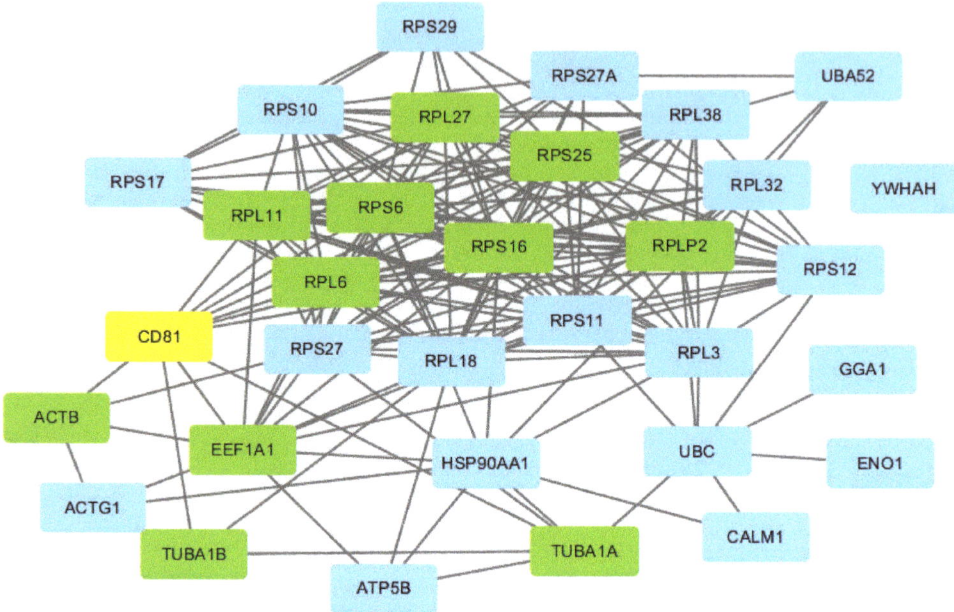

Figure 3. CD81 (yellow) and its nearest neighbors in the PPI network labeled in green.

Importantly, CD81 is a transmembrane protein that has tissue specific expression in various tissues including: tonsil, cerebral cortex, lymph node, smooth muscle, and reproductive tissues [35]. Shown in mice, CD81 regulates neuron-induced astrocytic differentiation [36]. Three alleles of the mice CD81 in seven genetic backgrounds were associated to abnormal brain development, hematopoietic system, and/or immune cells development and behavior [37]. In CD81-null mice, astrocyte and microglia cell numbers were upregulated, which lead to a 30% increase in brain size [38]. However, in human patients, deficiency of CD81 proteins was only associated to defected B cells [39]. As shown in mice and rats, expression of CD81 gene was up-regulated after seizure in three hippocampal cell layers: DGCL, CA1pyr, and CA3pyr [40,41]. In addition, the gene coding for the CD81 protein was recognized as a tumor-suppressor gene [42].

The second most frequent protein in the Pareto ranking of Betti centrality nodes is HSP90AA1 (heat shock protein 90 alpha family class A member 1). It is a stress induced isoform of the molecular chaperone Hsp90. In human, the gene has ubiquitous expression in brain and testis and 25 other tissues [43]. The expression of HSP90AA1 is known to be elevated when cellular stress is present, including in leukemia, several types of cancer, T cell under certain stimulation, and individuals suffering from chronic obstructive pulmonary disease (COPD) [44]. In rats, it was found that HSP90 protein levels in piriform cortex decreased after status epilepticus. The degradation was related to neuronal vulnerability to status epilepticus insult [45].

The third most frequent protein, excluding ribosomal proteins, according to the Pareto ranking of Betti centrality nodes is EEF1A1 (eukaryotic translation elongation factor 1, alpha 1). It plays an important role in the cellular translation process. Although EEF1A1 was not associated with neurological diseases, its isoform EEF1A2 was identified to be related. The two proteins are 92% similar in their amino acid sequences but exhibit non-overlapping expression patterns [46]. The difference in functions that the two isoforms perform might be due to post-translational modifications [47]. In wild-type mice, expression of A2 takes over from A1 starting at 21 days after birth; at the same time, a deletion or biallelic mutation of EEF1A2 gene was found responsible for the early-onset neurological abnormalities and early death in mice [48,49]. In human patients, EEF1A2 missense mutation was associated to early-onset epilepsy, severe intellectual disabilities and specific subtle facial dysmorphic features [50–52].

TUBA1A (Tubulin Alpha 1a) is one of the components that make up the cytoskeleton. In particular, together with beta tubulin, it forms a stable tubulin heterodimer that is a building block of microtubules. Microtubules play key roles in mitosis of dividing cells where they form mitotic spindles. In non-dividing cells such as neurons, microtubules form parallel bundles in axons and dendrites providing pathways for axoplasmic transport. Tubulin's gene is overexpressed in the brain's spinal cord and other tissues [53]. In human patients, mutations of this gene are expressed over a wide spectrum of phenotypes including lissencephaly, microcephaly, and early-onset epileptic seizures caused by defective neuronal migration [54,55]. Beta tubulin isotypes, TUBB2A, TUBB3, and TUBB4B have all been found over-expressed in post-traumatic brain injury patients and it has been therefore suggested that these tubulins along with CD44 may be appropriate targets for treatment [9], especially since there are numerous tubulin-binding pharmacological agents available [56]. However, virtually all of the approved and investigational tubulin-binding agents interact with beta and not alpha tubulin. We should point out that the Lipponen et al. study [9] was on rats and the data are available from GSE80174.

Ribosomal Proteins

The following analysis of ribosomal proteins is by no means exhaustive. First, we note that in the entire population of 130 patients, we found 34 ribosomal proteins in the Gibbs-homology networks at energy threshold 35. Of those 34 proteins, 10 were found in the literature to be related in some way to epilepsy, seizures, synaptic transmission, voltage-gated channels, and/or cytoskeleton. These 10 are RPS1, RPS4, RPS10, RPS11, RPS15, RPS27, RPL10A, RPL18, RPL32 [57]. Marrone et al. [58] discussed RPL32 in regard to seizures and aberrant cellular homeostasis. RPS6 is involved in co-expression of cyclin D1 in hemimegalencephaly [59] and is a key player in the mTOR signaling pathway and a contributor to epilepsy [60]. RPS6 is also implicated in X-chromosome brain diseases, as is RPS4, RPS1, RPL10 [61]. Notably, RPL6 is involved in a molecular pathway has been linked to temporal lobe epilepsy in childhood [62].

5. Discussion

Several studies based on the analysis of microarrays of epileptic tissue (human and animal models) reported both transcriptional and epigenetic alterations. In particular, the altered molecular pathways result in a variety of modifications in voltage-gated and receptor gated ion channels that lead to a perturbation of dendritic excitability [12]. Neuroproteomic studies on epilepsy revealed a significant contribution of proteins involved in energy metabolism, oxidative stress, inflammation, and excitatory imbalance [63]. Here, we report the results of our systems-biology based investigation, which analyzed the Gibbs Homology network for protein–protein interaction epilepsy data at threshold 32. The results of our investigations show that proteins with the highest Betti centrality are mainly transmembrane proteins, heat shock proteins, as well as neuronal elongation and cytoskeleton component proteins (alpha and beta tubulin isotypes). Importantly, by using the Pareto ranking, we also found significant ribosomal protein overexpression involving a number of these proteins. In the Supplementary Material, an additional analysis is provided showing similar results for a control group of 55 subjects.

Pires et al. [64] examining the proteome of brain samples from epilepsy found an overexpression of ribosomal proteins indicating an increased translational machinery. Increased ribosomal activity from microglia has been linked to neurological inflammatory conditions [65]. Gliosis as inflammatory responses, is frequently found in epilepsy patients and animal models. Moreover, the main histopathological features of TLE are hippocampal neuronal cell loss and gliosis with extensive synaptic rearrangement ('mossy fiber' innervation). The question if inflammation is the effect or the cause of seizure has been potentially answered in a mice model. The genetic deletion of Beta1-integrin leads to the development of astrogliosis that lead to spontaneous seizures [66]. Changes in glial cell phenotype have also been shown in specimens from patients with pharmaco-resistant temporal lobe epilepsy. Interestingly, astroglia molecular abnormalities have been revealed with the altered expression, localization, and function of the K+ and water channels [67]. These findings suggest that astroglia cells play a pivotal role in epilepsy and should be considered as promising targets for new therapeutic strategies.

This study identified a number of proteins that appear to play major roles in epilepsy and can, therefore, become attractive targets for pharmacological inhibition. The main proteins of interest are ribosomal proteins RPS1, RPS4, RPS10, RPS11, RPS15, RPS27, and RPL10A, RPL18, and RPL32. Unfortunately, only two of these protein structures have been solved and can be found in the Protein Data Bank (https://www.rcsb.org/ accessed on 30 August 2021), namely, RPS15 (PDB id: 1G1X) and RPL18 (PDB id: 1ILY). To the best of our knowledge, there are no known specific and selective inhibitors of the above-listed protein. However, there exist broad inhibitors of ribosomal activity, such as antibiotic molecules hygromyacin B, tetracycline, and pactamycin [68]. Future pharmacological development of specific inhibitors of the identified ribosomal proteins of interest could lead to important advances in this field.

Supplementary Materials: The following are available online at https://www.mdpi.com/article/10.3390/app11178059/s1. Figure S1: Target identification results for a control group of 55 patients.

Author Contributions: E.A.R. and J.A.T. designed the methodology used for this study, C.Y. performed computational work under the supervision of H.T.S., E.A.R. and M.C. provided clinical background interpretation. All authors have read and agreed to the published version of the manuscript.

Funding: This research was partially funded by NSERC (Canada) through a Discovery Grant awarded to J.A.T.

Institutional Review Board Statement: Not applicable.

Informed Consent Statement: Not applicable.

Data Availability Statement: All data used in this study were obtained from a publicly available database: BioGrid (https://thebiogrid.org/) accessed on 30 August 2021.

Acknowledgments: This research was supported by the Office of Naval Research and we acknowledge award number N00014-15-1-2126.ONR. E.A.R. was partly funded by CSTS Healthcare, Toronto, ON, Canada.

Conflicts of Interest: The authors declare no conflict of interest.

References

1. Epilepsy. Available online: https://www.who.int/news-room/fact-sheets/detail/epilepsy (accessed on 27 June 2021).
2. Epilepsy Data and Statistics | CDC. Available online: https://www.cdc.gov/epilepsy/data/index.html (accessed on 27 June 2021).
3. Stafstrom, C.E.; Carmant, L. Seizures and Epilepsy: An Overview for Neuroscientists. *Cold Spring Harb. Perspect. Med.* **2015**, *5*, a022426. [CrossRef] [PubMed]
4. Cunliffe, V.T. Building a Zebrafish Toolkit for Investigating the Pathobiology of Epilepsy and Identifying New Treatments for Epileptic Seizures. *J. Neurosci. Methods* **2016**, *260*, 91–95. [CrossRef] [PubMed]
5. Majores, M.; Eils, J.; Wiestler, O.D.; Becker, A.J. Molecular Profiling of Temporal Lobe Epilepsy: Comparison of Data from Human Tissue Samples and Animal Models. *Epilepsy Res.* **2004**, *60*, 173–178. [CrossRef] [PubMed]
6. Kamel, H.F.M.; Al-Amodi, H.S.A.B. Exploitation of Gene Expression and Cancer Biomarkers in Paving the Path to Era of Personalized Medicine. *Genom. Proteom. Bioinform.* **2017**, *15*, 220–235. [CrossRef] [PubMed]
7. Bhalla, D.; Godet, B.; Druet-Cabanac, M.; Preux, P.-M. Etiologies of Epilepsy: A Comprehensive Review. *Expert Rev. Neurother.* **2011**, *11*, 861–876. [CrossRef]
8. Fukuma, G.; Oguni, H.; Shirasaka, Y.; Watanabe, K.; Miyajima, T.; Yasumoto, S.; Ohfu, M.; Inoue, T.; Watanachai, A.; Kira, R.; et al. Mutations of Neuronal Voltage-Gated Na+ Channel A1 Subunit Gene SCN1A in Core Severe Myoclonic Epilepsy in Infancy (SMEI) and in Borderline SMEI (SMEB). *Epilepsia* **2004**, *45*, 140–148. [CrossRef]
9. Lipponen, A.; Paananen, J.; Puhakka, N.; Pitkänen, A. Analysis of Post-Traumatic Brain Injury Gene Expression Signature Reveals Tubulins, Nfe2l2, Nfkb, Cd44, and S100a4 as Treatment Targets. *Sci. Rep.* **2016**, *6*, 31570. [CrossRef]

10. Ding, K.; Gupta, P.K.; Diaz-Arrastia, R. Epilepsy after Traumatic Brain Injury. In *Translational Research in Traumatic Brain Injury;* Laskowitz, D., Grant, G., Eds.; Frontiers in Neuroscience; CRC Press/Taylor and Francis Group: Boca Raton, FL, USA, 2016; ISBN 978-1-4665-8491-4.
11. Schubert-Bast, S.; Rosenow, F.; Klein, K.M.; Reif, P.S.; Kieslich, M.; Strzelczyk, A. The Role of MTOR Inhibitors in Preventing Epileptogenesis in Patients with TSC: Current Evidence and Future Perspectives. *Epilepsy Behav.* **2019**, *91*, 94–98. [CrossRef] [PubMed]
12. Devinsky, O.; Vezzani, A.; O'Brien, T.J.; Jette, N.; Scheffer, I.E.; de Curtis, M.; Perucca, P. Epilepsy. *Nat. Rev. Dis. Primer* **2018**, *4*, 1–24. [CrossRef]
13. Amzica, F. *Neurophysiology of Epilepsy;* Oxford University Press: Oxford, UK, 2012; ISBN 978-0-19-175136-3.
14. Shorvon, S.D.; Bermejo, P.E.; Gibbs, A.A.; Huberfeld, G.; Kälviäinen, R. Antiepileptic Drug Treatment of Generalized Tonic–Clonic Seizures: An Evaluation of Regulatory Data and Five Criteria for Drug Selection. *Epilepsy Behav.* **2018**, *82*, 91–103. [CrossRef] [PubMed]
15. Wu, C.; Sun, D. GABA Receptors in Brain Development, Function, and Injury. *Metab. Brain Dis.* **2015**, *30*, 367–379. [CrossRef]
16. Dedeurwaerdere, S.; Boets, S.; Janssens, P.; Lavreysen, H.; Steckler, T. In the Grey Zone between Epilepsy and Schizophrenia: Alterations in Group II Metabotropic Glutamate Receptors. *Acta Neurol. Belg.* **2015**, *115*, 221–232. [CrossRef]
17. Rietman, E.A.; Karp, R.L.; Tuszynski, J.A. Review and Application of Group Theory to Molecular Systems Biology. *Theor. Biol. Med. Model.* **2011**, *8*, 21. [CrossRef]
18. Breitkreutz, D.; Hlatky, L.; Rietman, E.; Tuszynski, J.A. Molecular Signaling Network Complexity Is Correlated with Cancer Patient Survivability. *Proc. Natl. Acad. Sci. USA* **2012**, *109*, 9209–9212. [CrossRef] [PubMed]
19. Hinow, P.; Rietman, E.A.; Omar, S.I.; Tuszyński, J.A. Algebraic and Topological Indices of Molecular Pathway Networks in Human Cancers. Math. *Biosci. Eng. MBE* **2015**, *12*, 1289–1302. [CrossRef]
20. Benzekry, S.; Tuszynski, J.A.; Rietman, E.A.; Lakka Klement, G. Design Principles for Cancer Therapy Guided by Changes in Complexity of Protein-Protein Interaction Networks. *Biol. Direct* **2015**, *10*, 32. [CrossRef]
21. Rietman, E.A.; Scott, J.G.; Tuszynski, J.A.; Klement, G.L. Personalized Anticancer Therapy Selection Using Molecular Landscape Topology and Thermodynamics. *Oncotarget* **2016**, *8*, 18735–18745. [CrossRef] [PubMed]
22. Rietman, E.A.; Platig, J.; Tuszynski, J.A.; Lakka Klement, G. Thermodynamic Measures of Cancer: Gibbs Free Energy and Entropy of Protein–Protein Interactions. *J. Biol. Phys.* **2016**, *42*, 339–350. [CrossRef] [PubMed]
23. Brant, E.J.; Rietman, E.A.; Lakka Klement, G.; Cavaglia, M.; Tuszynski, J.A. Personalized therapy design for systemic lupus erythematosus based on the analysis of protein-protein interaction networks. *PLoS ONE* **2020**, *15*, e0226883. [CrossRef]
24. Golas, S.M.; Nguyen, A.N.; Rietman, E.A.; Tuszynski, J.A. Gibbs free energy of protein-protein interactions correlates with ATP production in cancer cells. *J. Biol. Phys.* **2019**, *45*, 423–430. [CrossRef]
25. Rietman, E.A.; Taylor, S.; Siegelmann, H.T.; Deriu, M.A.; Cavaglia, M.; Tuszynski, J.A. Using the Gibbs Function as a Measure of Human Brain Development Trends from Fetal Stage to Advanced Age. *Int. J. Mol. Sci.* **2020**, *21*, 1116. [CrossRef] [PubMed]
26. McGuire, S.H.; Rietman, E.A.; Siegelmann, H.; Tuszynski, J.A. Gibbs free energy as a measure of complexity correlates with time within C. elegans embryonic development. *J. Biol. Phys.* **2017**, *43*, 551–563. [CrossRef] [PubMed]
27. Huang, S. On the Intrinsic Inevitability of Cancer: From Foetal to Fatal Attraction. *Semin. Cancer Biol.* **2011**, *21*, 183–199. [CrossRef]
28. Moris, N.; Pina, C.; Arias, A.M. Transition States and Cell Fate Decisions in Epigenetic Landscapes. *Nat. Rev. Genet.* **2016**, *17*, 693–703. [CrossRef] [PubMed]
29. Kim, M.S.; Pinto, S.M.; Getnet, D.; Nirujogi, R.S.; Manda, S.S.; Chaerkady, R.; Madugundu, A.K.; Kelkar, D.S.; Isserlin, R.; Jain, S.; et al. A draft map of the human proteome. *Nature* **2014**, *509*, 575–581. [CrossRef] [PubMed]
30. Wilhelm, M.; Schlegl, J.; Hahne, H.; Moghaddas Gholami, A.; Lieberenz, M.; Savitski, M.M.; Ziegler, E.; Butzmann, L.; Gessulat, S.; Marx, H.; et al. Mass-spectrometry-based draft of the human proteome. *Nature* **2014**, *509*, 582–587. [CrossRef]
31. Guo, Y.; Sheng, Q.; Li, J.; Ye, F.; Samuels, D.C.; Shyr, Y. Large scale comparison of gene expression levels by microarrays and RNAseq using TCGA data. *PLoS ONE* **2013**, *8*, e71462. [CrossRef] [PubMed]
32. The Cancer Genome Atlas Program—National Cancer Institute. Available online: https://www.cancer.gov/about-nci/organization/ccg/research/structural-genomics/tcga (accessed on 28 June 2021).
33. BioGRID | Database of Protein, Chemical, and Genetic Interactions. Available online: https://thebiogrid.org/ (accessed on 28 June 2021).
34. Johnson, M.R.; Behmoaras, J.; Bottolo, L.; Krishnan, M.L.; Pernhorst, K.; Santoscoy, P.L.M.; Rossetti, T.; Speed, D.; Srivastava, P.K.; Chadeau-Hyam, M.; et al. Systems Genetics Identifies Sestrin 3 as a Regulator of a Proconvulsant Gene Network in Human Epileptic Hippocampus. *Nat. Commun.* **2015**, *6*, 6031. [CrossRef] [PubMed]
35. Uhlén, M.; Fagerberg, L.; Hallström, B.M.; Lindskog, C.; Oksvold, P.; Mardinoglu, A.; Sivertsson, Å.; Kampf, C.; Sjöstedt, E.; Asplund, A.; et al. Tissue-Based Map of the Human Proteome. *Science* **2015**, *347*. [CrossRef]
36. Kelić, S.; Levy, S.; Suarez, C.; Weinstein, D.E. CD81 Regulates Neuron-Induced Astrocyte Cell-Cycle Exit. *Mol. Cell. Neurosci.* **2001**, *17*, 551–560. [CrossRef]
37. Cd81 Phenotype Annotations. Available online: http://www.informatics.jax.org/marker/phenotypes/MGI:1096398 (accessed on 28 June 2021).
38. Geisert, E.E.; Williams, R.W.; Geisert, G.R.; Fan, L.; Asbury, A.M.; Maecker, H.T.; Deng, J.; Levy, S. Increased Brain Size and Glial Cell Number in CD81-Null Mice. *J. Comp. Neurol.* **2002**, *453*, 22–32. [CrossRef] [PubMed]

39. Van Zelm, M.C.; Smet, J.; Adams, B.; Mascart, F.; Schandené, L.; Janssen, F.; Ferster, A.; Kuo, C.-C.; Levy, S.; van Dongen, J.J.M.; et al. CD81 Gene Defect in Humans Disrupts CD19 Complex Formation and Leads to Antibody Deficiency. *J. Clin. Investig.* **2010**, *120*, 1265–1274. [CrossRef] [PubMed]
40. Long, Q.; Upadhya, D.; Hattiangady, B.; Kim, D.-K.; An, S.Y.; Shuai, B.; Prockop, D.J.; Shetty, A.K. Intranasal MSC-Derived A1-Exosomes Ease Inflammation, and Prevent Abnormal Neurogenesis and Memory Dysfunction after Status Epilepticus. *Proc. Natl. Acad. Sci. USA* **2017**, *114*, E3536–E3545. [CrossRef]
41. Borges, K.; Shaw, R.; Dingledine, R. Gene Expression Changes after Seizure Preconditioning in the Three Major Hippocampal Cell Layers. *Neurobiol. Dis.* **2007**, *26*, 66–77. [CrossRef]
42. Yoo, T.-H.; Ryu, B.-K.; Lee, M.-G.; Chi, S.-G. CD81 Is a Candidate Tumor Suppressor Gene in Human Gastric Cancer. *Cell. Oncol.* **2013**, *36*, 141–153. [CrossRef] [PubMed]
43. Fagerberg, L.; Hallström, B.M.; Oksvold, P.; Kampf, C.; Djureinovic, D.; Odeberg, J.; Habuka, M.; Tahmasebpoor, S.; Danielsson, A.; Edlund, K.; et al. Analysis of the Human Tissue-Specific Expression by Genome-Wide Integration of Transcriptomics and Antibody-Based Proteomics. *Mol. Cell. Proteom.* **2014**, *13*, 397–406. [CrossRef] [PubMed]
44. Zuehlke, A.D.; Beebe, K.; Neckers, L.; Prince, T. Regulation and Function of the Human HSP90AA1 Gene. *Gene* **2015**, *570*, 8–16. [CrossRef]
45. Kim, Y.-J.; Kim, J.-Y.; Ko, A.-R.; Kang, T.-C. Reduction in Heat Shock Protein 90 Correlates to Neuronal Vulnerability in the Rat Piriform Cortex Following Status Epilepticus. *Neuroscience* **2013**, *255*, 265–277. [CrossRef]
46. Lee, S.; Francoeur, A.M.; Liu, S.; Wang, E. Tissue-Specific Expression in Mammalian Brain, Heart, and Muscle of S1, a Member of the Elongation Factor-1 Alpha Gene Family. *J. Biol. Chem.* **1992**, *267*, 24064–24068. [CrossRef]
47. Soares, D.C.; Abbott, C.M. Highly Homologous EEF1A1 and EEF1A2 Exhibit Differential Post-Translationalmodification with Significant Enrichment around Localised Sites of Sequence Variation. *Biol. Direct* **2013**, *8*, 29. [CrossRef]
48. Chambers, D.M.; Peters, J.; Abbott, C.M. The Lethal Mutation of the Mouse Wasted (Wst) Is a Deletion That Abolishes Expression of a Tissue-Specific Isoform of Translation Elongation Factor 1alpha, Encoded by the Eef1a2 Gene. *Proc. Natl. Acad. Sci. USA* **1998**, *95*, 4463–4468. [CrossRef] [PubMed]
49. Davies, F.C.J.; Hope, J.E.; McLachlan, F.; Nunez, F.; Doig, J.; Bengani, H.; Smith, C.; Abbott, C.M. Biallelic Mutations in the Gene Encoding EEF1A2 Cause Seizures and Sudden Death in F0 Mice. *Sci. Rep.* **2017**, *7*, 46019. [CrossRef] [PubMed]
50. Lam, W.W.K.; Millichap, J.J.; Soares, D.C.; Chin, R.; McLellan, A.; FitzPatrick, D.R.; Elmslie, F.; Lees, M.M.; Schaefer, G.B.; Abbott, C.M. Novel de Novo EEF1A2 Missense Mutations Causing Epilepsy and Intellectual Disability. *Mol. Genet. Genom. Med.* **2016**, *4*, 465–474. [CrossRef] [PubMed]
51. Nakajima, J.; Okamoto, N.; Tohyama, J.; Kato, M.; Arai, H.; Funahashi, O.; Tsurusaki, Y.; Nakashima, M.; Kawashima, H.; Saitsu, H.; et al. De Novo EEF1A2 Mutations in Patients with Characteristic Facial Features, Intellectual Disability, Autistic Behaviors and Epilepsy. *Clin. Genet.* **2015**, *87*, 356–361. [CrossRef] [PubMed]
52. de Ligt, J.; Willemsen, M.H.; van Bon, B.W.M.; Kleefstra, T.; Yntema, H.G.; Kroes, T.; Vulto-van Silfhout, A.T.; Koolen, D.A.; de Vries, P.; Gilissen, C.; et al. Diagnostic Exome Sequencing in Persons with Severe Intellectual Disability. *N. Engl. J. Med.* **2012**, *367*, 1921–1929. [CrossRef]
53. GTEx Portal. Available online: https://www.gtexportal.org/home/gene/TUBA1A (accessed on 29 June 2021).
54. Hikita, N.; Hattori, H.; Kato, M.; Sakuma, S.; Morotomi, Y.; Ishida, H.; Seto, T.; Tanaka, K.; Shimono, T.; Shintaku, H.; et al. A Case of TUBA1A Mutation Presenting with Lissencephaly and Hirschsprung Disease. *Brain Dev.* **2014**, *36*, 159–162. [CrossRef]
55. Kumar, R.A.; Pilz, D.T.; Babatz, T.D.; Cushion, T.D.; Harvey, K.; Topf, M.; Yates, L.; Robb, S.; Uyanik, G.; Mancini, G.M.S.; et al. TUBA1A Mutations Cause Wide Spectrum Lissencephaly (Smooth Brain) and Suggest That Multiple Neuronal Migration Pathways Converge on Alpha Tubulins. *Hum. Mol. Genet.* **2010**, *19*, 2817–2827. [CrossRef] [PubMed]
56. Dumontet, C. Mechanisms of action and resistance to tubulin-binding agents. *Expert Opin. Investig. Drugs* **2000**, *9*, 779–788. [CrossRef]
57. Laurén, H.B.; Lopez-Picon, F.R.; Brandt, A.M.; Rios-Rojas, C.J.; Holopainen, I.E. Transcriptome Analysis of the Hippocampal CA1 Pyramidal Cell Region after Kainic Acid-Induced Status Epilepticus in Juvenile Rats. *PLoS ONE* **2010**, *5*, e10733. [CrossRef]
58. Marrone, A.K.; Kucherenko, M.M.; Wiek, R.; Göpfert, M.C.; Shcherbata, H.R. Hyperthermic seizures and aberrant cellular homeostasis in Drosophila dystrophic muscles. *Sci. Rep.* **2011**, *1*, 1–9. [CrossRef]
59. Aronica, E.; Boer, K.; Baybis, M.; Yu, J.; Crino, P. Co-Expression of Cyclin D1 and Phosphorylated Ribosomal S6 Proteins in Hemimegalencephaly. *Acta Neuropathol.* **2007**, *114*, 287–293. [CrossRef] [PubMed]
60. Cho, C.-H. Frontier of Epilepsy Research—MTOR Signaling Pathway. *Exp. Mol. Med.* **2011**, *43*, 231–274. [CrossRef]
61. Laumonnier, F.; Cuthbert, P.C.; Grant, S.G.N. The Role of Neuronal Complexes in Human X-Linked Brain Diseases. *Am. J. Hum. Genet.* **2007**, *80*, 205–220. [CrossRef] [PubMed]
62. Moreira-Filho, C.A.; Bando, S.Y.; Bertonha, F.B.; Iamashita, P.; Silva, F.N.; Costa, L.; da, F.; Silva, A.V.; Castro, L.H.M.; Wen, H.-T. Community Structure Analysis of Transcriptional Networks Reveals Distinct Molecular Pathways for Early- and Late-Onset Temporal Lobe Epilepsy with Childhood Febrile Seizures. *PLoS ONE* **2015**, *10*, e0128174. [CrossRef] [PubMed]
63. do Canto, A.M.; Donatti, A.; Geraldis, J.C.; Godoi, A.B.; da Rosa, D.C.; Lopes-Cendes, I. Neuroproteomics in Epilepsy: What Do We Know so Far? *Front. Mol. Neurosci.* **2021**, *13*, 604158. [CrossRef] [PubMed]
64. Pires, G.; Leitner, D.; Drummond, E.; Kanshin, E.; Nayak, S.; Askenazi, M.; Faustin, A.; Friedman, D.; Debure, L.; Ueberheide, B.; et al. Proteomic Differences in the Hippocampus and Cortex of Epilepsy Brain Tissue. *BiorXiv* **2020**. [CrossRef]

65. Yang, Y.; Boza-Serrano, A.; Dunning, C.J.R.; Clausen, B.H.; Lambertsen, K.L.; Deierborg, T. Inflammation Leads to Distinct Populations of Extracellular Vesicles from Microglia. *J. Neuroinflamm.* **2018**, *15*, 168. [CrossRef]
66. Reactive Astrogliosis Causes the Development of Spontaneous Seizures. *J. Neurosci.* **2015**, *35*, 3330–3345. Available online: https://www.jneurosci.org/content/35/8/3330/ (accessed on 29 June 2021). [CrossRef]
67. Scihub.Se. Available online: https://scihub.se/10.1038/nrdp.2018.24 (accessed on 29 June 2021).
68. Long, K.S. Ribosomal Protein Synthesis Inhibitors. In *Encyclopedia of Molecular Pharmacology*; Offermanns, S., Rosenthal, W., Eds.; Springer: Berlin/Heidelberg, Germany, 2008. [CrossRef]

MDPI

Article

Sea Purslane as an Emerging Food Crop: Nutritional and Biological Studies

Arona Pires [1], Sílvia Agreira [1], Sandrine Ressurreição [1], Joana Marques [2], Raquel Guiné [3], Maria João Barroca [1,2] and Aida Moreira da Silva [1,2,*]

[1] Polytechnic of Coimbra, Coimbra Agriculture School, Bencanta, 3045-601 Coimbra, Portugal; arona@esac.pt (A.P.); sagreira@esec.pt (S.A.); sandrine@esac.pt (S.R.); mjbarroca@esac.pt (M.J.B.)
[2] Unidade de I&D Química-Física Molecular, Department of Chemistry, University of Coimbra, 3004-535 Coimbra, Portugal; marques.jt@uc.pt
[3] CERNAS Research Centre, Department of Food Industry, Polytechnic of Viseu, 3504-510 Viseu, Portugal; raquelguine@esav.ipv.pt
* Correspondence: aidams@esac.pt

Abstract: Halophyte plants are highly adapted to salt marsh ecosystems due to their physiological and ecological characteristics. *Halimione portulacoides* (L.) Aellen is one abundant halophyte shrub that belongs to a *Chenopodiaceae* family and Caryophyllales order and is found on sandy or muddy coastlines and salt marshes. In this study, the leaves of sea purslane (*H. portulacoides*) grown in Figueira da Foz (Portugal) were characterized at nutritional and mineral concentration. Moreover, different methanolic extracts were obtained from the leaves, and the antioxidant activity was assessed by several methods. From a nutritional point of view, this halophyte plant may be considered a good source of dietary fiber, protein, natural minerals such as calcium, magnesium, manganese, copper, and potassium. The primary sugar found in leaves of sea purslane is maltose, followed by sucrose, glucose, and fructose. Finally, leaves showed a high content of phenolic compounds and considerable antioxidant activity. The novel products butter and pasta enriched with powder dried leaves of *H. portulacoides* revealed the plant's potential to be used as a salt substitute and a good alternative to enhance the sensory characteristics of products, with additional health benefits. The nutritional characteristics and the phytochemical value highlight *H. portulacoides* as a potential candidate crop in saline agriculture and to be used as a new vegetable, especially as a premium food in the novel "salty veggies" market or as a kitchen salt substitute.

Keywords: halophyte; sea purslane; minerals; antioxidant activity; novel ingredient

Citation: Pires, A.; Agreira, S.; Ressurreição, S.; Marques, J.; Guiné, R.; Barroca, M.J.; Moreira da Silva, A. Sea Purslane as an Emerging Food Crop: Nutritional and Biological Studies. *Appl. Sci.* **2021**, *11*, 7860. https://doi.org/10.3390/app11177860

Academic Editors: Francesco Cappello and Magdalena Gorska-Ponikowska

Received: 9 July 2021
Accepted: 24 August 2021
Published: 26 August 2021

Publisher's Note: MDPI stays neutral with regard to jurisdictional claims in published maps and institutional affiliations.

1. Introduction

The increase of the world population leads to increased agricultural production to obtain the necessary amount of food to feed everyone on the planet. Furthermore, global warming and freshwater reduction lead to increased land salinity and dryness [1]. Thus, innovations related to agricultural practices, type of value chain, and products have been developed to increase food availability for the population and to shift toward more sustainable food systems, either in the dominant food system regime or in alternative niches. An example of a new strategy is the introduction of non-conventional plant-based foods [2]. Regarding this issue, the interest in natural ingredients with good nutritional and functional components properties that can replace synthetic ones has been increasing in order to develop promising functional foods [2]. On the other hand, the loss of agro-biodiversity has been an incentive to the introduction in agrosystems of innovative crops with high-value biochemical composition and adaptability to climate change and soil salinities. In this context, halophyte plants are extremely adapted to salt marsh ecosystems due to their physiological and ecological characteristics (support at least 11.7 g L^{-1} of NaCl), allowing them to live and grow in places with very high salt concentrations, where most plants are unable to survive [1,3]. Furthermore, the leaves of some halophyte plants are rich in bioactive molecules, such as lipophilic compounds and phenols, including flavonoids [3,4]. Hence, the scientific community has tried to understand the importance of halophyte plants for human consumption, creating a market-positioning strategy as added-value components for the pharmaceutical and nutraceutical industries and also for the gastronomic area through the improvement of foods' organoleptic properties [3,5].

The perennial *Halimione portulacoides* (L.) Aellen, sea purslane ("gramata branca" or "beldroega do mar" in Portuguese), is a halophyte shrub that belongs to a Chenopodiaceae family and Caryophyllales order. It is found on sandy and muddy coastlines and salt marshes around the coasts of North Africa, Southwest Asia, and Europe [5,6]. In Europe, it is one of the most productive and abundant species in salt marshes. This plant is dispersed along the coast of the Iberian Peninsula, and the prevalence in the Portuguese estuaries is at Tagus and Ria de Aveiro salt marches [7–10]. With a controversial taxon background, *Atriplex portulacoides* is also acknowledged as a senior synonym of *Halimione portulacoides* in the International Plant Names Index [11]. This plant is characterized as a shrub, reaching up to 1.5 m in height (Figure 1). It is a monoecious, protruding-ascending species, with a silvery-gray color and with stems that are often radiant. The leaves are decussate, speared, hollowed out, whole, and fleshy. The inflorescence features unisexual flowers, and the fruit is sessile [12].

Figure 1. *Halimione portulacoides* (L.) Aellen, sea purslane.

The potential of sea purslane in healthy and functional food products has recently been highlighted by the similarity of its fatty acid composition with the *S. ramosissima*, which is a promising functional food with a renewed interest as a food and pharmaceutical product [4,13]. The lipophilic fraction of *H. portulacoides* leaves from estuarine environments of Portugal is mainly composed of long-chain aliphatic acids (e.g., octacosanoic, triacontanoic, oleic, hexadecanoic, and linoleic) and alcohols (e.g., octacosanol, hexacosanol, and triancontanol) (both in the C16–C30 range) while containing smaller amounts of sterols, such as schottenol, sitosterol, and sitostanol. Furthermore, the environmental stresses induce in a plant the synthesis of a wide range of phenolic compounds, such as sulfated flavonoids, particularly derivatives of isorhamnetin-sulfate and carotenoids (such as zeaxanthin, β-carotene, lutein, auroxanthin, violaxanthin, and antheraxanthin) [11,14–18]. These confer important biological properties, such as antioxidant, anti-inflammatory, anti-trombotic, and anti-cancerogenic activities [4,11,18]. Furthermore, *H. portulacoides* leaves are a good source of protein and important dietary minerals, namely Mg, K, Ca, Fe, Mn, Cu, and Zn [9,11,19–21]. The nutritional and biochemical profiles of the plant responsible for its positive effects on human health increase its interest as a new vegetable, especially as a premium food in the novel "salty veggies" market [13]. Furthermore, its high productivity in saline conditions and even in arid lands and its resistance to different environmental stresses render *H. portulacoides* a good candidate for exploitation in sustainable and saline agriculture. However, similar to most halophytes, relevant issues related to their cultivation remain to be defined, such as the effects of the season, geographical location, and the morphology of the environment on their nutritional quality and salinity content [1]. An adaptative trait of *H. portulacoides* is the capacity to concentrate seawater metal cations beneficial to human health. However, in addition to naturally occurring metals, halophytes could accumulate heavy metals derived from the human contamination of salt marshes' sediments [11,22]. The high concentration of toxic heavy metals existent in some lagoon environments and European estuaries have an impact on the macro and micro nutritional composition of *H. portulacoides* [11]. Therefore, the location from which the plants are harvested for food use constitutes a crucial factor for their nutritional quality and eventual toxicity. However, by itself, this is no evidence of toxicity of *H. portulacoides*, since more than 90% of toxic metals are retained in the plant's below-ground organs [23,24]. According to Cabrita et al. [24], the low concentration of mercury in the aerial parts of the plant are attributed to metal release by leaves and stems, probably via stomata.

Although *H. portulacoides* is an almost forgotten traditional food, its use as food (raw or cooked) and forage dates back thousands of years. Indeed, the use of this plant in human culture comes from the Early Neolithic period, as evidenced by the finding of this plant amongst ancient carbonized remains of food in northern Holland [11]. In Italy, *H. portucaloides* is traditionally used raw in salads or cooked in some recipes based on fish [11]. Moreover, its buds can be preserved in vinegar [25]. Nowadays, the fresh leaves of this alimurgical wild edible species have been used in gourmet preparations [11]. Sea purslane's visually appealing aspect in terms of freshness and color are attributes that potentiate its usage in a broad range of foods, namely to garnish dishes. However, sea purslane may be used not only as a fresh product but also as a dried herb. Hence, powdered dried leaves, which improve the product's availability and shelf-life, could represent an excellent alternative to creative reinterpretations of traditional foods. Moreover, it can be incorporated into a broad range of foods, developing functional foods that are a trend in the food industry driven by consumer's acceptance and awareness of their positive health effects. The dry halophyte leaves can be a natural salt substitute and a good alternative to develop innovative traditional products with peculiar flavor and color traits enriched with antioxidants compounds.

As referred above, the chemical, mineral, and bioactive compounds related to the composition of the salt marsh sediments where the plant lives are not yet fully known. Hence, to use *H. portucaloides* as food in safe conditions, a nutritional, biochemical, and mineral characterization of the plant leaves is essential. To the best of our knowledge, the nutritional, biochemical, and mineral composition of *H. portucaloides* of the salt marsh of Figueira da Foz (Portugal) has not yet been characterized. Moreover, this is the first time that it is developed into enriched products with sea purslane dried leaves. In this context, the present study aimed to evaluate the macro and micro nutritional compounds and phytochemical value of *H. portulacoides* concerning its potential to be used in the development of new functional food ingredients. Furthermore, to incentivize the cultivation of *H. portulacoides*, we proposed two novel products, pasta and butter, enriched with powdered dried leaves. In addition, a sensory evaluation of the products was performed by a set of consumer panelists.

2. Materials and Methods

2.1. Plant Material

H. portulacoides (L.) Aellen leaves were collected at Armazéns de Lavos (40°06′43″ N 8°49′59″ W), Figueira da Foz salterns, Portugal, in July 2019. The nutritional and mineral profile was evaluated in the fresh leaves, and the biological profile was evaluated in grinded freeze-dried (in a CoolSafe 100-9 Pro Freeze Dryer, Labogene, Denmark) leaves.

The powdered samples were stored at room temperature and protected from light until further use.

2.2. Chemicals

2,2′-Azino bis(3 ethylbenzothiazoline 6 sulfonic acid) diammonium salt (≥98%), 2,2′-azobis(2-methylpropionamidine) dihydrochloride (97%), 2,2-diphenyl-1-picrylhydrazyl, 2,4,6-tris(2-pyridyl)-s-triazine (≥99%), 5,5′-dithiobis(2-nitrobenzoic acid) (99%), acetylcholinesterase from Electrophorus electricus (electric eel), acetylthiocholine iodide (≥99.0%), aluminum chloride ($AlCl_3$, for synthesis), ammonium acetate (≥98%), butylated hydroxytoluene (≥99%), copper(II) chloride ($CuCl_2$, for synthesis), ethylenediaminetetraacetic acid (≥98.5%), ferrozine (97%), galantamine hydrobromide, gallic acid (≥98.0%), iron(II) chloride ($FeCl_2{\cdot}4(H_2O)$, ≥99%), linoleic acid (≥99%), neocuproine (≥98%), quercetin (≥95%), sodium carbonate (≥99.5%), sodium dihydrogen phosphate ($NaH_2PO_4{\cdot}2(H_2O)$, ≥98%), thiobarbituric acid (≥98%), trichloroacetic acid, TRIS (≥99%), Trolox (97%), tryptic soy agar, Tween®® 80, β-carotene (≥93%), as well as solvents (of analytical grade) were obtained from Merck (Oeiras, Portugal). Acetic acid (glacial p.a.) was purchased from Pronalab (Sintra, Portugal), the Folin–Ciocalteu's reagent, HCl (35%) and iron(III) chloride ($FeCl_3{\cdot}6(H_2O)$) (≥98%) from Panreac (Barcelona, Spain), and potassium persulfate (99%) and sodium phosphate dibasic (Na_2HPO_4, ≥99%) from Honeywell (Carnaxide, Portugal).

2.3. Nutritional Composition Analysis

The Association of Official Analytical Chemists (AOAC, 1997) methodologies were used to determine the chemical properties of *H. portulacoides*. Moisture content (method 930.04), ashes (method 930.05), crude protein (method 978.04) using a nitrogen conversion factor of 6.25, total lipids (method 930.09), and dietary fiber (AOAC 985.29) and crude fiber (method 930.10) were determined. The carbohydrate content was determined from the difference between 100 and the sum of the percentages of moisture, ashes, crude protein, dietary fiber, and total lipid contents.

The Regulation (EU) No. 1169/2011 of the European Parliament and of the Council of 25 October 2011 was used for calculation of the energy values (expressed in kcal/100 g and kJ/100 g) [26].

Quantification of sugars was performed by high-performance liquid chromatography with refractive index detection (HPLC-RI), consisting of an LC1110 high-pressure pump (GBC, Australia), LC-100 oven (Perkin-Elmer, USA), refraction 830-RI (Jasco, Japan), and an HC-75 Ca ++ 305 × 7.8 mm column (Hamilton, Energy Way, Reno, NV, USA). The mobile phase used was ultrapure water (Direct-pure, 10 Uv) with traces of sodium azide, with a flow rate of 0.6 mL/minute, at 80 °C. The quantification was performed by BioUltra standards (Sigma-Aldrich, St Louis, MO, USA). The data were collected by an Interface Hercule Lite (JMBS) and processed by the software Borwin Chromatography Software, version 1.5, build 16 by Jasco-Borwin (Japan).

2.4. Mineral and Heavy Metal Composition Analysis

For minerals and heavy metals analyses of *H. portulacoides*, a PerkinElmer PinAAcle 900 T Atomic Absorption Spectrometer (USA) was used. The contents of calcium, copper, iron, magnesium, manganese, potassium, sodium, and zinc were quantified by flame atomic absorption spectrometry (FAAS) (ISO 6869:2000) [27]. Cadmium, lead, and chromium were analyzed by graphite furnace atomic absorption spectrometry (GFAAS) (EN 14082:2003) [28]. A Thermo X series II inductively coupled plasma mass spectrometer (ICP-MS) (Thermo Fisher Scientific, Waltham, MA, USA) was used to determine the iodine content in *H. portulacoides* (EN 15111:2007). The phosphorus content was determined by spectrophotometry (ISO 6491:1998) [29] with a PG instruments T80+ UV/VIS spectrophotometer (UK). Traces of mercury were analyzed by an AMA254 Mercury Analyzer (Leco, USA).

2.5. Color Coordinates

The color coordinates of the *H. portulacoides* leaves were assessed in the top and bottom sides using a colorimeter (Chroma Meter—CR-400, Konica Minolta, Tokyo, Japan), and they were registered in the CIE Lab color space. The axis for L^* corresponds to brightness and varies from 0 (black) until 100 (absolute white). The chromatic coordinate a^* ranges from green (negative values) to red (positive values), and the b^* coordinate ranges from blue (negative) to yellow (positive). The total color difference (TCD) was calculated using Equation (1), which allows quantifying the overall color difference between dried and fresh leaves, which is this case was the control sample:

$$TCD = \sqrt{\left(L^* - L_0^*\right)^2 + \left(a^* - a_0^*\right)^2 + \left(b^* - b_0^*\right)^2} \quad (1)$$

where L_0*, a_0*, and b_0* are the color coordinates for the control sample [30].

For evaluation of color, 20 measurements were made in the fresh leaves (top and bottom sides) and in the dried powder.

2.6. Extraction Procedure

H. portulacoides powdered leaves samples (5 g) were extracted with 100 mL of ethanol for 3 h, at room temperature, in a magnetic stirrer. The extract was centrifuged for 10 min at 2500 rpm; then, the supernatant was filtered and stored at a concentration of 5 mg/mL at 4 °C until further analysis.

2.7. Chemical Composition and Antioxidant and Enzymatic Activities

The total phenolic and flavonoid contents (TPC and TFC, respectively) and the determination of the enzymatic (cholinesterase inhibition) and antioxidant activities (DPPH and ABTS radical scavenging methods, β-carotene/linoleic acid bleaching method, lipid peroxidation inhibition, metal chelating ability, reducing power—FRAP and CUPRAC) and enzymatic (cholinesterase inhibition) activities were performed according to modified versions of literature reported methods [31].

2.8. Drying

The drying of *H. portulacoides* fresh leaves (5 kg) was performed in a laboratory scale tray dryer. The drying unit consists of a chamber equipped with six trays (45 × 90 cm), heating elements, and a flow fan with adjustable speed yielding. The leaves were dried at a temperature of 65 °C and air velocity of 1.5 m/s for 22 h. The dried leaves were grinded using a homogenizer and reduced to powder.

2.9. Food Usage Suggestions

2.9.1. Pasta

A basic pasta recipe was prepared (control) with the following ingredients: 100 g of wheat flour T65, one egg, and 1.6 g of salt (control). In enriched pasta, the content of salt was replaced by 7.7 g of powdered dried *H. portulacoides* (equivalent to 1.6 g of salt, assuming that all Na of the dried leaves is linked to NaCl). In both pastas, the ingredients were mixed and kneaded by hand for 15 min and left to rest in the cold, at approximately 4 °C, for 30 min. Then, the paste was prepared and shaped with a home pasta-shaping machine (Figure 2). The machine was adjusted to produce pasta with 1 mm of thickness.

Figure 2. Ingredients and pasta preparation.

The pasta cooking time was 10 min in boiling water (for 100 g of pasta, 2 L of water was used, to which 6 g of salt was added). Cooking was carried out with precision by respecting the standardized protocol ISO 7304-1:2016 [32] in order to make it possible to compare both samples (control and enriched pasta). After cooking, the pasta was poured into a sieve and drained.

2.9.2. Butter

Pasteurized cream (with 40% of fat) was cooled at ±6 °C. After that, cream was mixed in a tank, and the biological ripening was performed at 15 ± 2 °C until reaching a pH between 4.8 and 5.2 by adding a dairy starter culture (*Lactococcus lactis* subsp. cremoris, *Leuconostoc, Lactococcus lactis* subsp. Lactis, and *Lactococcus lactis* subsp. lactis biovar diacetylactis) (FS-DVS Flora Danica). The mixture was introduced in a churning machine. In the churning process, the cream was violently agitated to break down fat globules, allowing the fat to coagulate into butter grains. During churning at a temperature between 8 and 12 °C, the buttermilk was drained off, and butter began to appear in the form of grains.

The grains were washed with pasteurized water to remove any residual buttermilk and milk solids, controlling the butter's moisture content up to 16%. Next, to a portion of butter was added salt to produce butter with 1% of NaCl salt, and to another part was added 1% of dried powdered sea purslane. After salting, the butter was worked vigorously to ensure an even distribution of the salt or sea purslane powder, depending on the case. The butter was packed in vegetal paper and refrigerated at 3 °C until further use.

2.10. Sensory Evaluation

Consumer panelists were recruited from the Coimbra Agriculture School (ESAC) in order to evaluate possible changes in the organoleptic characteristics of pasta and butter resulting from the addition of *H. portulacoides*. To assess the preference for a given product by the tasters, a Product Preference Test was used. Samples of products non-fortified with dried powder of *H. portulacoides* (control) and fortified were presented randomly to 40 tasters. Panelists were placed randomly at room temperature, and water was served to clean their palates prior to proceeding to the next sample.

The organoleptic attributes (color, flavor, texture, appearance, and overall acceptance) of pasta and butter samples (control and enriched with *H. portulacoides*) were evaluated using a 9-point hedonic scale (1 = dislike extremely to 9 = like extremely). Tasters were also asked to rank products according to their purchasing preference and enquired about the regularity of consumption of pasta and butter.

2.11. Statistical Analysis

Statistical analyses were performed to evaluate if the differences between mean values (color and sensorial attributes) were statistically significant. For comparison of mean values between two groups (samples), the independent samples T-test was used. The results of biological activity were analyzed using one-way ANOVA (for three or more groups) followed by Tukey's post hoc test for statistical comparison between the experimental data. In all cases, the level of significance considered was 5%. The statistical analyses were performed using the software SPSS V26 and GraphPad Prism (GraphPad Software, USA)

The IC50 values were calculated for each extract fitting the results using nonlinear regression analysis in sigmoidal dose–response curves (variable slope).

3. Results and Discussion

3.1. Nutritional Composition

The *H. portulacoides* plant can become important for application in new food products due to its nutritional composition and health-beneficial properties. Table 1 shows the average values of the main chemical components for sea purslane grown in Figueira da Foz (Portugal), expressed in raw and dry matter.

Table 1. Nutritional composition of *Halminione portulacoides*.

Composition	Raw Matter	Dry Matter
Energy (kcal/100 g)	48.03 ± 0.06	218.59 ± 0.30
Moisture (g/100 g)	78.03 ± 0.01	-
Ash (g/100 g)	6.09 ± 0.02	27.70 ± 0.09
Dietary fiber (g/100 g)	8.90 ± 0.01	40.49 ± 0.06
Crude fiber (g/100 g)	4.54 ± 0.01	20.64 ± 0.07
Protein (g/100 g)	2.08 ± 0.02	9.47 ± 0.07
Lipids (g/100 g)	0.46 ± 0.01	2.07 ± 0.05
Carbohydrates * (g/100 g)	4.45 ± 0.01	20.26 ± 0.06

* excluding fiber.

The moisture content of wild *H. portulacoides* evaluated in this work was 78.03 g/100 g. However, the moisture content of the same plant growing in hydroponics conditions was reported as being around 90% [9].

In general, halophyte plants such as *H. portulacoides* have higher ash contents than other edible plants. As an example, *Sarcocornia ambigua* has an ash content of 24.98% (dry matter) [33], contrasting with *Sarcocornia perennis* (43.62%, dry matter) [2], *Arthrocnemum macrostachyum* (31.6%, dry matter), and *Salicornia ramosissima* (29.2%, dry matter) [4]. The ash content is related to the total concentration of minerals. Therefore, the high concentration of ash observed in these sea plants is probably related to their ability to retain the minerals of the seacoast saline soils [4].

The total mineral content (ash) of sea purslane leaves (27.70%, dry matter) was very similar to that reported by Briens et al. (around 28%, dry matter), which were both collected in salt marshes [34]. Sea purslane that grows in saline hydroponic conditions, with different nutrient solutions, presented a leaf mineral content of 36.67% (dry weight) [9].

Plants have a variety of lipids with important biological functions involving plant metabolism. These lipids play structural and signaling roles that are significant in the metabolic regulation, protection, and homeostasis of the cell [13].

The lipid content extractable with petroleum ether was 0.46 g/100 g of raw matter. Custódio et al. [9] reported that the lipid content in leaves was 0.33 g/100 g, and in another study, the same author presented values in the range 0.74–0.94 g/100 g [3]. The lipids in plants act as signalers, energy storage compounds, and hydrophobic barriers for the membrane [35]. In human health, lipids are essential for promoting the absorption of some vitamins and helping build some tissues [36]. Maciel et al. (2018) presented nineteen different fatty acids in *H. portulacoides* leaves characterized by a high percentage of polyunsaturated fatty acids (PUFA) (approximately 60%) and omega-3 (n-3) (approximately 45%). The percentage of saturated fatty acids (SFA) was around 27%, and monounsaturated fatty acids (MUFAs) represented approximately 12%. In addition, it was referred that the first and second most abundant SFA were C18:3 (n-3) (approximately 43.5%) and C16:0 (19.2%), respectively. The MUFA fatty acid present in the highest amount was C18:1 (n-9)—oleic acid (approximately 10%). The ratio of n-6/n-3 fatty acids (0.32) and the presence of phospholipids and glycolipids of high biological value [13] increase the nutritional value of sea purslane, enabling it to be used as gourmet food with potential health benefits.

Leaves' protein content, 9.47 g/100 g of dry matter, was similar to values reported for halophyte plants such as Sarcocornia perennis subsp. alpini (8.10 g/100 g) and *Sarcocornia perennis* subsp. perennis (6.90 g/100 g) [4]. The recommended intake for adults is 0.8 g of protein per kilogram of body weight [36].

The total carbohydrates of leaves were 4.45 g/100 g (fresh weight) or 20.26 g/100 g (dry matter). The carbohydrates in plants are the main sources of energy and constitute carbon skeletons for organic compounds and storage components [37,38]. In addition, they help to maintain glycemic homeostasis and gastrointestinal integrity. The lowest amount of carbohydrates that humans should consume per day is 130 g [36,39].

H. portulacoides presented higher inorganic matter (ash), lipids, and protein contents than, for example, *Salicornia* spp., which is a halophyte plant suitable for human consumption and considered as a promising functional food [40]. The results highlight that, as other commercially available halophytes, *H. portulacoides* has the potential to be consumed fresh, processed, or used in novel food products with health benefits.

Figure 3 illustrates the four sugars (maltose, glucose, fructose, and sucrose) identified in these plant leaves. The major sugar was maltose with 3.01 $\pm$ 0.08 g/100 g, followed by sucrose with 0.49 $\pm$ 0.02 g/100 g, glucose with 0.30 $\pm$ 0.02 g/100 g, and fructose with 0.21 $\pm$ 0.01 g/100 g, as expressed in raw matter. Among the main soluble sugars, maltose was predominant in leaves of sea purslane, and glucose and fructose were not accumulated significantly in this plant. Total sugars represented 90% of the carbohydrates found in *H. portulacoides* leaves. Custódio et al. [9] reported a content of 0.3 g/100 g (raw matter) of total sugars in sea purslane leaves. According to Briens et al. [34], the amount of carbohydrates in *H. portucaloides* leaves (dry matter) was 127 µmol/g, corresponding to 50 µmol/g of sucrose, 41 µmol/g of fructose, 23 µmol/g of glucose, and 13 µmol/g of other carbohydrates. Based on the capacity to accumulate carbohydrates and (or) nitrogenous solutes, *H. portulacoides* is a species that produces more nitrogenous solutes than soluble carbohydrates under saline stress.

Based on the ratio between carbohydrates (20.26%) and ash (27.70%), which are both expressed in dry matter, *H. portulacoides* could be considered a plant with a high level of inorganic ions and a low content of sugars. Other halophyte plants with this behavior are *Atriplex, Aster, Salicornia,* and *Suaeda* [34]. Moreover, 100 g of fresh leaves provide 48.03 $\pm$ 0.06 kcal of energy, which is higher than the value of 18.5 kcal described by Custodio et al. [9].

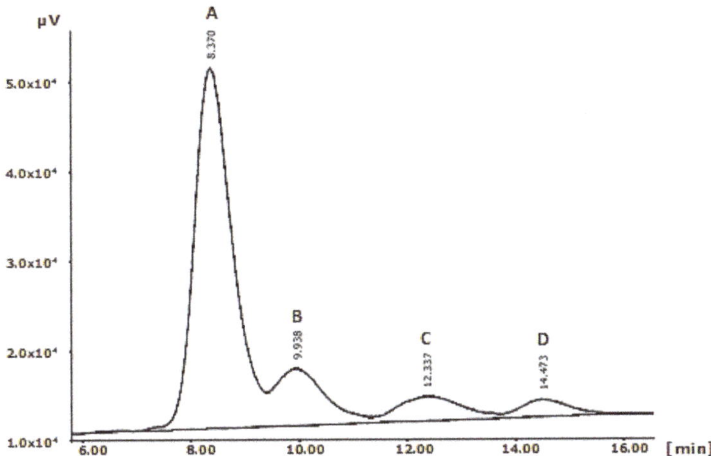

Figure 3. Chromatogram of *Halimione portulacoides* sugars. A: maltose, B: sucrose, C: glucose, and D: fructose.

3.2. Color

Since the sea purslane leaves have a slightly different color on the top and bottom, the color coordinates were evaluated on both sides. The values for the L^*, a^*, and b^* coordinates on the top fresh leaves were 49.74, −7.20, and 6.48, respectively. The values for the bottom of the fresh leaves were similar: 49.98, −7.36, and 7.32, respectively for L^*, a^*, and b^*. The fresh leaves showed a silvery–gray color.

The color parameters (L^*, a^*, and b^*) for fresh leaves and dried powder of *H. portulacoides* are presented in Figure 4. Comparing the color parameters of the dried samples with those obtained for fresh leaves, it was possible to conclude that drying induced a rise in L^* and b^* color parameters, indicating an increase in the lightness and yellowness of the dehydrated plant. These differences were statistically significant for the color coordinates L^* and b^* ($p < 0.0005$ in both cases) but not significant for a^* ($p = 0.832$).

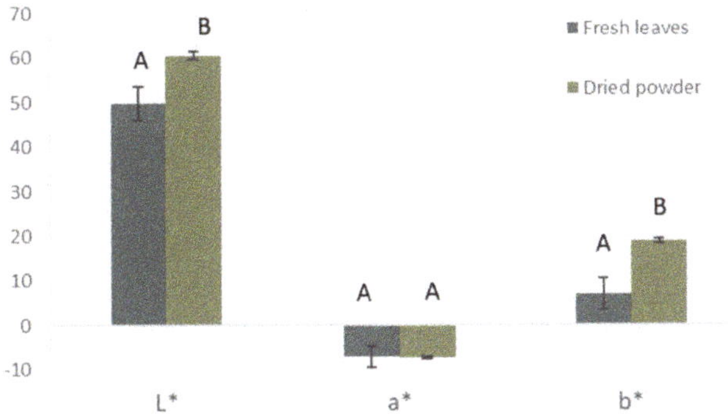

Figure 4. Color parameters of fresh leaves and dried powder of *H. portulacoides*. Bars with the same letter are not statistically different for each of the color coordinates. The color used for each series in the graph is the estimated real color of sea purslane in the fresh and dried states.

The total color difference ΔE, which is a combination of the L^*, a^*, and b^* values, is a colorimetric parameter extensively used to characterize the variation of color in food during processing. The color difference had a value of 16.1 for the leaves dried at 65 °C, being mainly attributed to differences in lightness and yellowness parameters.

3.3. Mineral and Heavy Metal Composition

Due to the saline environment, halophytes, in general, have higher contents in minerals than other edible plants [41]. The mineral content of *H. portulacoides* leaves, obtained by atomic absorption spectrometry, is shown in Table 2.

Table 2. Mineral composition (mg/100 g raw matter) and the mineral intake (%) provided by 100 g of fresh leaves of *H. portulacoides*.

Composition	Raw Matter mg/100 g	Intake Provided by 100 Fresh Leaves (%)
Sodium, Na	1799.38 ± 13.98	78.0
Potassium, K	314.97 ± 0.89	9.5
Calcium, Ca	168.02 ± 0.63	24.0
Magnesium, Mg	67.24 ± 0.29	28.0
Phosphorus, P	40.41 ± 1.58	7.3
Iron, Fe	2.18 ± 0.15	7.6
Manganese, Mn	1.51 ± 0.04	73.7
Zinc, Zn	0.64 ± 0.05	7.8
Copper, Cu	0.21 ± 0.02	16.8
Iodine, I	0.011 ± 0.004	7.3

The intake of minerals provided by 100 g of fresh leaves of *H. portulacoides* was estimated from the average RNIs (Recommended Nutrient Intakes) for adult females and males in the European Union (when applicable) (adapted from [42–45]).

The most abundant minerals were sodium, potassium, calcium, magnesium, and phosphorus, followed by iron, manganese, zinc, copper, and iodine with minor concentration. Phosphorus, calcium, magnesium, sodium, potassium, and iron are essential minerals for human health [37]. Zinc, copper, manganese, chromium, and nickel are necessary in residual concentrations in the human diet [37].

In general, a distinctive property of the halophyte plant is its exceptionally high sodium content. In fact, the amount of sodium in fresh leaves of *H. portucaloides* is around 1.8% (raw matter). The sodium consumption per day should not exceed 2300 mg [46] and, consequently, the intake of 100 g of fresh *H. portulacoides* corresponds to 78% of the daily value recommended for sodium. Sodium is considered an essential nutrient, but its excessive consumption is associated with several pathologies such as hypertension and cardiovascular disease [4]. Hence, the sodium concentration in leaves is acceptable for human intake, but special care must be taken to not exceed the daily dose recommended by the FDA [46].

The content of sodium (8.19 g/100 g, dry matter) was similar to the value 7.82 g/100 g (dry matter) found by Custódio et al. [9] for leaves of *H. portulacoides* and *Salicornia* species (Table 3), which are succulent shoots highly appreciated in gourmet cuisine due to their salty taste. The level of sodium accumulation in plant tissues depends on the availability of elemental nutrients concentrations, namely nitrogen, in saline environments [22,47].

Potassium and calcium are other minerals in high concentration in this plant, respectively, 1433.64 and 764.77 mg/100 g (dry matter). In dry matter, the potassium concentration in leaves of *H. portulacoides* was similar to that in *Salicornia bigelovii* but much higher than *Sarcocornia perennis alpini* or *Salicornia ramosissima* [48]. Leaves of *H. portulacoides* presented much higher calcium contents than *Sarcocornia* and *Salicornia* species (Table 3). Amongst the halophyte plants, the *H. portulacoides* plant is a good source of calcium and potassium.

Copper, zinc, and iron are essential micronutrients necessary in chloroplast reactions, enzyme systems, protein synthesis, and hormone growth [49]. Iron is an essential mineral and cofactor in the synthesis of neurotransmitters, as well as an important constituent of proteins involved in oxygen transport and metabolism [19]. Concentrations of iron (9.92 mg/100 g of dried matter) in leaves of sea purslane are similar to the halophyte *Salicornia bigelovii* Torr [48].

Zinc concentration in *H. portulacoides* leaves (2.93 mg/100 g in dry matter) is higher than the value reported by Reboredo et al. (1.94 mg/100 g in dry matter) [50].

Iodine is a vital nutrient for human health, since it regulates thyroid function. Low iodine intake is responsible for thyroid disorders [51]. The iodine content (0.05 mg/100 g of dry matter) presented in sea purslane leaves was higher than in cereals and grains (ranging from 0.0016 to 0.039 g/100 g), meat (0.0034–0.034 g/100 g), dairy products (0.047–0.069 g/100 g), fresh fruit (0.00018 g/100 g), fresh vegetables (0.0036 g/100 g), leafy vegetables (salad) (0.0236 g/100 g), mushrooms (0.021 g/100 g), or nuts (0.0218 g/100 g), which are all expressed in dry matter [52].

Although foods of marine environment such as marine fish and seaweed are the major suppliers of iodine [53], *H. portucloides* represents an important source of iodine. The oral intake of iodine recognized for adequate nutrition in human adolescents and adults is 150 mg per day [52]. The ingestion of 100 g of fresh leaves corresponds to 7.3% of the daily values recommended for iodine [43].

In general, *H. portulacoides* leaves are a good source of minerals such as Ca, Mg, Mn, and Cu. It should be noted that 100 g of fresh leaves provide values of 24% and 28%, respectively, for calcium and magnesium, 74% for manganese, and 17% for copper.

Table 3. Mineral composition (dry matter) of *Halimione portulacoides* and other halophyte plants.

Composition	*H. portulacoides* Present Study	*Sarcocornia perennis alpini* [4]	*Salicornia ramosissima* [4]	*Salicornia bigelovii* Torr [43]
Na (mg/100 g)	8190.18 ± 35.10	6430 ± 90	8990 ± 50	8618 ± 613
K (mg/100 g)	1433.64 ± 4.07	1030 ± 10	892 ± 23	1520 ± 69
Ca (mg/100 g)	764.77 ± 2.86	263 ± 1	486 ± 5	535 ± 17
Mg (mg/100 g)	306.06 ± 1.32	703 ± 4	943 ± 8	1019 ± 52
P (mg/100 g)	183.93 ± 2.68	-	-	155 ± 9
Fe (mg/100 g)	9.92 ± 0.67	128 ± 5	153 ± 2	8.64 ± 0
Mn (mg/100 g)	6.87 ± 0.16	6.52 ± 0.03	20.4 ± 0. 4	-
Zn (mg/100 g)	2.93 ± 0.22	2.52 ± 0.01	6.87 ± 0.01	3.5 ± 0.12
Cu (mg/100 g)	0.94 ± 0.08	-	-	0.79 ± 0.12
I (mg/100 g)	0.05 ± 0.02	-	-	-
Cd (μg/100 g)	89.02 ± 0.47	19 ± 0.00	nd	8.63 ± 0.00
Pb (μg/100 g)	17.91 ± 0.55	131 ± 2	145 ± 2	17.27 ± 8.64
Hg (μg/100 g)	6.98 ± 0.10	-	-	-
Cr (μg/100 g)	-	492 ± 11	524 ± 5	-

nd: not detected; -: not presented. Adapted from [4,48,54].

Cadmium, lead, and mercury, considered human carcinogens [55], were detected in *H. portulacoides*. According to the Commission Regulation (EC) n° 466/2001 of 8 March 2001, the maximum content of cadmium in leafy vegetables, fresh herbs, celery, and all cultured mushrooms is 0.2 mg/kg (20 μg/100 g of fresh weight), and the maximum lead content in brassica, leafy vegetables, fresh herbs, and all mushrooms is 0.3 mg/kg (30 μg/100 g of fresh weight) [56]. Thus, the concentrations of toxic metals were below the legislated values and there was a much lower concentration of lead than in *Sarcocornia perennis alpini* and *Salicornia ramosissima* from Castro Marim (Algarve, south of Portugal) (Table 3) [4]. The different amounts of these contaminants are dependent on the level of contamination of sediments in salt marshes, since halophytes have the ability to accumulate metals such as Zn, Cr, Pb, Ni, and Cd, among others. However, metal concentrations found in the above-ground tissues of halophyte plants such as *Salicornia fruticosa* and *Salicornia maritima* from Tagus and Guadiana estuaries (Portugal) were up to four orders of magnitude lower than in below-ground parts, confirming metal retention in their roots and a residual upward translocation [54,57,58].

Mercury is a dangerous pollutant due to its high toxicity, making it a major threat to coastal ecosystems [24]. As other metals, the mobility of mercury is greater in the roots, and only a small part is translocated to the above-ground parts of the plant [10]. Hence, the mercury concentration of above-ground tissues can be 174 to 545-fold times lower than that of the roots [10].

In fact, the content of mercury in *H. portulacoides* leaves (1.53 μg/100 g of raw matter) is around 33 times lower than the limit values defined for fishery products such as fish (50 μg/100 g) [56]. Moreover, the content of mercury in *H. portulacoides* leaves (under study) collected in Figueira da Foz (Portugal) is lower than the values reported for the same plant tissues (6–14 μg/100 g) collected in Laranjo Bay salt marsh (Ria de Aveiro, Portugal), during April 2003 and April 2004 [10]. The higher concentration of mercury in sea purslane leaves of Ria Formosa can be attributed to its high level of mercury contamination, being one the most mercury-contaminated systems in Europe [10].

The content of heavy metals (cadmium, lead, and mercury) found in *H. portulacoides* leaves was much lower than the values that could be considered dangerous to human health, making the leaves a safe product.

3.4. Total Phenolic and Flavonoid Content

Halophytes are rich in highly bioactive phytochemicals [15], particularly in phenolic compounds, which are abundantly present in the human diet, and to which are attributed important antioxidant properties, their intake being associated with a decreased risk of development of oxidative stress-related diseases [14]. Flavonoids are one of the most important families of phenolic compounds, thus contributing to their health benefits, namely as anticancer and chemopreventive agents [17]. Therefore, the measurement of the phytochemical composition of an extract in terms of total phenolic and flavonoid content can be used to estimate its antioxidant potential. Regarding the TPC of the *H. portulacoides* extract, the results presented in Table 4 support the advantage of using ethanol as extraction solvent, since a higher content of phenolic compounds was obtained when compared with reported extractions using hexane, chloroform, methanol [16], ethyl acetate, and water [59]. Moreover, the TPC for the *H. portulacoides* extract is in agreement with previous results obtained for this plant harvested in the same region of Portugal [18]. The TFC is higher than those reported for other halophytes, namely *Ipomoea pes-caprae* [60], which can be attributed to the rich content of sulfated flavonoids, particularly derivatives of isorhamnetin-sulfate [18], which proved to possess higher antioxidant activity than α-tocopherol [61].

Table 4. Chemical composition and antioxidant and enzymatic activities of the *H. portulacoides* leaves extract.

Assay	*H. portulacoides* Extract
Chemical composition	
TPC (mg GAE/g extract)	16.10 ± 0.20
TFC (mg QCE/g extract)	26.60 ± 0.80
Antioxidant activity	
DPPH (IC_{50} mg/mL)	3.70 ± 0.40
ABTS (IC_{50} mg/mL)	>5
β-carotene/linoleic acid (IC_{50} mg/mL)	0.15 ± 0.03
Lipid peroxidation (IC_{50} mg/mL)	>5
Metal chelating ability (IC_{50} mg/mL)	2.30 ± 0.50
FRAP (mg TE/g extract)	19.90 ± 1.90
CUPRAC (mg TE/g extract)	44.00 ± 2.30
Enzymatic activity	
AChE inhibition (IC_{50} mg/mL)	>5

GAE, gallic acid equivalents; QCE, quercetin equivalents; TE, Trolox equivalents.

3.5. Antioxidant and Enzymatic Activities

The ethanolic extract of *H. portulacoides* exhibits a higher DPPH activity (IC50 = 3.70 ± 0.40 mg/mL) when compared to data obtained for extracts using other solvents (IC50 > 10 mg/mL) and analogous ABTS scavenging capacity as reported hexane, methanol, and water extracts [16]. When compared to the inhibitory potential of other halophytes, *H. portulacoides* (IC50 = 0.15 mg/mL) shows higher activity than *Suaeda pruinose, Suaeda maritima*, and *Suaeda mollis* (IC50 values of 0.54, 1.42, and 0.54 mg/mL, respectively) [62]. Nevertheless, it is still considerably less active than the antioxidant BHT (IC50 = 0.005 mg/mL) [31]. Regarding the TBARS assay, the *H. portulacoides* extract did not present the capacity to inhibit lipid peroxidation in the range of concentrations tested.

Antioxidants are able to chelate and reduce prooxidant metal ions responsible for the production of ROS, e.g., the ferrous ions that produce free radicals via the Fenton reaction [63]. The *H. portulacoides* extract is more effective in reducing copper (CUPRAC—44.00 mg TE/g extract) than iron (FRAP—19.90 mg TE/g extract). This was already reported for water, methanol, and ethyl acetate extracts of *H. portulacoides*, with FRAP values ranging from 31.59 to 50.06 mg TE/g extract and CUPRAC from 51.83 to 71.21 mg TE/g extract [60].

The acetylcholinesterase inhibition was tested for the *H. portulacoides* extract, though no significant activity was observed in the range of concentrations tested. Previous studies had already reported a very modest activity for water, methanol, and ethyl acetate extracts of *H. portulacoides* against AChE [60]. Nevertheless, these results contrast with those reported for ethanolic extracts of the leaves of other halophytes, such as *Armeria pungens*, with a high activity against AChE (IC50 = 90.3 µg/mL) [64].

The results for the β-carotene bleaching assay were obtained after 2 h of reaction. Values represent the mean ± standard deviation of three independent experiments.

3.6. Sensory Evaluation

The assessors, 30% men and 60% women, were aged between 18 and 63 years old, being on average of 31.25 ± 14.12 years.

The test is presented with two coded samples to evaluate the product attributes in the hedonic scale. Figure 5 shows the control and enriched pasta with *H. portulacoides*.

Figure 5. Control (non-enriched) and enriched pasta with *H. portulacoides*.

Figure 6 shows the results of the sensory evaluation of four pasta attributes: color, flavor, texture, and overall acceptance; all were expressed in the scale from 1 (dislike extremely) to 9 (like extremely). Among the sensorial parameters, texture showed the best results (7.2 and 7.7, respectively, for enriched pasta and control pasta). The overall acceptance also had results very close to 7.6 for the pasta without *H. portulacoides* and close to 6.8 for the pasta with *H. portulacoides*. The attributes of color and flavor had, respectively, scores of 8 and 6.5 for the enriched pasta and around 7.7 and 7.5 for the control pasta. Although the lower scores were obtained for the enriched pasta, its acceptance was reasonable. Statistical analysis showed that the differences between the mean scores for the control and enriched pastas were significant for attributes flavor ($p = 0.004$) and overall acceptance ($p = 0.003$), while for the other attributes (color and texture), there were no significant differences ($p > 0.05$). Figure 6 shows the results of the sensory evaluation of four pasta attributes: color, flavor, texture, and overall acceptance, which are all expressed in the scale from 1 (dislike extremely) to 9 (like extremely). Among the sensorial parameters, texture showed the best results (7.2 and 7.7, respectively, for enriched pasta and control pasta). The overall acceptance also had results very close to 7.6 for the pasta without *H. portulacoides* and close to 6.8 for the pasta with *H. portulacoides*. The attributes of color and flavor had, respectively, scores of 8 and 6.5 for the enriched pasta and around 7.7 and 7.5 for the control pasta. Although the lower scores were obtained for the enriched pasta, its acceptance was reasonable. Statistical analysis showed that the differences between the mean scores for the control and enriched pastas were significant for attributes flavor ($p = 0.004$) and overall acceptance ($p = 0.003$), while for the other attributes (color and texture), there were no significant differences ($p > 0.05$).

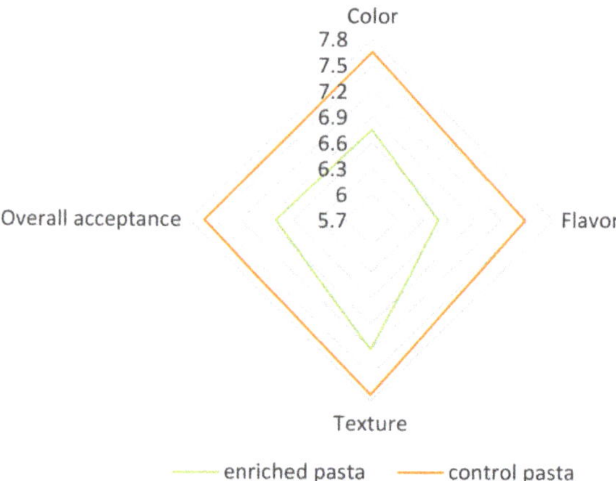

Figure 6. Sensory analysis for control (non-enriched) and enriched pasta with *H. portulacoides*.

When asked about the purchasing preference, 37.5% of the panel members preferred pasta enriched with *H. portulacoides* and 62.5% preferred control pasta.

Figure 7 shows the butter enriched with *H. portulacoides*.

Figure 7. Butter enriched with *H. portulacoides*.

Figure 8 presents the results of the sensory evaluation of six parameters: appearance, color, flavor, scent, texture, and overall acceptance to enriched and control butter. It should be noted that the enriched butter had the higher scores in all organoleptic characteristics. Texture had the higher scores for both butters, but the enriched butter presented the highest value. The overall acceptance and appearance of fortified butter also scored with values of 8.15 in both cases, while the values were, respectively, 7.48 and 7.60 for the butter without *H. portulacoides*. Overall, the tasters preferred the butter with *H. portulacoides* to the control butter. Statistical analysis revealed that significant differences were found between the control and enriched butter samples only for the mean scores for appearance ($p = 0.003$), while for all other attributes, the differences were not significant ($p > 0.05$).

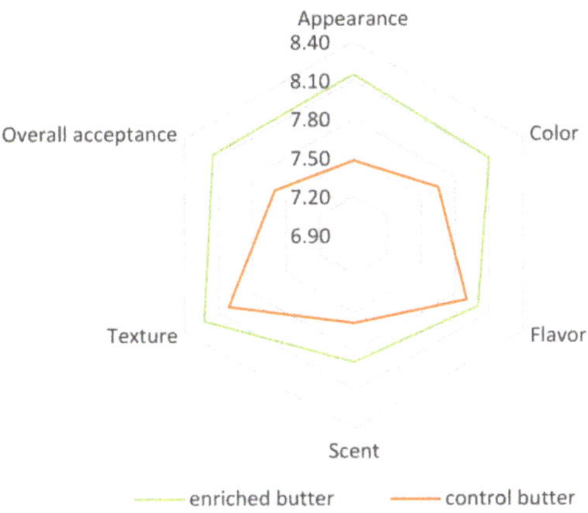

Figure 8. Sensory analysis for control (non-enriched) and enriched butter with *H. portulacoides*.

Moreover, when asked about the purchasing preference, only 7.5% of the panel members referred that they would not buy enriched butter, resulting in a good market acceptance if this product would be available in the market. Although both products had been enriched with dried powder leaves of *H. portucaloides*, their acceptability is different, depending strongly on the type of product.

4. Conclusions

The present work determined the nutritional and mineral profile of *H. portulacoides* leaves collected in the salterns of Figueira da Foz (Portugal) and their biological activity. The exploitation of powder dried leaves as a salt substitute and enhancer of sensory characteristics of foods (pasta and butter) was also assessed.

The halophyte sea purslane plant may be considered a good source of dietary fiber, protein, and lipids, presenting higher concentration of these nutrients than some Salicornia species that are halophyte plants suitable for human consumption and considered as promising functional foods. Moreover, high concentration of minerals such as sodium, potassium, calcium, magnesium, and phosphorous were found in sea purslane leaves. Although they have low concentration of manganese, the ingestion of 100 g of fresh leaves provides 74% of the daily dose recommended for adults.

The *H. portulacoides* leaves extract with the green solvent (ethanol) yielded more phenolic compounds than extractions using other organic solvents and higher content in flavonoids when compared to other halophytes such as Ipomoea pes-caprae. Moreover, an increased antioxidant potential measured by the DPPH and ABTS radical scavenging assays was found compared to the use of other solvents and when compared to other halophytes such as *Suaeda* species. In addition, the *H. portulacoides* leaves extract was more effective in reducing copper than iron, as assessed by the CUPRAC and FRAP assays.

In terms of its use as a novel ingredient, butter and pasta enriched with powder dried leaves of *H. portulacoides* revealed the plant's potential to be used as a salt substitute that enhances the sensory characteristics of products, providing health benefits to the consumers.

Author Contributions: Conceptualization, A.M.d.S. and M.J.B.; methodology, A.P., S.A., S.R. and J.M.; software, J.M. and R.G.; validation, A.M.d.S., M.J.B. and R.G.; formal analysis, A.M.d.S., M.J.B. and R.G.; investigation, A.P., S.R.; resources, A.P., S.A., S.R., M.J.B., J.M., R.G. and A.M.d.S.; data curation, A.M.d.S., M.J.B. and R.G.; writing—A.P., M.J.B. and A.M.d.S.; original draft preparation, A.P., M.J.B. and A.M.d.S.; writing—review and editing, A.M.d.S., M.J.B. and R.G.; visualization, A.M.d.S., M.J.B. and R.G.; supervision, A.M.d.S. and M.J.B.; project administration, A.M.d.S. and M.J.B.; funding acquisition, A.M.d.S. and M.J.B. All authors have read and agreed to the published version of the manuscript."

Funding: This research was funded by the Portuguese Foundation for Science and Technology (UID/MULTI/00070/2019) and from the European Regional Development Fund, through Portugal 2020-POCI-01-0145-FEDER-029305, IDEAS4life—Novos IngreDiEntes Alimentares de Plantas MarítimaS; and Centro 2020-Centro-01-0145-FEDER-000007, Project ReNATURE—Valorization of the Natural Endogenous Resources of the Centro Region.

Institutional Review Board Statement: Not applicable.

Informed Consent Statement: Not applicable.

Acknowledgments: The authors would like to thank Gilda Saraiva for the halophyte supply. A very special thanks is owed to David Gomes, Maria Adélia Vaz, and Maria Lurdes Pires for assisting in the manufacture of butter and in the incorparation of dried powder of *H. portulacoides* leaves.

Conflicts of Interest: The authors declare no conflict of interest.

References

1. Martins-Noguerol, R.; Cambrollé, J.; Mancilla-Leytón, J.; Puerto-Marchena, A.; Muñoz-Vallés, S.; Millán-Linares, M.; Millán, F.; Martínez-Force, E.; Figueroa, M.; Pedroche, J.; et al. Influence of soil salinity on the protein and fatty acid composition of the edible halophyte *Halimione portulacoides*. *Food Chem.* **2021**, *352*, 129370. [CrossRef]
2. Barroca, M.J.; Guiné, R.P.F.; Amado, A.M.; Ressurreição, S.; Da Silva, A.M.; Marques, M.P.M.; De Carvalho, L.A.E.B. The drying process of *Sarcocornia perennis*: Impact on nutritional and physico-chemical properties. *J. Food Sci. Technol.* **2020**, *57*, 4443–4458. [CrossRef]
3. Custódio, M.; Maciel, E.; Domingues, M.R.; Lillebø, A.I.; Calado, R. Nutrient availability affects the polar lipidome of *Halimione portulacoides* leaves cultured in hydroponics. *Sci. Rep.* **2020**, *10*, 1–13. [CrossRef] [PubMed]
4. Barreira, L.; Resek, E.; Rodrigues, M.J.; Rocha, M.I.; Pereira, H.; Bandarra, N.; da Silva, M.M.; Varela, J.; Custódio, L. Halophytes: Gourmet food with nutritional health benefits? *J. Food Compos. Anal.* **2017**, *59*, 35–42. [CrossRef]
5. Cambrollé, J.; Mancilla-Leytón, J.M.; Muñoz-Vallés, S.; Cambrón-Sena, A.; Figueroa, M.E. Advances in the use of *Halimione portulacoides* stem cuttings for phytoremediation of Zn-polluted soils. *Estuarine. Coast. Shelf Sci.* **2016**, *175*, 10–14. [CrossRef]
6. Andrades-Moreno, L.; Cambrollé, J.; Figueroa, M.; Mateos-Naranjo, E. Growth and survival of *Halimione portulacoides* stem cuttings in heavy metal contaminated soils. *Mar. Pollut. Bull.* **2013**, *75*, 28–32. [CrossRef]
7. Anjum, N.A.; Israr, M.; Duarte, A.C.; Pereira, M.E.; Ahmad, I. *Halimione portulacoides* (L.) physiological/biochemical characterization for its adaptive responses to environmental mercury exposure. *Environ. Res.* **2014**, *131*, 39–49. [CrossRef]
8. Brito, P.; Ferreira, R.A.; Martins-Dias, S.; Azevedo, O.M.; Caetano, M.; Caçador, I. Cerium uptake, translocation and toxicity in the salt marsh halophyte *Halimione portulacoides*(L.), Aellen. *Chemosphere* **2020**, *266*, 128973. [CrossRef]
9. Custódio, M.; Villasante, S.; Calado, R.; Lillebø, A.I. Testing the hydroponic performance of the edible halophyte *Halimione portulacoides*, a potential extractive species for coastal integrated multi-trophic aquaculture. *Sci. Total Environ.* **2020**, *766*, 144378. [CrossRef]
10. Válega, M.; Lillebø, A.; Caçador, I.; Pereira, M.; Duarte, A.; Pardal, M. Mercury mobility in a salt marsh colosined by *Halimione portulacoides*. *Chemosphere* **2008**, *72*, 1607–1613. [CrossRef]
11. Zanella, L.; Vianello, F. Functional food from endangered ecosystems: *Atriplex portulacoides* as a case study. *Foods* **2020**, *9*, 1533. [CrossRef] [PubMed]
12. Marins, V.C.M. Comunidade bacteriana endofítica cultivável de *Halimione portulacoides*. Master's Thesis, Universidade de Aveiro, Aveiro, Portugal, 2011.
13. Maciel, E.; Lillebø, A.; Domingues, P.; da Costa, E.; Calado, R.; Domingues, M.R.M. Polar lipidome profiling of *Salicornia ramosissima* and *Halimione portulacoides* and the relevance of lipidomics for the valorization of halo-phytes. *Phytochemistry* **2018**, *153*, 94–101. [CrossRef] [PubMed]

14. Abbasi, A.M.; Shah, M.H.; Khan, M.A. *Wild Edible Vegetables of Lesser Himalayas: Ethnobotanical and Nutraceutical Aspects*; Springer International Publishing: Berlin, Germany, 2015; Volume 1, pp. 1–360.
15. Faustino, M.A.; Pinto, D.C.G.A. Halophytic grasses, a new source of nutraceuticals? A review on their secondary metabolites and biological activities. *Int. J. Mol. Sci.* **2019**, *20*, 1067. [CrossRef]
16. Odrigues, M.J.; Gangadhar, K.N.; Vizetto-Duarte, C.; Wubshet, S.G.; Nyberg, N.T.; Barreira, L.; Varela, J.; Custódio, L. Maritime halophyte species from southern Portugal as sources of bioactive molecules. *Mar. Drugs* **2014**, *12*, 2228–2244. [CrossRef] [PubMed]
17. Valavanidis, A.; Vlachogianni, T. *Plant Polyphenols: Recent Advances in Epidemiological Research and Other Studies on Cancer Prevention*, 1st ed.; Elsevier BV: Amsterdam, The Netherlands, 2013; Volume 39, pp. 269–295.
18. Vilela, C.; Santos, S.; Coelho, D.; Silva, A.; Freire, C.; Neto, C.; Silvestre, A. Screening of lipophilic and phenolic extractives from different morphological parts of *Halimione portulacoides*. *Ind. Crop. Prod.* **2014**, *52*, 373–379. [CrossRef]
19. Aberoumand, A.; Deokule, S.S. Determination of elements profile of some wild edible plants. *Food Anal. Methods* **2008**, *2*, 116–119. [CrossRef]
20. Everest, A.; Ozturk, E. Focusing on the ethnobotanical uses of plants in Mersin and Adana provinces (Turkey). *J. Ethnobiol. Ethnomed.* **2005**, *1*. [CrossRef] [PubMed]
21. Joint, F.; World Health Organization. *Evaluation of Certain Food Additives and Contaminants: Seventy-Third [73rd] Report of the Joint FAO/WHO Expert Committee on Food Additives*; World Health Organization: Geneva, Switzerland, 2011.
22. Kant, S.; Kant, P.; Lips, H.; Barak, S. Partial substitution of NO3− by NH4+ fertilization increases ammonium assimilating enzyme activities and reduces the deleterious effects of salinity on the growth of barley. *J. Plant Physiol.* **2007**, *164*, 303–311. [CrossRef]
23. Castro, R.; Pereira, S.; Lima, A.; Corticeiro, S.; Valega, M.; Pereira, E.; Duarte, A.; Figueira, E. Accumulation, distribution and cellular partitioning of mercury in several halophytes of a contaminated salt marsh. *Chemosphere* **2009**, *76*, 1348–1355. [CrossRef]
24. Cabrita, M.T.; Duarte, B.; Cesário, R.; Mendes, R.; Hintelmann, H.; Eckey, K.; Dimock, B.; Caçador, I.; Canário, J. Mercury mobility and effects in the salt-marsh plant *Halimione portulacoides*: Uptake, transport, and toxicity and tolerance mechanisms. *Sci. Total Environ.* **2018**, *650*, 111–120. [CrossRef]
25. Anoè, N.; Calzavara, D.; Salviato, L.; Zanaboni, A. *Flora E Vegetazione delle Barene. Gli Ambienti Salmastri della Laguna di Venezia*; Soc Veneziana di Scienze Naturali: Venezia, Italy, 2001; Volume 26, pp. 9–84.
26. European Union. *Regulation (eu) no 1169/2011 of the European Parliament and of the Council*; European Union: Bruxelles, Belgium, 2011; No 1169/2011.
27. ISO:6869. *Animal Feeding Stuffs–Determination of the Contents of Calcium, Copper, Iron, Magnesium, Manganese, Potassium, Sodium and Zinc–Method Using Atomic Absorption Spectrometry*; ISO: Geneva, Switzerland, 2000.
28. European Union. *Foodstuffs—Determination of Trace Elements—Determination of Lead, Cadmium, Zinc, Copper, Iron and Chromium by Atomic Absorption Spectrometry (AAS) after Dry Ashing*; European Union: Bruxelles, Belgium, 2003; Volume EN 14082:2003.
29. International Organization for Standardization. *Animal Feeding Stuffs—Determination of Phosphorus Content—Spectrometric Method*; ISO: Geneva, Switzerland, 1998; Volume ISO 6491.
30. de Guiné, R.P.F.; Barroca, M.J. Mass Transfer Properties for the Drying of Pears. In *Transactions on Engineering Technologies*; Springer: Berlin, Germany, 2014; pp. 271–280.
31. Marques, J.M.D.; Amado, A.M.; Lysenko, V.; Osório, N.; Batista de Carvalho, L.A.E.; Marques, M.P.M.; Barroca, M.J.; Moreira da Silva, A. Novel insights into *Corema album* berries: Vibrational profile and biological activity. *Plants* **2021**, *10*, 1761, under revision. [CrossRef]
32. ISO:7304-1. *Durum Wheat Semolina and Alimentary Pasta—Estimation of Cooking Quality of Alimentary Pasta by Sensory Analysis—Part 1: Reference Method*; ISO: Geneva, Switzerland, 2016.
33. Bertin, R.L.; Gonzaga, L.V.; Borges, G.d.S.C.; Azevedo, M.S.; Maltez, H.F.; Heller, M.; Micke, G.A.; Tavares, L.B.B.; Fett, R. Nutrient composition and, identification/quantification of major phenolic compounds in *Sarcocornia ambigua* (Amaranthaceae) using HPLC–ESI-MS/MS. *Food Res. Int.* **2014**, *55*, 404–411. [CrossRef]
34. Briens, M.; Larher, F. Osmoregulation in halophytic higher plants: A comparative study of soluble carbohydrates, polyols, betaines and free proline. *Plant Cell Environ.* **1982**, *5*, 287–292.
35. Baeza-Jiménez, R.; López-Martínez, L.X.; García-Varela, R.; García, H.S. *Lipids in Fruits and Vegetables: Chemistry and Biological Activities, Fruit and Vegetable Phytochemicals: Chemistry and Human Health*; Yahia, E.M., Ed.; Wiley: Hoboken, NJ, USA, 2017; pp. 423–449.
36. Lupton, J.R.; Brooks, J.; Butte, N.; Caballero, B.; Flatt, J.; Fried, S. *Dietary Reference Intakes for Energy, Carbohydrate, Fiber, Fat, Fatty Acids, Cholesterol, Protein, and Amino Acids*; National Academy Press: Washington, DC, USA, 2002; Volume 5, pp. 589–768.
37. Pedreiro, S.; da Ressurreição, S.; Lopes, M.; Cruz, M.T.; Batista, T.; Figueirinha, A.; Ramos, F. *Crepis vesicaria* L. subsp. taraxacifolia leaves: Nutritional profile, phenolic composition and biological properties. *Int. J. Environ. Res. Public Health* **2021**, *18*, 151. [CrossRef]
38. Zhu, J.; Qi, J.; Fang, Y.; Xiao, X.; Li, J.; Lan, J.; Tang, C. Characterization of sugar contents and sucrose metabolizing enzymes in developing leaves of Hevea brasiliensis. *Front. Plant Sci.* **2018**, *9*, 58. [CrossRef]
39. Trouvelot, S.; Héloir, M.-C.; Poinssot, B.; Gauthier, A.; Paris, F.; Guillier, C.; Combier, M.; Trdá, L.; Daire, X.; Adrian, M. Carbohydrates in plant immunity and plant protection: Roles and potential application as foliar sprays. *Front. Plant Sci.* **2014**, *5*, 592. [CrossRef]

40. Loconsole, D.; Cristiano, G.; De Lucia, B. Glassworts: From wild salt marsh species to sustainable edible crops. *Agriculture* **2019**, *9*, 14. [CrossRef]
41. Borah, S.; Baruah, A.M.; Das, A.K.; Borah, J. Determination of mineral content in commonly consumed leafy vegetables. *Food Anal. Methods* **2008**, *2*, 226–230. [CrossRef]
42. Authority, E.F.S. *Tolerable Upper Intake Levels for Vitamins and Minerals*; EFSA: Parma, Italy, 2006.
43. WHO. *Trace Elements in Human Nutrition and Health*; World Health Organization: Geneva, Switzerland, 1996.
44. WHO. *Vitamin and Mineral Requirements in Human Nutrition*; World Health Organization: Geneva, Switzerland, 2004.
45. Institute of Medicine. *Dietary Reference Intakes for Vitamin A, Vitamin K, Arsenic, Boron, Chromium, Copper, Iodine, Iron, Manganese, Molybdenum, Nickel, Silicon, Vanadium, and Zinc*; National Academy Press: Washington, DC, USA, 2001.
46. FDA. Food Labeling: Revision of the Nutrition and Supplement Facts Labels. Available online: https://s3.amazonaws.com/public-inspection.federalregister.gov/2016-11867.pdf (accessed on 6 June 2021).
47. Hessini, K.; Gandour, M.; Megdich, W.; Soltani, A.; Abdely, C. How does ammonium nutrition influence salt tolerance in *Spartina alterniflora* Loisel? In *Salinity and Water Stress*; Springer: Berlin, Germany, 2009; pp. 91–96.
48. Lu, D.; Zhang, M.; Wang, S.; Cai, J.; Zhou, X.; Zhu, C. Nutritional characterization and changes in quality of *Salicornia bigelovii* Torr. during storage. *LWT* **2010**, *43*, 519–524. [CrossRef]
49. Milić, D.; Luković, J.; Ninkov, J.; Zeremski-Škorić, T.; Zorić, L.; Vasin, J.; Milić, S. Heavy metal content in halophytic plants from inland and maritime saline areas. *Cent. Eur. J. Biol.* **2012**, *7*, 307–317. [CrossRef]
50. Reboredo, F. Zinc compartmentation in *Halimione portulacoides* (L.) Aellen and some effects on leaf ultrastructure. *Environ. Sci. Pollut. Res.* **2012**, *19*, 2644–2657. [CrossRef] [PubMed]
51. Opazo, M.C.; Coronado-Arrázola, I.; Vallejos, O.P.; Moreno-Reyes, R.; Fardella, C.; Mosso, L.; Kalergis, A.M.; Bueno, S.M.; Riedel, C.A. The impact of the micronutrient iodine in health and diseases. *Crit. Rev. Food Sci. Nutr.* **2020**, 1–14. [CrossRef]
52. Haldimann, M.; Alt, A.; Blanc, A.; Blondeau, K. Iodine content of food groups. *J. Food Composit. Anal.* **2005**, *18*, 461–471. [CrossRef]
53. Müssig, K. Iodine-induced toxic effects due to seaweed consumption. In *Comprehensive Handbook of Iodine: Nutritional, Biochemical, Pathological and Therapeutic Aspects*; Preedy, V.R., Burrow, G.N., Watson, R.R., Eds.; Academic Press: Cambridge, MA, USA, 2009; pp. 897–908.
54. Caetano, M.; Vale, C.; Cesário, R.; Fonseca, N. Evidence for preferential depths of metal retention in roots of salt marsh plants. *Sci. Total Environ.* **2008**, *390*, 466–474. [CrossRef] [PubMed]
55. Tchounwou, P.B.; Yedjou, C.G.; Patlolla, A.K.; Sutton, D.J. Heavy metal toxicity and the environment. *Mol. Clin. Environ. Toxicol.* **2012**, *101*, 133–164.
56. EC Commission. *Setting Maximum Levels for Certain Contaminants in Foodstuffs*; European Commission: Brussels, Belgium, 2001; No 466/2001.
57. Redondo-Gómez, S.; Mateos-Naranjo, E.; Figueroa, M.; Davy, A. Salt stimulation of growth and photosynthesis in an extreme halophyte, Arthrocnemum macrostachyum. *Plant Biol.* **2010**, *12*, 79–87. [CrossRef] [PubMed]
58. Silva, M.M.D. *Metals and Butyltins in Sediments of Ria Formosa—The Role of Spartina maritima and Sarcocornia fruticosa*; University of Porto: Porto, Portugal, 2008.
59. Zengin, G.; Aumeeruddy-Elalfi, Z.; Mollica, A.; Yilmaz, M.A.; Mahomoodally, M.F. In vitro and in silico perspectives on biological and phytochemical profile of three halophyte species—A source of innovative phytopharmaceuticals from nature. *Phytomedicine* **2018**, *38*, 35–44. [CrossRef] [PubMed]
60. Banerjee, D.; Hazra, A.K.; Chakraborti, S.; Ray, J.; Mukherjee, A.; Mukherjee, B. Variation of total phenolic con-tent, flavonoid and radical scavenging activity of Ipomoea pes-caprae with respect to harvesting time and lo-cation. *Ind. J. Mar. Sci.* **2013**, *42*, 106–109.
61. Yagi, A.; Uemura, T.; Okamura, N.; Haraguchi, H.; Imoto, T.; Hashimoto, K. Antioxidative sulphated flavonoids in leaves of Polygonum hydropiper. *Phytochemistry* **1994**, *35*, 885–887. [CrossRef]
62. Oueslati, S.; Trabelsi, N.; Boulaaba, M.; Legault, J.; Abdelly, C.; Ksouri, R. Evaluation of antioxidant activities of the edible and medicinal Suaeda species and related phenolic compounds. *Ind. Crop. Prod.* **2012**, *36*, 513–518. [CrossRef]
63. Zengin, G.; Sarikurkcu, C.; Aktumsek, A.; Ceylan, R. Sideritis galatica Bornm: A source of multifunctional agents for the management of oxidative damage, Alzheimer's and diabetes mellitus. *J. Funct. Foods* **2014**, *11*, 538–547. [CrossRef]
64. Rodrigues, M.J.; Pereira, C.A.; Oliveira, M.; Neng, N.R.; Nogueira, J.M.; Zengin, G.; Mahomoodally, M.F.; Custódio, L. Sea rose (Armeria pungens (Link) Hoffmanns. & Link) as a potential source of innovative industrial products for anti-ageing applications. *Ind. Crop. Prod.* **2018**, *121*, 250–257. [CrossRef]

Article

Association between Serum Heat Shock Proteins and Gamma-Delta T Cells—An Outdated Clue or a New Direction in Searching for an Anticancer Strategy? A Short Report

Dorota Pawlik-Gwozdecka [1,*], Justyna Sakowska [2], Maciej Zieliński [2], Magdalena Górska-Ponikowska [3], Francesco Cappello [4,5], Piotr Trzonkowski [2] and Maciej Niedźwiecki [1]

1 Department of Pediatrics, Hematology and Oncology, Medical University of Gdańsk, 80-211 Gdańsk, Poland; maciej.niedzwiecki@gumed.edu.pl
2 Department of Medical Immunology, Medical University of Gdańsk, 80-211 Gdańsk, Poland; justynas@gumed.edu.pl (J.S.); mzielinski@gumed.edu.pl (M.Z.); piotr.trzonkowski@gumed.edu.pl (P.T.)
3 Department of Medical Chemistry, Medical University of Gdańsk, 80-211 Gdańsk, Poland; magdalena.gorska-ponikowska@gumed.edu.pl
4 Euro-Mediterranean Institute of Science and Technology, 90127 Palermo, Italy; francesco.cappello@unipa.it
5 Department of Biomedicine, Neurosciences and Advanced Diagnostics (BiND), University of Palermo, 90127 Palermo, Italy
* Correspondence: dorota.pawlik-gwozdecka@gumed.edu.pl

Citation: Pawlik-Gwozdecka, D.; Sakowska, J.; Zieliński, M.; Górska-Ponikowska, M.; Cappello, F.; Trzonkowski, P.; Niedźwiecki, M. Association between Serum Heat Shock Proteins and Gamma-Delta T Cells—An Outdated Clue or a New Direction in Searching for an Anticancer Strategy? A Short Report. *Appl. Sci.* **2021**, *11*, 7325. https://doi.org/10.3390/app11167325

Academic Editor: Giuseppina Andreotti

Received: 4 July 2021
Accepted: 4 August 2021
Published: 9 August 2021

Publisher's Note: MDPI stays neutral with regard to jurisdictional claims in published maps and institutional affiliations.

Abstract: HSPs demonstrate a strong association with gamma-delta (γδ) T cells. Most of the studies regarding interactions between the parameters were conducted in the 1990s. Despite promising results, the concept of targeting γδ T cells by HSPs seems to be a forgotten direction due to potent non-peptidic phosphoantigens rather than HSPs have been found to be the essential stimulatory components for human γδ cells. Currently, with greater knowledge of lymphocyte diversity, and more accurate diagnostic methods, we decided to study the correlation once again in the neoplastic condition. Twenty-one children with newly diagnosed acute lymphoblastic leukaemia (ALL) were enrolled on the study. Serum HSP90 concentrations were evaluated by an enzyme-linked immunosorbent assay (ELISA), subsets of γδ T cells (CD3+ γδ, CD3+ γδ HLA/DR+, CD4+ γδ and CD8+ γδ) by flow cytometry. We have shown statistically relevant correlations between serum HSP90 and CD3+ HLA/DR+ γδ T cells in paediatric ALL at diagnosis ($R = 0.53$, $p < 0.05$), but not after induction chemotherapy. We also have demonstrated decreased levels of both serum HSP90 and CD3+ HLA/DR+ γδ T cells before treatment, which may indirectly indicate dose-dependent unknown interaction between the parameters. The results of our study may be a good introduction to research on the association between HSPs and CD3+ HLA/DR+ γδ T cells, which could be an interesting direction for the development of anti-cancer strategies, not just for childhood ALL.

Keywords: serum HSP90; gamma-delta T cells; acute lymphoblastic leukaemia

1. Introduction

Heat-shock proteins (HSPs) have recently been extensively studied in the context of anticancer properties, especially their extracellular form. Serum HSPs have been found to elicit antitumour immunity by acting as tumour-specific antigens, and adjuvants that facilitate uptake, processing, and presentation [1,2].

According to many reports, HSPs demonstrate a strong association with gamma-delta (γδ) T cells by different mechanisms including direct recognition of specific epitopes in their free form or as peptide-HSPs complexes [3].

Human γδ T cells represent a small subset of CD3+ T lymphocytes (1–10%), however, these cells have been gaining the interest of scientists' and clinicians' as they demonstrate both innate and adaptive immune properties. Their primary functions include phagocytosis and the presentation of soluble antigens to alpha-beta (αβ) T cells, induction of dendritic cells (DC), maturation and the production of cytokines [4]. The key advantage of γδ T cells is their ability to identify antigens out of the context of the classical major histocompatibility complex (MHC) and the natural tropism of γδ T cells for the tumour microenvironment [5].

Numerous reports have confirmed the safety of using γδ T cells in adoptive immunotherapy [6]. Unfortunately, the efficacy of γδ T cell immunotherapy has been limited. This is hypothesized to be due to the ambiguous effects of specific γδ T cell subsets on cancer cells. Furthermore, energy or exhaustion of the effector γδ T cells has been observed after induction by ligands such as n-aminobisphosphonates or phosphorylated antigens [7]. The targeting of γδ T cells by HSPs seems to be a forgotten direction.

Most of the studies regarding interactions between γδ T cells and HSPs were conducted in the 1990s but, despite promising results, the concept was abandoned. Currently, with better technical capabilities, greater knowledge of lymphocyte diversity, and more accurate diagnostic methods we decided to study the correlation once again in the neoplastic condition.

Due to the limited reports concerning extracellular HSP90—one of the most investigated proteins of the HSP family and the correlation with the frequency of specific subunits of γδ T cells in cancers, we examined the relationship between these parameters in the peripheral blood of 21 paediatric patients with B-cell acute lymphoblastic leukaemia (ALL)—the most common cancer in children. In the past, the prognosis was distressingly poor with only a 31% chance of a five-year survival. New diagnostic and treatment modalities have contributed to a drastic improvement in patient outcomes [8,9]. However, despite the relatively satisfying results of conventional chemotherapy, leukaemia remains the leading cause of cancer-related death among children [10].

2. Materials and Methods

Twenty-one patients (10 male and 11 female) aged 1 to 18 years were enrolled in the study. The diagnosis of acute lymphoblastic B-cell leukaemia was verified in accordance with the therapeutic protocol (ALL IC BFM 2009). The most important clinical data concerning patients are summarized in Table 1. Blood samples were collected in two time points (before and after induction chemotherapy—on the 0 and 33rd day of therapy). Data regarding haematological parameters were obtained from the medical records. As a control, blood samples from twenty-two healthy children were collected once. This study was approved by the local Research Ethics Committee. All samples were obtained following written informed consent.

Table 1. Patient characteristics.

Age	1–4 Years = 13	5–7 Years = 6	8–18 years = 2
Gender	Male = 10	Female = 11	
Steroid sensitivity	Good = 20	Poor = 1	
BM on day 15	M1 = 15	M2 = 5	M3 = 1
MRD	<0.1% = 6	0.1–10% = 13	>10% = 2
BM on day 33	M1 = 21	M2 = 0	M3 = 0
Risk group	SR = 1	IR = 15	HR = 5

Abbreviations: BM—bone marrow, M—bone marrow status (% blasts in bone marrow), MRD—minimal residual disease, SR—standard risk (group), IR—intermediate risk (group), HR—high risk (group).

Serum HSP90 concentrations were evaluated by an enzyme-linked immunosorbent assay (ELISA) (human serum HSP90 ELISA Kit, Cloud Clone Corp., Wuhan, China) according to the manufacturer's instructions. The minimum detectable dose of HSP90 in serum—less than 1.22 ng/mL. Intra-assay coefficient variation < 10%, interassay coefficient variation < 12%. Serum samples were diluted to 1:2 by PBS.

Freshly obtained EDTA whole blood was stained using antibody cocktails: TCRγδ FITC (clone, IMMU510)/TCRαβ PE (clone, IP26A)/CD4 APC (clone, 13B8.2)/CD8 AF700 (clone, B9.11)/HLA-DR PC5 (clone, B8.12.2)/CD3 Krome Orange (clone, UCHT1). Samples were then lysed with an Immunoprep Reagent Kit and TQPrep Workstation (Beckman Coulter, IN, USA). Finally, fluorescence beads for absolute counting were used. For the sample readout, a Navios flow cytometer was used and the data were analysed with Kaluza software (all from Beckman Coulter, IN, USA). The data were interpreted according to the fluorescence minus one approach.

Statistical analysis was performed using IBM SPSS 25 and Statistica 13. The association between serum HSP90 and $\gamma\delta$ T cells and in leukaemic patients was analysed by the correlation (R Spearman, Pearson). Wilcoxon test was used to indicate alterations in HSP90 serum level/level of subsets $\gamma\delta$ T cells in the research group (before and after chemotherapy). U Mann–Whitney test to compare patients with the control group. A value of $p < 0.05$ indicated statistical significance.

3. Results

We demonstrated no correlations between serum HSP90 and CD3+ $\gamma\delta$ T cells before and after induction chemotherapy (before R = 0.01, after treatment R = −0.41), CD4+ $\gamma\delta$ T cells (before R = 0.21, after treatment R = −0.07) and CD8+ $\gamma\delta$ (before R = 0.31, after treatment R = 0.09).

The same investigations also showed no statistically relevant differences among healthy controls (serum HSP90 and CD3+ $\gamma\delta$ T cells R = 0.45; CD4+ $\gamma\delta$ T cells R = 0.2; CD8+ $\gamma\delta$ T cells R = 0.12).

Analysis showed a strong association between serum HSP90 and CD3+ HLA/DR+ $\gamma\delta$ T cells in ALL patients before treatment (R = 0.53, $p < 0.05$) vs after induction protocol (R = 0.13); after removal of the outlier (23.85 ng/mL) the correlation between serum HSP90 and CD3+ HLA/DR+ $\gamma\delta$ T cells remained still strong (R = 0.53, $p < 0.05$) (Figure 1). The same patient presented extremely high level of serum HSP90 after treatment (52.51 ng/mL) as well. The difference among the healthy controls was R = −0.2 (Table 2).

In our previous report, we demonstrated that patients before and on the 33rd day of therapy showed decreased serum HSP90 levels compared to healthy controls with a higher difference on the day of diagnosis than after 33rd day [11].

Interestingly, CD3+ HLA/DR+ $\gamma\delta$ T cells were also decreased before chemotherapy relative to the moment after induction protocol.

Table 2. The results of the correlation coefficient between serum HSP90 and $\gamma\delta$ T cells in the research group and controls.

	Serum HSP90		
	Before Chemotherapy	**After Chemotherapy**	**Controls**
Diagnosis			
CD3 + $\gamma\delta$ T cells	0.01	0 *	0.45
CD3 + HLA/DR + $\gamma\delta$ T cells	0.53	0.06 *	−0.20
CD4 + $\gamma\delta$ T cells	0.21	−0.08 *	0.20
CD8 + $\gamma\delta$ T cells	0.31	0.09 *	0.12
After chemotherapy			
CD3 + $\gamma\delta$ T cells	−0.41 *	0.02	
CD3 + HLA/DR + $\gamma\delta$ T cells	0.13 *	−0.27	
CD4 + $\gamma\delta$ T cells	−0.07 *	0.13	
CD8 + $\gamma\delta$ T cells	0.09 *	0	

* results not described in the main text.

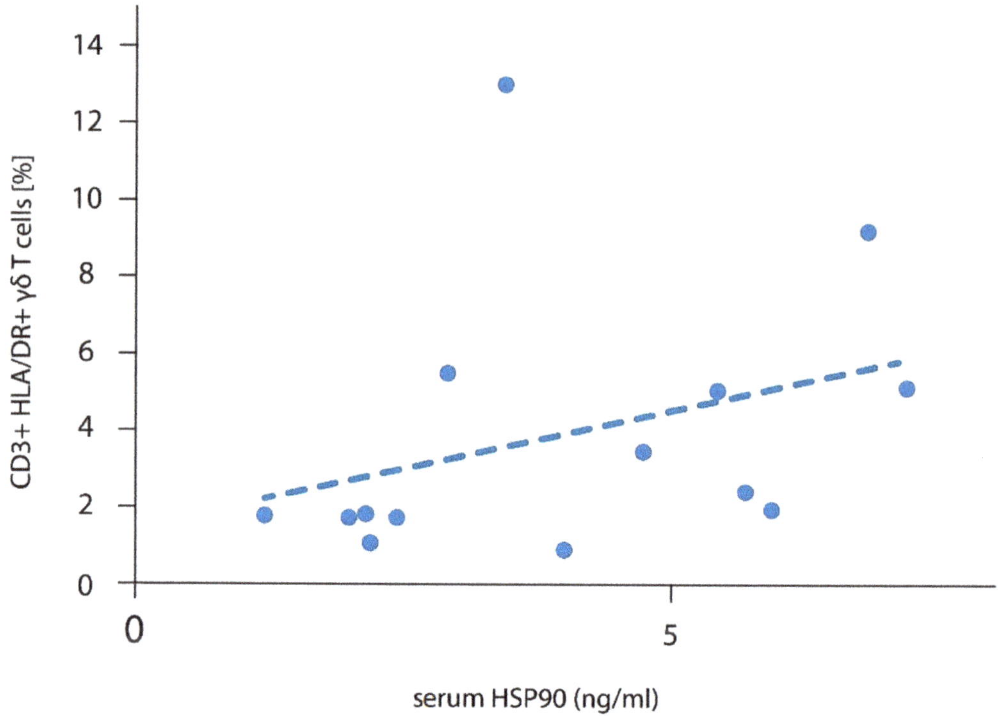

Figure 1. Correlation of activated serum HSP90 and CD3 + γδ T cells among children with ALL at diagnosis.

The median CD3+ HLA/DR+ γδ T cells on the day of the diagnosis was 2.36% (range 0.91–13.00%), and 4.96% (range 0.88–55.17%) on the 33rd day of therapy. Median CD3+ HLA/DR+ γδ T cells among the control group was 5.24% (range 1.6%-13.88%). We found statistical differences between CD3+ HLA/DR+ γδ T cells in ALL patients in two time points ($p = 0.029$, Wilcoxon signed-rank test). Children at disease presentation showed a decreased level of CD3+ HLA/DR+ γδ T cells compared to the healthy controls ($p = 0.026$, U Mann–Whitney test). However, there was no statistical significance between the level of the lymphocytes among patients on the 33rd day of therapy in comparison to the healthy children ($p = 0.627$, U Mann-Whitney test) (Figure 2).

We found no statistical differences before and after chemotherapy (Wilcoxon signed-rank test) in the levels of CD3 + γδ T cells ($p = 0.87$), CD4 + γδ T cells ($p = 0.57$), and CD8 + γδ T cells ($p = 0.39$) (Table 3).

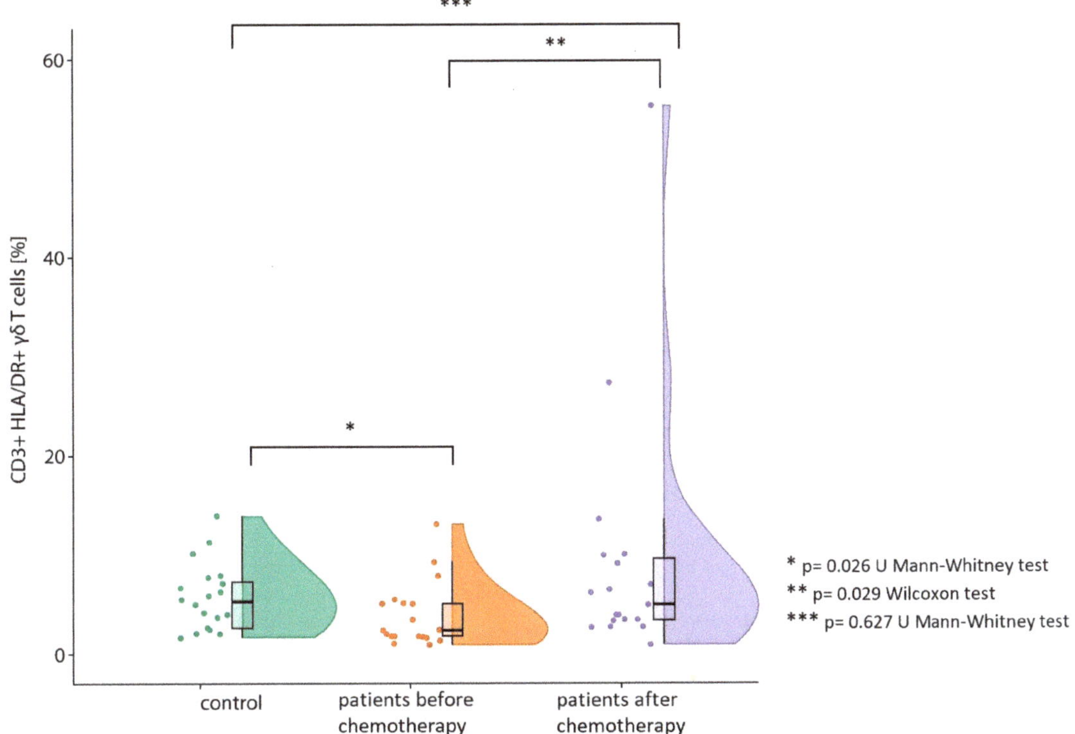

Figure 2. Decreased concentration of CD3 + HLA/DR+ γδ T cells among patients at disease presentation compared to the concentration of the lymphocytes after induction.

Table 3. The results of statistical tests regarding the comparison of γδ T cells in two time points in the research group (Wilcoxon test) and between the research and study group (U Mann–Whitney test).

Research Group before Chemotherapy	**Research Group after Chemotherapy**				**Control Group**			
	CD3 + γδ T cells	CD3 + HLA/DR + γδ T cells	CD4 + γδ T cells	CD8 + γδ T cells	CD3 + γδ T cells	CD3 + HLA/DR + γδ T cells	CD4 + γδ T cells	CD8 + γδ T cells
CD3 + γδ T cells	0.87				0.08 *			
CD3 + HLA/DR + γδ T cells		0.029				0.026		
CD4 + γδ T cells			0.57				0.51 *	
CD8 + γδ T cells				0.39				0.93 *

* results not described in the main text.

4. Discussion

Currently, HSPs are under intensive investigation. In clinical models heat shock proteins, alone or in a complex with tumour-derived peptides, have been shown to elicit an anti-tumour response in cancer patients. HSPs can act as common tumour-specific antigens as well as adjuvants that facilitate the uptake, processing and presentation of antigens. Due to their immunogenic properties, they are used in autologous tumour-derived HSP peptide-based vaccines [1].

γδ T cells represent a small subset of the T cells in peripheral blood. Despite this, they are considered to be good candidates for effective antitumor therapy [4]. γδ T cells can recognize a wide variety of structurally different ligands including phosphoantigens, aminobisphosphonates, alkylamines and several self-proteins such as HSPs, that can be detected without a presentation by other cells or molecules [12].

Activated (CD3+ HLA/DR+) $\gamma\delta$ T cells secrete cytokines, which influence the tumour microenvironment and involve IFN-γ inhibiting tumour growth, blocking angiogenesis and macrophage stimulation [4].

In the 90s the correlation between HSPs and $\gamma\delta$ T cells was extensively studied following the report demonstrating that murine $\gamma\delta$ T cells could be stimulated with HSP65 from mycobacterial extracts, which results in the induction of cytotoxic immune response against affected host cells. It has been reported that a similar mechanism can also take part in the elimination of cancer cells [13,14]. Laad et al. showed that Vγ9Vδ2 T cells recognize HSPs on oral tumour cells and Thomas et al. on oesophageal tumour targets [15,16]. Increased cytotoxicity of $\gamma\delta$ T lymphocytes has been demonstrated relative to cell lines expressing HSPs [17].

Despite years of research, the specific mechanisms of interaction remain enigmatic, while the targeting of $\gamma\delta$ T cells by HSPs seems to be a forgotten direction because potent non-peptidic phosphoantigens rather than HSPs have been found to be the essential stimulatory components of mycobacterial extracts for human Vγ9Vδ2 [18].

The most popular trend in cancer immunotherapy based on $\gamma\delta$ T cells is focused on the stimulation of cells by the systemic administration of phosphoantigens, nitrogen-containing bisphosphonates (N-bis) or synthetic phosphoagonist bromohydrin pyrophosphate (BrH-PP). Despite the proven safety of $\gamma\delta$ T cells immunotherapy, its clinical benefit remains an issue. This could be the effect of $\gamma\delta$ T cell anergy, decreased number of peripheral blood $\gamma\delta$ T cells after the infusion of stimulants or the dual nature of $\gamma\delta$ T cells, because it has been reported they could also promote cancer progression through inhibiting antitumour responses and enhancing cancer angiogenesis [7].

In this report we have shown statistically relevant correlations between serum HSP90 and CD3+ HLA/DR+ $\gamma\delta$ T cells in paediatric ALL at diagnosis ($R = 0.53$, $p < 0.05$) (Figure 1), but not after chemotherapy ($R = 0.13$). Our team have demonstrated, that the correlation of serum HSP90 with $\gamma\delta$ T cells may depend on lymphocytes immunophenotype rather than chains.

We also have noticed that serum HSP90 and CD3+ HLA/DR+ $\gamma\delta$ T cells are both decreased before chemotherapy relative to the moment after induction protocol (Figure 2), which indirectly indicate unknown dose-dependent interaction between the parameters in cancer conditions.

The results of our study may be a good introduction to research on the activation of $\gamma\delta$ T cells by HSPs which could be an interesting direction for the development of adjuvant anti-cancer strategies, not just for childhood ALL [19].

5. Conclusions

Summing up the correlations between serum HSP90 and activated CD3+ $\gamma\delta$ T cells provide a promising suggestion that these cells may enhance the effect of conventional chemotherapy by supporting the immune system. Further studies, including in-vitro experiments, are needed to determine the clinical importance of our findings.

Author Contributions: D.P.-G., M.Z., M.G.-P., F.C., M.N. and P.T. contributed to the concept development and study design. D.P.-G. obtained approvals and collected patient samples and wrote the manuscript. J.S. and M.N. performed statistical analyses and prepared figures. M.G.-P., M.Z. and J.S. performed laboratory analyses. D.P.-G. and M.N. coordinated the study. All authors critically reviewed the manuscript. All authors have read and agreed to the published version of the manuscript.

Funding: The study was supported by a grant for young researchers (MN 01-0262/08/152) from the Medical University of Gdańsk, Poland.

Institutional Review Board Statement: The study was conducted according to the guidelines of the Declaration of Helsinki, and approved by the Independent Bioethics Committee for Scientific Research at Medical University of Gdańsk (NKBBN/251/2016, decision issued on 28 June 2016).

Informed Consent Statement: Informed consent was obtained from all subjects involved in the study.

Data Availability Statement: The statistical analysis (as Excel and Statistica files) used to support the findings of this study are available from the corresponding author upon request.

Acknowledgments: M.G.-P. acknowledges the support of ST46 funding from the Medical University of Gdańsk (Gdańsk, Poland) and the Polish Ministry of Health and Higher Education. Our team would like to thank GetResponse Cares Foundation for research equipment.

Conflicts of Interest: The authors declare that there is no conflict of interest regarding the publication of this paper.

Abbreviations

$\gamma\delta$ T cells	gamma-delta T cells
$\alpha\beta$ T cells	alpha-beta T cells
ALL	acute lymphoblastic leukaemia
ALL IC BFM 2009	acute lymphoblastic leukaemia intercontinental Berlin-Frankfurt-Munchen
BM	bone marrow
BrH-PP	phosphoagonist bromohydrin pyrophosphate
CD	cluster differentiation antigen
DC	dendritic cells
EDTA	ethylenediaminetetraacetic acid
ELISA	enzyme-linked immunosorbent assay
HR	high risk (group)
HSPs	heat shock proteins
IR	intermediate-risk (group)
M	bone marrow status (% blasts in bone marrow)
MHC	major histocompatibility complex
MRD	minimal residual disease
N-bis	nitrogen-containing bisphosphonates
SR	standard risk (group)
WBC	white blood cells

References

1. Shevtsov, M.; Multhoff, G. Heat Shock Protein–Peptide and HSP-Based Immunotherapies for the Treatment of Cancer. *Front. Immunol.* **2016**, *7*, 171. [CrossRef]
2. Calderwood, S.K.; Gong, J.; Murshid, A. Extracellular HSPs: The complicated roles of extracellular HSPs in immunity. *Front. Immunol.* **2016**, *7*, 159. [CrossRef] [PubMed]
3. Cao, W.; He, W. The recognition pattern of gammadelta T cells. *Front. Biosci.* **2005**, *10*, 2676–2700. [CrossRef] [PubMed]
4. Wu, Y.L.; Ding, Y.P.; Tanaka, Y.; Shen, L.W.; Wei, C.H.; Minato, N.; Zhang, W. $\gamma\delta$ T cells and their potential for immunotherapy. *Int. J. Biol. Sci.* **2014**, *10*, 119–135. [CrossRef] [PubMed]
5. Morandi, F.; Yazdanifar, M.; Cocco, C.; Bertaina, A.; Airoldi, I. Engineering the Bridge between Innate and Adaptive Immunity for Cancer Immunotherapy: Focus on $\gamma\delta$ T and NK Cells. *Cells* **2020**, *9*, 1757. [CrossRef]
6. Buccheri, S.; Guggino, G.; Caccamo, N.; Li Donni, P.; Dieli, F. Efficacy and safety of gammadeltaT cell-based tumor immunotherapy: A meta-analysis. *J. Biol. Regul. Homeost. Agents* **2014**, *28*, 81–90. [PubMed]
7. Zhao, Y.; Niu, C.; Cui, J. Gamma-delta (gammadelta) T cells: Friend or foe in cancer development? *J. Transl. Med.* **2018**, *16*, 3. [CrossRef] [PubMed]
8. Hunger, S.P.; Lu, X.; Devidas, M.; Camitta, B.M.; Gaynon, P.S.; Winick, N.J.; Reaman, G.H.; Carrollet, W.L. Improved survival for children and adolescents with acute lymphoblastic leukemia between 1990 and 2005: A report from the children's oncology group. *J. Clin. Oncol.* **2012**, *30*, 1663–1669. [CrossRef] [PubMed]
9. Hunger, S.P.; Mullighan, C.G. Acute lymphoblastic leukemia in children. *N. Engl. J. Med.* **2015**, *373*, 1541–1552. [CrossRef] [PubMed]
10. Kato, M.; Manabe, A. Treatment and biology of pediatric acute lymphoblastic leukemia. *Pediatrics Int.* **2018**, *60*, 4–12. [CrossRef] [PubMed]
11. Pawlik-Gwozdecka, D.; Górska-Ponikowska, M.; Adamkiewicz-Drożyńska, E.; Niedźwiecki, M. Serum heat shock protein 90 as a future predictive biomarker in childhood acute lymphoblastic leukemia. *Centr. Eur. J. Immunol.* **2021**, *46*, 63–67. [CrossRef] [PubMed]
12. Deseke, M.; Prinz, I. Ligand recognition by the $\gamma\delta$ TCR and discrimination between homeostasis and stress conditions. *Cell Mol. Immunol.* **2020**, *17*, 914–924. [CrossRef] [PubMed]

13. Kabelitz, D.; Bender, A.; Schondelmaier, S.; Schoel, B.; Kaufmann, S.H. A large fraction of human peripheral blood gamma/delta + T cells is activated by Mycobacterium tuberculosis but not by its 65-kD heat shock protein. *J. Exp. Med.* **1990**, *171*, 667–679. [CrossRef]
14. Hirsh, M.I.; Junger, W.G. Roles of heat shock proteins and gamma delta T cells in inflammation. *Am. J. Respir. Cell Mol. Biol.* **2008**, *39*, 509–513. [CrossRef]
15. Laad, A.D.; Thomas, M.L.; Fakih, A.R.; Chiplunkar, S.V. Human gamma delta T cells recognize heat shock protein-60 on oral tumor cells. *Int. J. Cancer* **1999**, *80*, 709–714. [CrossRef]
16. Thomas, M.L.; Samant, U.C.; Deshpande, R.K.; Chiplunkar, S.V. Gammadelta T cells lyse autologous and allogenic oesophageal tumours: Involvement of heat-shock proteins in the tumour cell lysis. *Cancer Immunol. Immunother.* **2000**, *48*, 653–659. [CrossRef]
17. Zhang, H.; Hu, H.; Jiang, X.; He, H.; Cui, L.; He, W. Membrane HSP70: The molecule triggering gammadelta T cells in the early stage of tumorigenesis. *Immunol. Invest.* **2005**, *34*, 453–468. [CrossRef]
18. Champagne, E. $\gamma\delta$ T cell receptor ligands and modes of antigen recognition. *Arch. Immunol. Ther. Exp.* **2011**, *59*, 117–137. [CrossRef] [PubMed]
19. Rane, S.S.; Dearman, R.J.; Kimber, I.; Uddin, S.; Bishop, S.; Shah, M.; Podmore, A.; Pluen, A.; Derrick, J.P. Impact of a Heat Shock Protein Impurity on the Immunogenicity of Biotherapeutic Monoclonal Antibodies. *Pharm. Res.* **2019**, *36*, 51. [CrossRef]

Article

Changes in Elderberry (*Sambucus nigra* L.) Juice Concentrate Polyphenols during Storage

Cláudia M. B. Neves [1], António Pinto [1,2,3], Fernando Gonçalves [1,3] and Dulcineia F. Wessel [1,2,4,*]

1 Agrarian School of Viseu, Polytechnic Institute of Viseu, 3500-606 Viseu, Portugal; cmneves@esav.ipv.pt (C.M.B.N.); apinto@esav.ipv.pt (A.P.); goncalves7ster@gmail.com (F.G.)
2 CITAB, University of Trás-os-Montes e Alto Douro, 5001-801 Vila Real, Portugal
3 CERNAS-IPV, Agrarian School of Viseu, 3500-606 Viseu, Portugal
4 LAQV-REQUIMTE, Department of Chemistry, University of Aveiro, 3810-193 Aveiro, Portugal
* Correspondence: ferdulcineia@esav.ipv.pt

Abstract: Elderberry (*Sambucus nigra* L.) juice concentrate is highly rich in polyphenols, particularly anthocyanins and flavonols, which have been associated with a wide range of health-promoting properties. Phenolic compounds, in particular anthocyanins, are unstable and may change during storage, which might influence the product color quality and its potential health effects. The aim of this study was to evaluate the changes in the polyphenols profile of elderberry juice concentrate produced at an industrial scale during seven months of storage at 5 °C and at room temperature. The total phenolic content, the total monomeric anthocyanins, the percent polymeric color, and the $ABTS^{\bullet+}$ scavenging activity were monitored over time. In addition, the profile and content of the main individual phenolic compounds were also assessed by HPLC-DAD. The results show that cyanidin-3-*O*-sambubioside, cyanidin-3-*O*-glucoside, cyanidin-3-*O*-sambubioside-5-*O*-glucoside, cyanidin-3,5-*O*-diglucoside, chlorogenic acid, rutin, and quercetin-3-*O*-glucoside were the main phenolic compounds identified. Storage at room temperature resulted in a strong reduction in total monomeric anthocyanin content accompanied by an increase in percent polymeric color values. Cyanidin-3-*O*-sambubioside and cyanidin-3-*O*-glucoside degraded faster than cyanidin-3,5-*O*-diglucoside and cyanidin-3-*O*-sambubioside-5-*O*-glucoside. Concentration of chlorogenic acid also decreased over storage, whereas rutin and quercetin-3-*O*-glucoside were quite stable. Storage at 5 °C caused a lower impact on the contents of anthocyanins and chlorogenic acid and the percent polymeric color was not affected. The total phenolic content and the in vitro antioxidant activity remained quite similar over the time, for both temperatures, suggesting that elderberry concentrates still preserve their health benefits of antioxidant capacity after seven months of storage.

Keywords: elderberry; *Sambucus nigra* L.; juice concentrate; storage effect; phenolic compounds; berries; anthocyanins' stability; polymeric color; $ABTS^{\bullet+}$ antioxidant capacity

Citation: Neves, C.M.B.; Pinto, A.; Gonçalves, F.; Wessel, D.F. Changes in Elderberry (*Sambucus nigra* L.) Juice Concentrate Polyphenols during Storage. *Appl. Sci.* **2021**, *11*, 6941. https://doi.org/10.3390/app11156941

Academic Editors: Francesco Cappello and Magdalena Gorska-Ponikowska

Received: 13 July 2021
Accepted: 25 July 2021
Published: 28 July 2021

1. Introduction

Elderberry (*Sambucus nigra* L.) is a widespread shrub that grows in most parts of Europe and North Africa [1]. It has for a long time been used in folk medicine as a diuretic agent and in the treatment of colds, influenza, and herpes. In recent years, elderberry fruits have received great attention due to the presence of large amounts of anthocyanin pigments and other polyphenols [2–4], which are known for their antioxidant activity and health benefits [5,6]. Several studies have shown that elderberries exhibit anti-inflammatory, antiviral, anti-proliferative, anti-diabetic, and immunostimulatory activities [7]. Due to their health-promoting properties and high content of anthocyanin pigments, they are frequently processed to juice concentrate to be used as food colorant, and in pharmaceutical and nutraceutical fields. Some examples of predominant polyphenols that can be found in elderberry juices and concentrates are presented in Figure 1. The main anthocyanins have been identified as cyanidin-3-*O*-glucoside, cyanidin-3-*O*-sambubioside, cyanidin-3,5-*O*-diglucoside, and cyanidin-3-*O*-sambubioside-5-*O*-glucoside [8–10]. Phenolic acids such as chlorogenic acid and the flavonols quercetin-3-*O*-glucoside and rutin have also been detected in elderberry juice [11,12].

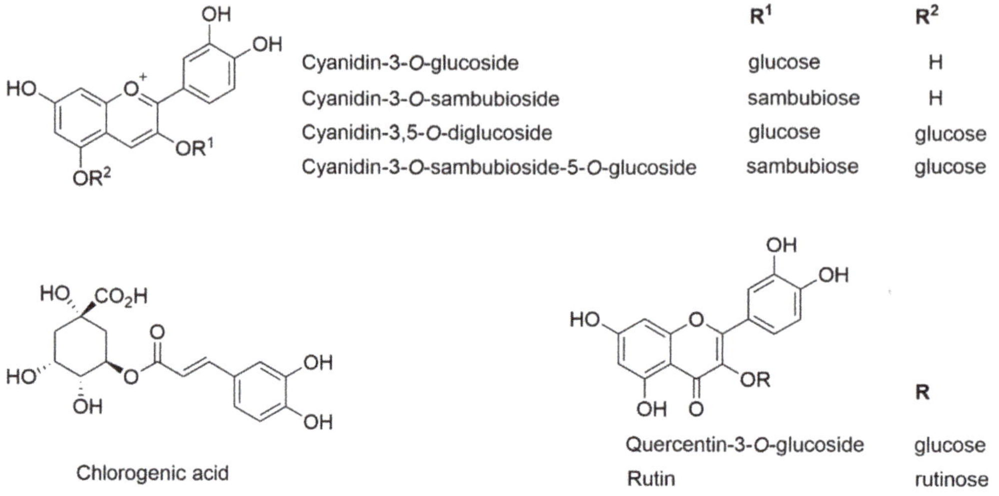

	R^1	R^2
Cyanidin-3-*O*-glucoside	glucose	H
Cyanidin-3-*O*-sambubioside	sambubiose	H
Cyanidin-3,5-*O*-diglucoside	glucose	glucose
Cyanidin-3-*O*-sambubioside-5-*O*-glucoside	sambubiose	glucose

Chlorogenic acid

	R
Quercentin-3-*O*-glucoside	glucose
Rutin	rutinose

Figure 1. Some polyphenols identified in elderberry (*Sambucus nigra* L.) juices and concentrates [8–12].

During storage, the anthocyanins in berry juices may undergo several reactions to form more stable compounds, which typically involve oxidation, polymerization, co-pigmentation with other phenolic compounds, and cleavage reactions [13–15]. The cleavage of anthocyanins results in colorless compounds, polymerization is accompanied by browning, and co-pigmentation might result in various colored compounds [16]. The transformation of these compounds influences not only the product color quality and consumer acceptance [17], but also the potential health effects [18].

The main objective of the present work was the evaluation of the effect of storage on the polyphenolic composition of elderberry juice concentrate and the possible consequences on its color pigments and antioxidant capacity. Two elderberry juice concentrates produced at an industrial scale were monitored during seven months of storage at 5 °C and at room temperature.

2. Materials and Methods

2.1. Chemicals

The HPLC standards chlorogenic acid, quercetin, quercetin-3-*O*-glucoside, and cyanidin-3-*O*-sambubioside chloride were purchased from Sigma-Aldrich (St. Louis, MO, USA), quercetin-3-*O*-rutinoside (rutin) and cyanidin-3,5-*O*-diglucoside chloride from Phytolab (Vestenbergsgreuth, Germany), cyanidin-3-*O*-sambubioside-5-*O*-glucoside chloride from Extrasynthese (Lyon, France), and cyanidin-3-*O*-glucoside chloride from USP (Rockville, MD, USA). Folin–Ciocalteu phenol reagent was acquired from Fisher Scientific (Hampton, NH, USA), and Trolox ((±)-6-hydroxy-2,5,7,8-tetramethylchromane-2-carboxylic acid) and ABTS (2,2-azinobis(3-ethyl-benzothiazoline-6-sulfonic acid) from Sigma-Aldrich. Gallic acid 1-hydrate (99%), aluminum chloride 6-hydrate (pure, pharma grade), sodium carbonate anhydrous, sodium acetate anhydrous, and potassium chloride were obtained from Panreac (Barcelona, Spain) and potassium metabisulfite from LabChem (Zelienople, PA, USA). Methanol, ethanol, and formic acid (analytical grade) were purchased from Fisher Scientific and hydrochloric acid ≥37% from Fluka/Honeywell (Porto Salvo, Portugal). Acetonitrile HPLC far UV gradient grade wasobtained from J.T.Baker (Waltham, MA, USA).

2.2. *Equipments*

All UV-Vis spectrophotometric measurements were carried out using a Shimadzu UV-1280 (Izasa Scientific, Barcelona, Spain).

Total soluble solids were recorded on an ATAGO digital refractometer (ATAGO USA Inc., Belleuve, WA, USA) and pH values measured on a Consort C1010 multi-parameter pH analyzer (Turnhout, Belgium).

For the HPLC analysis, an Ultimate 3000 HPLC (Dionex, Waltham, MA, USA) equipped with an Ultimate 3000 pump, Ultimate 3000 autosampler, Ultimate 3000 column compartment, and a Thermo Scientific Dionex Ultimate 3000 diode array detector was used.

2.3. *Elderberry Juice Concentrate Samples*

The elderberry fruits (*Sambucus nigra* L.) used in the production of juice concentrate were harvested in August 2020 from plants cultivated in Varosa Valley in northern Portugal. The fruits were from three cultivars, "Sabugueira", "Bastardeira", and "Sabugueiro". After harvest, the berries were separated from the branches and crushed in a fruit-processing factory (Régifrutas Company, Tarouca, Portugal). The obtained elderberry mash was immediately transported in isothermal tank trucks at a temperature between −2 and 0 °C to a fruit juice factory (Indumape Company, Pombal, Portugal) where it was processed into juice concentrate. Two samples of juice concentrate (I and II) were collected and sent to the Agrarian School of Viseu where they were stored at −18 °C until use. The technological process to produce the juice concentrates involved the following steps: crushing, mash pressing, depectinization, ultrafiltration, and concentration. The process to obtain concentrate I, in addition to the steps described above, also involved a prior heating step of the mash at 80 °C.

Concentrate I and II were used in the studies of chemical and microbiological stability over time at two different storage temperatures (5 °C and room temperature). For each concentrate, two separate aliquots of 200 mL were transferred under aseptic conditions to sterilized bottles duly identified to remain in the dark, under two different conditions, at room temperature and at 5 °C. For the chemical analyses, "stock solutions" of the concentrates were prepared in ultrapure water (1:25, *v*:*v*), hereinafter referred to as Solutions A. For the preparation of these solutions, 4 mL of each sample was measured into a 100 mL flask and dissolved in ultrapure water. Solutions A were kept at 5 °C for analysis, for no more than two days.

2.4. *Microbiological Analysis*

The groups of microorganisms evaluated were the mesophiles at 30 °C and fungi (mold and yeasts). The enumeration of the mesophiles was carried out by counting colonies in plates, using the culture medium PCA (Plate Count Agar), based on the "Norma Portuguesa" NP 4405:2002 and ISO 4833:2003 standards, using the pour plate technique for inoculation. The enumeration of fungi was carried out by counting colonies, in plates with culture medium of Dichloran Rose-Bengal Agar, with chloramphenicol, based on the ISO 21527-1:2008 standard. The results are expressed in CFU mL^{-1} (colony-forming units per mL).

2.5. *Measurement of Total Soluble Solids*

Total soluble solid content in elderberry juice concentrates was determined from Solutions A using a digital refractometer. The auto-zero was carried out with distilled water and each analysis was performed in triplicate. The results were obtained considering the dilution factor used to obtain Solutions A and are expressed as °Brix.

2.6. Total Phenolic Content

The content of total phenolic compounds was determined by the Folin–Ciocalteu method with some modifications [19]. For this analysis, Solutions A were previously diluted with distilled water in the ratio 2:25. To 125 µL of appropriately diluted juice concentrate solution A, or standard solution, 1 mL of distilled water and 125 µL of Folin-Ciocalteu reagent were added. The mixture was homogenized with a vortex and, after a pause of 6 min, 2 mL of 5% sodium carbonate (*w*/*v*) was added. The above mixture was homogenized with a vortex and allowed to stand in the dark for 60 min at room temperature. The absorbance of the samples at 760 nm was read in triplicate. The total phenolic content of the juice concentrate samples was determined in triplicate from a standard curve of gallic acid and expressed as g gallic acid equivalents per kg (g GAE kg^{-1}) of concentrate.

2.7. Total Monomeric Anthocyanins Content

The content of total monomeric anthocyanins was determined by the pH differential method according to Giusti and Wrolstad [20]. The appropriate dilution factor (DF) for sample analysis was first defined by diluting the juice concentrate Solutions A with pH 1 potassium chloride buffer to absorbance < 1.2 at $\lambda_{510\ nm}$. The dilution factor determined was 50. For the analyses, Solutions A were diluted by separately adding the potassium chloride buffer (pH 1) or the potassium acetate buffer (pH 4.5) according to the dilution factor determined above. The mixtures were allowed to equilibrate for 15 min, and the absorbance was read at 510 and 700 nm. The content of total monomeric anthocyanins was determined according to the following equation:

$$\text{Total monomeric anthocyanins } (\text{g L}^{-1}) = \frac{A \times MW \times DF}{\varepsilon \times l} \quad (1)$$

where *MW* is the molecular mass of cyanidin-3-*O*-glucoside (449.2 g mol^{-1}), *DF* is the dilution factor used, ε is the molar absorptivity coefficient of cyanidin-3-*O*-glucoside (26,900 L mol^{-1} cm^{-1}), *l* is optical path length in cm, and *A* was calculated using Equation (2):

$$A = (A_{510\ nm} - A_{700\ nm})_{pH1} - (A_{510\ nm} - A_{700\ nm})_{pH4.5} \quad (2)$$

The results are expressed in g cyanidin-3-*O*-glucoside equivalents per kg (g cy3gluE kg^{-1}) of concentrate.

2.8. Color Density, Polymeric Color, and Percent Polymeric Color

The color density (CD), the polymeric color (PC), and the percent polymeric color (%PC) of the juice concentrates were determined according to Giusti and Wrolstad [20]. Solutions A were diluted in distilled water with the same dilution factor previously used for the quantification of total monomeric anthocyanins. In two test tubes was added 2.8 mL of diluted sample; in one of the tubes was added 0.2 mL of distilled water and in the other was added 0.2 mL of the aqueous potassium metabisulfite solution (200 g L^{-1}). After 15 min, the absorbance was measured at 420, 510, 620, and 700 nm (to correct for haze).

Color density was quantified in samples treated with distilled water and polymeric color in samples with potassium metabisulfite according to Equation (3).

$$\text{CD or PC} = [(A_{420\ nm} - A_{700\ nm}) + (A_{510\ nm} - A_{700\ nm}) + (A_{620\ nm} - A_{700\ nm})] \times \text{DF} \quad (3)$$

The percent polymeric color of the juices was calculated using the following equation:

$$\%\text{PC} = \frac{\text{PC}}{\text{CD}} \times 100 \quad (4)$$

2.9. Quantification of Flavan-3-ols

The flavan-3-ol content was determined by the vanillin method according to Sun et al. [21], which consists of the reaction of vanillin with compounds that contain the flavan3-3-ol nucleus in acidic medium. Briefly, the method consisted of the following steps. In test tubes was placed 500 µL of solution A diluted in methanol p.a. grade in the ratio 1:25, or standard to which 1.25 mL of 9 N HCl methanolic solution and then 1.25 mL of 1% vanillin methanolic solution were added. The mixture was homogenized with a vortex and the tubes were placed at 30 °C in the dark for 15 min. The absorbance of the samples at 500 nm was read in triplicate. To avoid the interference of anthocyanins, which, despite not reacting with vanillin, have a maximum absorption at 490–540 nm in an acidic medium, which coincides with that of the colored product of the vanillin method, in parallel, blank tests were prepared replacing the vanillin solution with an equal volume of methanol. Flavan-3-ol content was expressed as g of epicatechin equivalents per kg (g EpiE kg^{-1}) of concentrate, using a standard curve of epicatechin.

2.10. Quantification of Flavonols

The flavonol content was determined according to a spectrophotometric method based on the formation of aluminum–flavonoid complex [22]. In test tubes, 1 mL of solution A diluted 1:10 in methanol p.a. grade or standard, 0.5 mL of distilled water and 0.5 mL of 2% aluminum chloride aqueous solution were placed. The mixtures were homogenized with a vortex and the formation of a yellow complex was observed. The samples were left in the dark for 10 min at room temperature and the absorbance at 425 nm was read in triplicate. The flavonol content was expressed as g quercetin equivalents per kg (g QE kg^{-1}) of concentrate, using a standard curve of quercetin.

2.11. Determination of Antioxidant Activity

The antioxidant activity of elderberry juice concentrate was assessed by the $ABTS^{\bullet+}$ scavenging method, according to Ozgen et al. [23]. $ABTS^{\bullet+}$ was prepared by dilution of ABTS to a final concentration of 7 mM with 2.45 mM of aqueous potassium persulfate solution. This mixture was allowed to stand at room temperature in the dark for 12–16 h and then was stored at −18 °C until use. On the day of analysis, the $ABTS^{\bullet+}$ solution was diluted with ethanol to an absorbance of 0.700 at 734 nm. Assays were conducted by mixing 2 mL of the prepared $ABTS^{\bullet+}$ solution with 100 µL of solution A previously diluted 1:100 in distilled water. After 15 min in the dark at room temperature, the absorbance of the samples at 734 nm was measured in triplicate. Simultaneously, a blank solution was prepared by replacing the 100 µL of sample with distilled water. The percent inhibition of $ABTS^{\bullet+}$ was determined according to Equation (5).

$$\%\ ABTS^{\bullet+}\ \text{inibition} = \frac{A_{(\text{blank solution})} - A_{(\text{sample})}}{A_{(\text{blank solution})}} \times 100 \quad (5)$$

The antioxidant activity of the juice concentrates was expressed in mmol Trolox equivalents per kg (mmol TE kg^{-1}) of concentrate, using a Trolox calibration curve.

2.12. HPLC-DAD Analysis and Quantification of Polyphenols

The polyphenol profile of elderberry juice concentrates was analyzed by reversed phase HPLC. Samples A were diluted in 5% (*v*/*v*) formic acid in ultrapure water (0.2 mL of sample in a final volume of 1.5 mL) and filtered through a syringe filter before analysis (0.2 µm nylon membrane filter, Whatman, Amadora, Portugal). The compounds were separated on a Macherey-Nagel C18 Nucleodur column, 250 mm of length, 4 mm of internal diameter, and 5 µm particle size, using as eluent A 5% (*v*/*v*) formic acid in ultrapure water and as eluent B acetonitrile, a flow rate of 1 mL min^{-1}, and column temperature stabilized at 25 °C, according to the following gradient elution program: 0–10 min—5% B; 10–20 min—20% B; 20–27 min—20% B; 27–37 min—70% B; 37–45 min—70% B; 45–55 min—

5% B; 55–65 min—5% B. The eluents were prepared daily and filtered through a 0.22 μm nylon membrane filter (Filter-Lab). The UV-Vis detection was performed at wavelengths between 190 and 800 nm and the injection volume was 20 μL. Instrument control and data acquisition were performed using the software Chromeleon version 6.80 (Dionex, Sunnyvale, CA, USA). Compounds were identified according to the retention times of the peaks and their UV-Vis spectra, in comparison with those of commercial standards.

The quantification of the polyphenolic compounds was carried out using a calibration curve in the range of 2.5–13.0 mg L^{-1} for chlorogenic acid, 4.0–24.0 mg L^{-1} for cyanidin-3-*O*-glucoside, 4.0–44.0 mg L^{-1} for cyanidin-3-*O*-sambubioside, 4.0–44.0 mg L^{-1} for cyanidin-3-*O*-sambubioside-5-*O*-glucoside, 1.0–22.0 mg L^{-1} for quercetin-3-glucoside, and 2.5–28.0 mg L^{-1} for rutin. The results are presented in g kg^{-1} of concentrate. Cyanidin-3-*O*-sambubioside-5-*O*-glucoside and cyanidin-3,5-*O*-diglucoside were quantified as the sum of the two compounds, since they appear at the same retention time, and are expressed as g cyanidin-3-*O*-sambubioside-5-*O*-glucoside per kg.

3. Results and Discussion

3.1. Characterization of the Initial Eldeberry Concentrates

The concentrates were obtained by an industrial process that involved crushing, mash pressing, depectinization, ultrafiltration, and concentration. The production of concentrate I, in addition to the steps described above, also involved a thermal treatment of the mash at 80 °C, prior to pressing. The main objective of this step is the inactivation of oxidoreductases such as polyphenol oxidase and peroxidase, which are assumed to be involved in the degradation of anthocyanins [24,25]. Some chemical characteristics of the elderberry juice concentrates before storage are presented in Table 1.

Table 1. Analytic data of the initial elderberry juice concentrates [1].

	Concentrate I	Concentrate II
Thermal processing step	Heating of mash at 80 °C	No
Total soluble solids (°Brix)	55.8 ± 0.8	80.6 ± 1.0
Total phenolic content (g GAE kg^{-1})	37.7 ± 0.4	54.7 ± 1.9
Total monomeric anthocyanins (g Cy3gluE kg^{-1})	10.6 ± 0.1	15.8 ± 0.4
Flavonols (g QE kg^{-1})	3.1 ± 0.04	4.8 ± 0.2
Flavan-3-ols (g EpiE kg^{-1})	1.0 ± 0.6	1.5 ± 0.3
Polymeric color (%)	26.2 ± 1.6	28.3 ± 1.2
Antioxidant activity (mmol TE kg^{-1})	352.3 ± 2.6	455.0 ± 19.4

[1] Results are expressed as mean values ± standard deviation (n = 3). GAE—gallic acid equivalents, QE—quercetin equivalents, EpiE—epicatechin equivalents, Cy3gluE—cyanidin-3-*O*-glucoside equivalents, TE—Trolox equivalents.

Concentrate I and concentrate II presented an average content of total soluble solids of 55.8 and 80.6 °Brix, respectively. Concentrate II contained 54.7 g GAE kg^{-1} of total phenolic compounds, 15.8 g kg^{-1} of total monomeric anthocyanins, 4.8 g QE kg^{-1} of flavonols, and 1.5 g EpiE kg^{-1} of flavan-3-ols. The ratio between the concentrates in all these parameters was around 1.5, which indicates that concentrate II is 1.5-fold more concentrated compared to concentrate I. Therefore, for these analyzed parameters, the lower contents observed in concentrate I are explained by the concentration process, which results in different values of °Brix, and are not related to the heating treatment. The results obtained allow us to infer that the initial juices have similar phenolic characteristics. The percentage of polymeric color (%PC) was also similar; 26.2 ± 1.6% and 28.3 ± 1.2% for concentrate I and concentrate II, respectively. These values of %PC are within the expected values for fruit juices that have been subjected to industrial processing [20]. The concentrates I and II showed an $ABTS^{\bullet+}$ antioxidant activity equal to 352.3 and 455.0 mmol TE kg^{-1}, respectively. The ratio between these values reveals the same trend observed for the content of phenolic compounds, where juice II is about 1.5 times more concentrated. Overall, the values presented in Table 1 are in agreement with those obtained by Bermúdez-Soto and Tomás-Barberán [9] for elderberry concentrate.

The most important phenolic compounds in the two concentrates were identified and quantified by HPLC-DAD. The concentrations of the phenolic compounds identified are presented in Table 2.

Table 2. Contents of individual phenolic compounds in the initial elderberry concentrates quantified by HPLC in g kg^{-1}.

Compounds	Concentrate I	Concentrate II
Cyanidin-3-*O*-sambubioside	3.30 ± 0.06	4.69 ± 0.04
Cyanidin-3-*O*-glucoside	1.99 ± 0.07	2.86 ± 0.02
Cyanidin-3,5-diglucoside + cyanidin-3-*O*-sambubioside-5-*O*-glucoside	4.39 ± 0.15	6.49 ± 0.08
Rutin	2.51 ± 0.04	3.70 ± 0.04
Quercentin-3-*O*-glucoside	0.79 ± 0.01	0.96 ± 0.01
Chlorogenic acid	0.78 ± 0.02	1.11 ± 0.01

The phenolic profile was similar in the two concentrates. As an example, the chromatograms of the concentrate I before storage are shown in Figure 2a,b.

In the chromatogram at 325 nm, the peak at about 18.0 min (**1**) was identified as chlorogenic acid and the peaks at 24.8 (**5**) and 25.9 min (**6**) as rutin and quercetin-3-*O*-glucoside, respectively. In the chromatogram at 520 nm, three peaks corresponding to anthocyanins can be observed. The peaks at 20.1 (**3**) and 20.3 min (**4**) were identified as cyanidin-3-*O*-sambubioside and cyanidin-3-*O*-glucoside, respectively. Cyanidin-3-*O*-sambubioside and cyanidin-3-*O*-glucoside corresponded, respectively, to around 33% and 20% of the total individual anthocyanins in both concentrates. Cyanidin-3-*O*-sambubioside-5-*O*-glucoside and cyanidin-3,5-*O*-diglucoside are other anthocyanins described to be present in high amounts in elderberries and elderberry products [3,8,9,12,26]. The peak at 18.2 min (**2**) was attributed to these two anthocyanins that eluted at the same retention time and appeared overlaid. Therefore, they were quantified together, the result being expressed in g of cyanidin-3-*O*-sambubioside-5-*O*-glucoside equivalents per kg. The sum of cyanidin-3-*O*-sambubioside-5-*O*-glucoside and cyanidin-3,5-*O*-diglucoside accounted for 45% of the total analyzed anthocyanins. The total anthocyanins content determined by HPLC was 9.68 and 14.04 g kg^{-1}, for concentrate I and II, respectively, which is slightly lower than the total monomeric anthocyanins obtained using the pH differential method (Table 1). These differences may be explained by the presence of some compounds that were not quantified by HPLC.

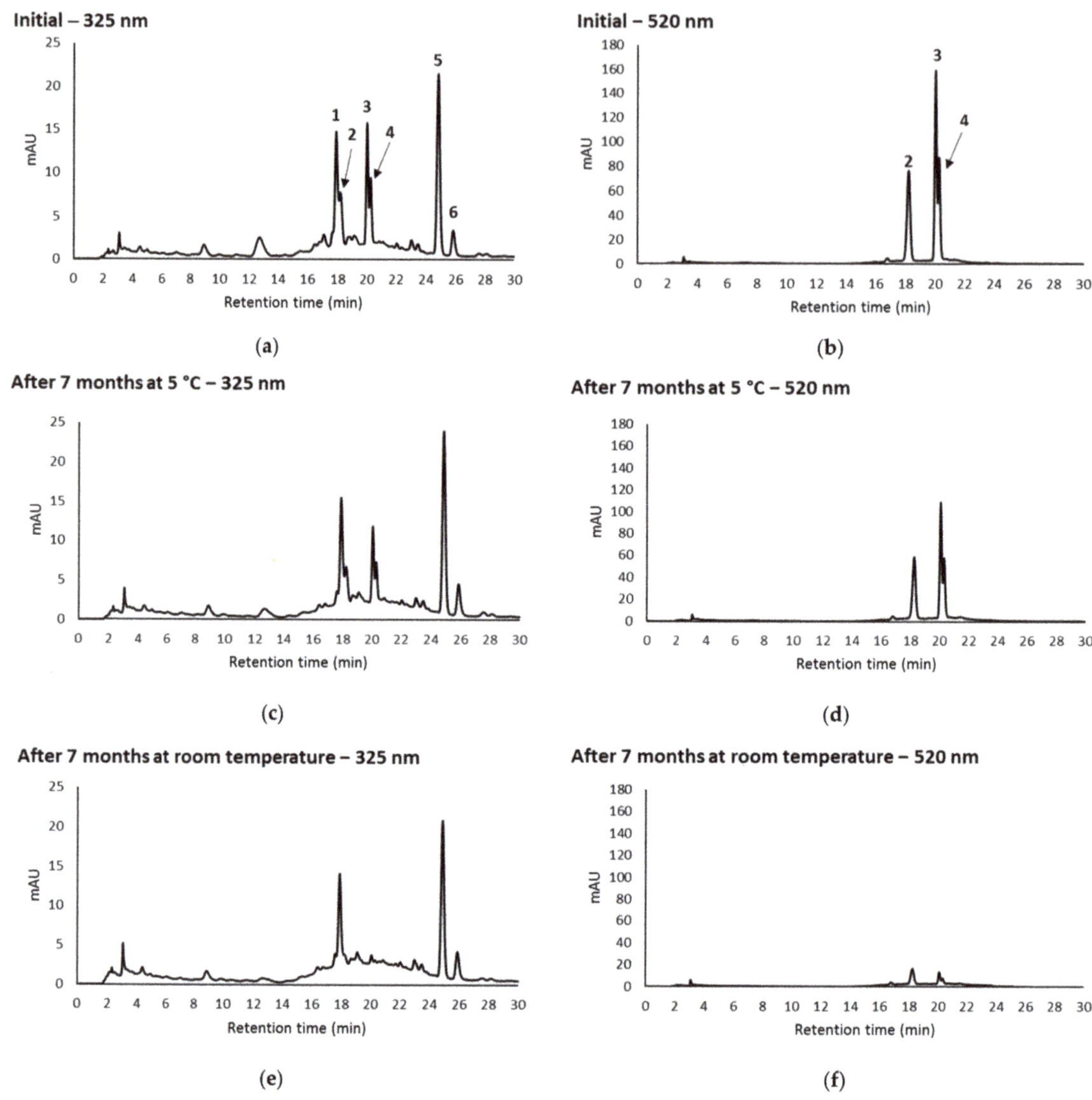

Figure 2. HPLC chromatograms of concentrate I (**a**) before storage with detection at 325 nm, (**b**) before storage with detection at 520 nm, (**c**) after 7 months of storage at 5 °C with detection at 325 nm, (**d**) after 7 months of storage at 5 °C with detection at 520 nm, (**e**) after 7 months of storage at room temperature with detection at 325 nm, and (**f**) after 7 months of storage at room temperature with detection at 520 nm. (**1**) Chlorogenic acid, (**2**) cyanidin-3-*O*-sambubioside-5-*O*-glucoside + cyanidin-3,5-*O*-diglucoside, (**3**) cyanidin-3-*O*-sambubioside, (**4**) cyanidin-3-*O*-glucoside, (**5**) rutin, and (**6**) quercetin-3-*O*-glucoside.

3.2. Effect of Storage on Total Phenolic Content, Total Monomeric Anthocyanins, Percent Polymeric Color, and Antioxidant Activity

The changes in total phenolic content (TPC), total monomeric anthocyanins (TMA), percentage of polymeric color (%PC), and antioxidant activity (AA) were evaluated during storage at 5 °C and at room temperature (Figure 3).

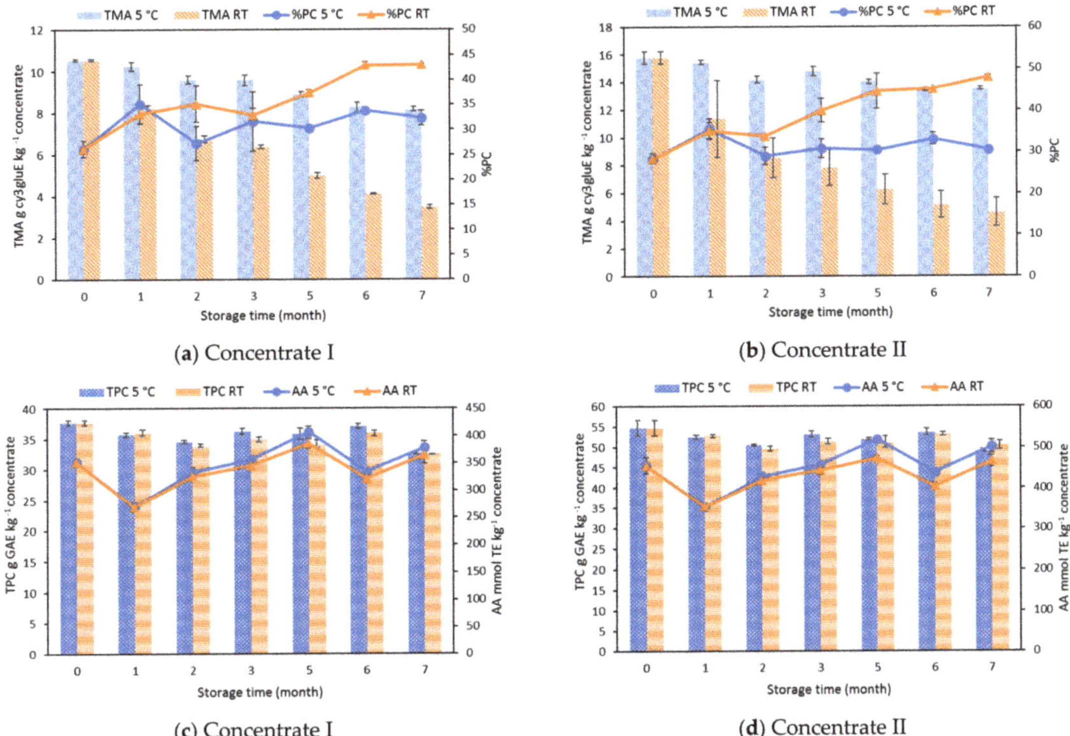

Figure 3. Evolution during storage at 5 °C and at room temperature (RT) for total monomeric anthocyanins (TMA) and percent polymeric color (%PC) in (**a**) concentrate I, (**b**) concentrate II; and total phenolic content (TPC) and antioxidant activity (AA) in (**c**) concentrate I and (**d**) concentrate II.

After seven months of storage at 5 °C, the content of total monomeric anthocyanins decreased from 10.6 to 8.2 g cy3gluE kg^{-1} in concentrate I and from 15.8 to 13.5 g cy3gluE kg^{-1} in concentrate II, representing losses of 22% and 14%, respectively. Moreover, storage at room temperature resulted in a strong reduction in total monomeric anthocyanins in both concentrates. After the same seven months, their content was 3.5 g cy3gluE kg^{-1} in concentrate I and 4.6 g cy3gluE kg^{-1} in concentrate II, corresponding to losses of 67% and 71%. These results demonstrate the greater stability of monomeric anthocyanins at lower temperatures, which is in accordance with studies conducted by Buckow et al. [27] in which they observed that degradation of anthocyanins in blueberry juices was significantly accelerated with increasing storage temperatures. At the same time, the percentage of polymeric color was slightly affected at 5 °C for both juices, while at room temperature increased from 26.2% to 43.0% in concentrate I and from 28.3% to 47.9% in concentrate II. Similar observations were made by Brownmiller et al. [28] in blueberry juices with 64–68% losses of anthocyanins after six months of storage at 25 °C accompanied by a linear increase in percent polymeric color values. In addition, losses up to 75% in monomeric anthocyanins and a marked increase in percent polymeric color in blackberry [29] and in raspberry juices [30] after six months of storage at 25 °C were reported. Our results are also consistent with the kinetic studies developed by Casati et al. [31] that predict a reduction of 50% of the monomeric anthocyanins in elderberry juice after 120 days of storage at 25 °C.

The total phenolic content of the juice concentrates measured by the Folin–Ciocalteu method, whether stored at room temperature or at refrigerated temperature, does not follow the decrease in total monomeric anthocyanin content, and small variations over time were observed. The higher stability observed in the total phenolic content may be explained by the formation of polymerized phenolic compounds that are detected by the Folin–Ciocalteu method [32]. This hypothesis is corroborated by the increase in the percent polymeric color. In fact, the decrease in monomeric anthocyanins content and the increase in percent polymeric color indicates that anthocyanins were transformed into polymeric compounds. For total phenolic compounds, after seven months of storage losses of about 13% and 14% were obtained for concentrate I and 10% and 8% of losses were found for concentrate II, at room temperature and at 5 °C, respectively. The results demonstrate that the total phenolic content is not much affected by the storage temperature, contrary to what happens with the monomeric anthocyanin content. The antioxidant activity, in general, follows the same profile of variation of total phenolic content and remained approximately stable over the seven months of storage despite the losses of total monomeric anthocyanins (Figure 3). Other authors [28,31] have shown no significant changes in antioxidant capacity of juices during storage, despite a marked loss of anthocyanins. This may be related to the presence of polymeric compounds which have been associated with antioxidant capacity [33].

3.3. *Effect of Storage on Individual Phenolic Compounds*

The content of individual phenolic compounds in the juice concentrates was followed by HPLC-DAD over the seven months of storage. The chromatograms of concentrate I after seven months of storage are presented in Figure 2c–f. A decrease in the areas of anthocyanin peaks at refrigerated temperature and an almost disappearance of anthocyanin peaks at room temperature were observed.

Figure 4 shows the evolution of the content of individual phenolic compounds in concentrate I and concentrate II over storage.

After seven months, at 5 °C the concentrates presented a composition of anthocyanins that represented 61–76% of the initial concentration, while for room temperature, the values represented 3–21%—Figure 4a–c. Chlorogenic acid showed a slight decrease around 10% over the time—Figure 4d, and the amount of rutin and quercetin-3-*O*-glucoside remained quite stable, with a tendency to increase after five months—Figure 4e,f.

For both concentrate I and concentrate II there is a marked loss of anthocyanin content at room temperature over the storage time—Figure 4a–c. Degradation was more pronounced for cyanidin-3-*O*-sambubioside and cyanidin-3-*O*-glucoside—Figure 4a,b, respectively—than for cyanidin-3,5-*O*-diglucoside and cyanidin-3-*O*-sambubioside-5-*O*-glucoside—Figure 4c. In the case of concentrate I, after seven months of storage at room temperature, 21.5% of the initial concentration of cyanidin-3,5-*O*-diglucoside + cyanidin-3-*O*-sambubioside-5-*O*-glucoside was found, whereas only 9.0% of cyanidin-3-*O*-sambubioside and 9.5% of cyanidin-3-*O*-glucoside were maintained. A similar result was obtained for concentrate II with retained concentrations of 19.4%, 2.5%, and 3.4%, respectively for cyanidin-3,5-*O*-diglucoside + cyanidin-3-*O*-sambubioside-5-*O*-glucoside, cyanidin-3-*O*-sambubioside, and cyanidin-3-*O*-glucoside, respectively. It has been described that anthocyanins' stability is influenced by the glycosylation site, and by the type and number of glycosyl moieties attached [34]. Our findings are in line with other studies that showed lower stability of 3-*O*-glycosides compared to the 3,5-*O*-diglycosides counterparts [35,36]. The lower stability of 3-*O*-glycosides can be explained by the mesomeric effect of the hydroxyl group at C-5 position that favors the electrophilic attack at C-6 and C-8 positions. The glycosyl substitution at C-5 reduces the nucleophilicity of C-6 and C-8 positions which makes 3,5-*O*-diglycosides more stable to electrophilic attack [37].

The contents of chlorogenic acid also decreased substantially during storage at room temperature in both concentrates. After seven months, 65.0% and 68.4% of its initial concentration was obtained for concentrate I and II, respectively. This indicates a possible involvement of chlorogenic acid in anthocyanins polymerization, which agrees with the observations of other authors [24].

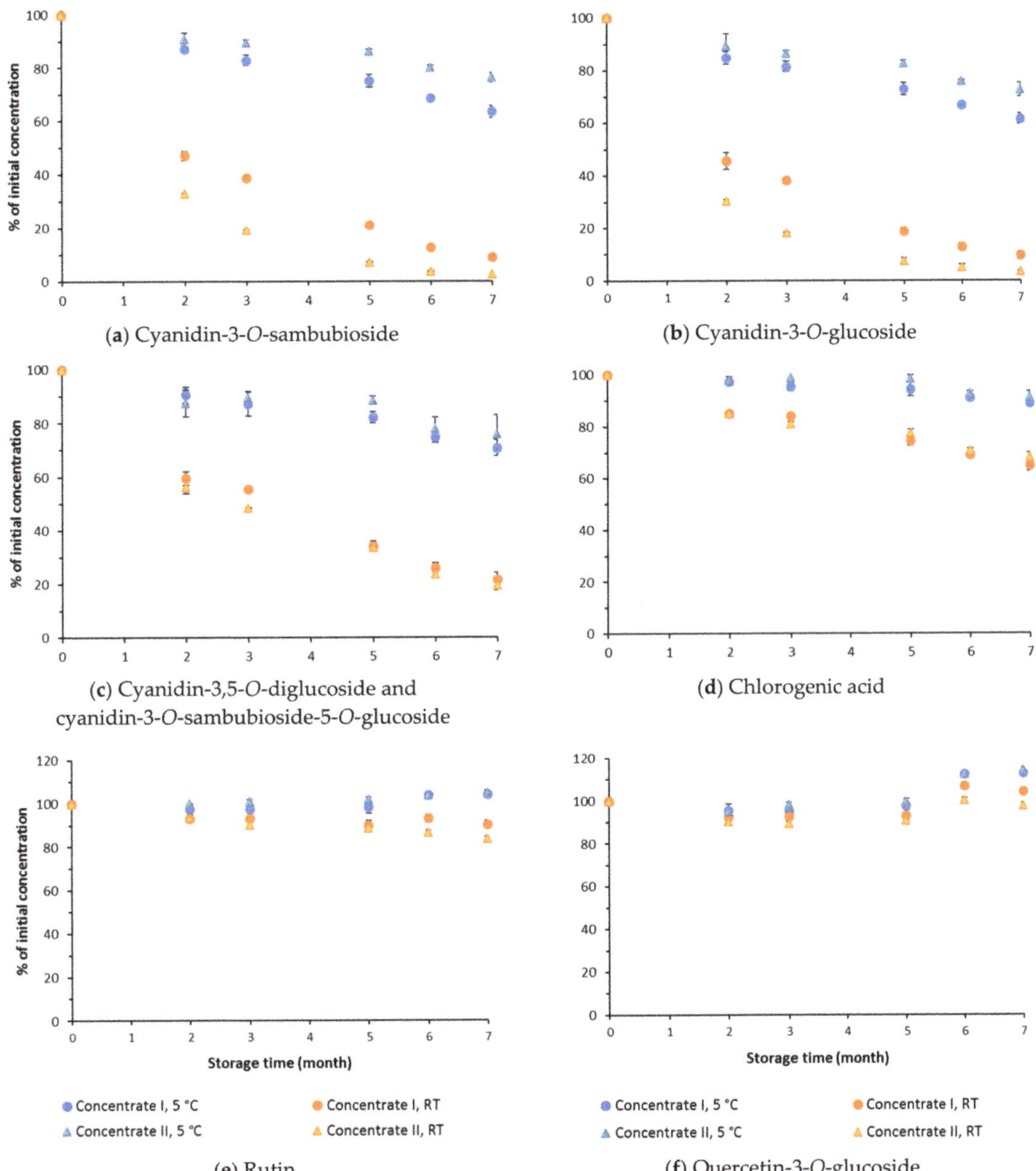

Figure 4. Changes in (**a**) cyanidin-3-*O*-sambubioside, (**b**) cyanidin-3-*O*-glucoside, (**c**) cyanidin-3,5-*O*-diglucoside + cyanidin-3-*O*-sambubioside-5-*O*-glucoside, (**d**) chlorogenic acid, (**e**) rutin, and (**f**) quercetin-3-*O*-glucoside of elderberry concentrates I and II during storage at 5 °C and at room temperature (RT).

In the present study, concentrate II, with about 80.6 °Brix, showed a faster degradation of some compounds at room temperature when compared with concentrate I, which had 55.8 °Brix. For instance, after three months of storage, concentrate I presented 39% of the initial concentration of cyanidin-3-*O*-sambubioside, while in concentrate II only 19% was retained. A similar result was obtained for cyanidin-3-*O*-glucoside. Conversely, Kirca et al. [38] showed that the degradation of monomeric anthocyanins in black carrot juice decreased with increasing soluble solid content during storage at 20 °C. It should be noted that concentrate II was not subjected to the heat treatment at the initial stage of processing and the native enzymes may be present in concentrate II, favoring the polymerization reactions between anthocyanins and other phenolic compounds. In addition, after five months at room temperature, concentrate II presented higher microbial activity than concentrate I (Table 3), which may be another reason for the greater degradation of the compounds. In the storage at refrigerated temperature, concentrate I and concentrate II showed low microbial growth and, in this case, the stability of anthocyanins was similar in both concentrates.

Table 3. Results of microbial counting obtained for concentrates before storage and after five months of storage.

Sample	Mesophiles at 30 °C (CFU mL^{-1})	Fungi (CFU mL^{-1})
Concentrate I before storage	<1	<1
Concentrate I stored at 5 °C	<1	<1
Concentrate I stored at room temperature	2.1×10^2	<1
Concentrate II before storage	<1	<1
Concentrate II stored at 5 °C	2.2×10	1.0×10
Concentrate II stored at room temperature	1.1×10^3	1.1×10^4

4. Conclusions

The storage at low temperature (5 °C) was shown to be important to minimize anthocyanin degradation and maintain the color attributes of elderberry juice concentrates. At 5 °C there was a loss of 22% and 14% (concentrate I and II) in total monomeric anthocyanins, while at room temperature, losses of 67% and 71% were observed. At the same time, the percent polymeric color increased, especially at room temperature, which indicates that anthocyanins may be involved in polymerization reactions. The analysis of individual phenolic compounds showed that cyanidin-3-*O*-sambubioside and cyanidin-3-*O*-glucoside degraded in higher proportion than cyanidin-3,5-*O*-diglucoside and cyanidin-3-*O*-sambubioside-5-*O*-glucoside. At room temperature, cyanidin-3-*O*-sambubioside and cyanidin-3-*O*-glucoside degraded faster in the concentrate that was not subjected to the additional heating step during the technological processing. Thus, the thermal treatment of the mash at 80 °C prior to pressing appears to have a beneficial effect on the conservation of anthocyanins during storage at room temperature.

The total phenolic content and the in vitro $ABTS^{\bullet+}$ antioxidant capacity were relatively stable over storage, suggesting that elderberry concentrates may still preserve their health benefits after seven months of storage.

Author Contributions: Conceptualization, C.M.B.N., A.P., F.G. and D.F.W.; methodology, C.M.B.N., A.P. and D.F.W.; investigation, C.M.B.N., F.G. and D.F.W.; writing—original draft preparation, C.M.B.N.; writing—review and editing, A.P., F.G. and D.F.W.; supervision, D.F.W. All authors have read and agreed to the published version of the manuscript.

Funding: This research was funded by the FEDER, within the PT2020 Partnership Agreement and Compete 2020, grant number POCI-01-0247-FEDER-033558.

Institutional Review Board Statement: Not applicable.

Informed Consent Statement: Not applicable.

Acknowledgments: Thanks are due to the Polytechnic Institute of Viseu and to FCT/MEC for the financial support to CERNAS-IPV, LAQV-REQUIMTE (UIDB/50006/2020) and CITAB (UIDB/04033/2020) research units, through national funds, and the co-funding by the FEDER, within the PT2020 Partnership Agreement and Compete 2020. The authors thank the financial support of the Project BagaConValor—Criação de valor no processo tecnológico de produção de sumo concentrado de baga de sabugueiro (POCI-01-0247-FEDER-033558).

Conflicts of Interest: The authors declare no conflict of interest.

References

1. Schmitzer, V.; Veberic, R.; Stampar, F. European elderberry (*Sambucus nigra* L.) and American Elderberry (*Sambucus canadensis* L.): Botanical, chemical and health properties of flowers, berries and their products. In *Berries: Properties, Consumption and Nutrition*; Tuberoso, C., Ed.; Nova Science Publishers Inc.: New York, NY, USA, 2012; ISBN 9781614702573.
2. Lee, J.; Finn, C.E. Anthocyanins and Other Polyphenolics in American Elderberry (*Sambucus Canadensis*) and European Elderberry (*S. nigra*) Cultivars. *J. Sci. Food Agric.* **2007**, *87*, 2665–2675. [CrossRef] [PubMed]
3. Veberic, R.; Jakopic, J.; Stampar, F.; Schmitzer, V. European Elderberry (*Sambucus nigra* L.) Rich in Sugars, Organic Acids, Anthocyanins and Selected Polyphenols. *Food Chem.* **2009**, *114*, 511–515. [CrossRef]
4. Ferreira, S.S.; Silva, P.; Silva, A.M.; Nunes, F.M. Effect of Harvesting Year and Elderberry Cultivar on the Chemical Composition and Potential Bioactivity: A Three-Year Study. *Food Chem.* **2020**, *302*, 125366. [CrossRef] [PubMed]
5. Sidor, A.; Gramza-Michałowska, A. Advanced Research on the Antioxidant and Health Benefit of Elderberry (*Sambucus nigra*) in Food—A Review. *J. Funct. Foods* **2015**, *18*, 941–958. [CrossRef]
6. Przybylska-Balcerek, A.; Szablewski, T.; Szwajkowska-Michałek, L.; Świerk, D.; Cegielska-Radziejewska, R.; Krejpcio, Z.; Suchowilska, E.; Tomczyk, Ł.; Stuper-Szablewska, K. *Sambucus nigra* Extracts–Natural Antioxidants and Antimicrobial Compounds. *Molecules* **2021**, *26*, 2910. [CrossRef] [PubMed]
7. Młynarczyk, K.; Walkowiak-Tomczak, D.; Łysiak, G.P. Bioactive Properties of *Sambucus nigra* L. as a Functional Ingredient for Food and Pharmaceutical Industry. *J. Funct. Foods* **2018**, *40*, 377–390. [CrossRef] [PubMed]
8. Goiffon, J.-P.; Mouly, P.P.; Gaydou, E.M. Anthocyanic Pigment Determination in Red Fruit Juices, Concentrated Juices and Syrups Using Liquid Chromatography. *Anal. Chim. Acta* **1999**, *382*, 39–50. [CrossRef]
9. Bermúdez-Soto, M.J.; Tomás-Barberán, F.A. Evaluation of Commercial Red Fruit Juice Concentrates as Ingredients for Antioxidant Functional Juices. *Eur. Food Res. Technol.* **2004**, *219*, 133–141. [CrossRef]
10. Vlachojannis, C.; Zimmermann, B.F.; Chrubasik-Hausmann, S. Quantification of Anthocyanins in Elderberry and Chokeberry Dietary Supplements. *Phytother. Res.* **2015**, *29*, 561–565. [CrossRef]
11. Kaack, K.; Fretté, X.C.; Christensen, L.P.; Landbo, A.-K.; Meyer, A.S. Selection of Elderberry (*Sambucus nigra* L.) Genotypes Best Suited for the Preparation of Juice. *Eur. Food Res. Technol.* **2008**, *226*, 843–855. [CrossRef]
12. Senica, M.; Stampar, F.; Veberic, R.; Mikulic-Petkovsek, M. Processed Elderberry (*Sambucus nigra* L.) Products: A Beneficial or Harmful Food Alternative? *LWT Food Sci. Technol.* **2016**, *72*, 182–188. [CrossRef]
13. Patras, A.; Brunton, N.P.; O'Donnell, C.; Tiwari, B.K. Effect of Thermal Processing on Anthocyanin Stability in Foods; Mechanisms and Kinetics of Degradation. *Trends Food Sci. Technol.* **2010**, *21*, 3–11. [CrossRef]
14. Buvé, C.; Kebede, B.T.; de Batselier, C.; Carrillo, C.; Pham, H.T.T.; Hendrickx, M.; Grauwet, T.; van Loey, A. Kinetics of Colour Changes in Pasteurised Strawberry Juice during Storage. *J. Food Eng.* **2018**, *216*, 42–51. [CrossRef]
15. Deng, J.; Yang, H.; Capanoglu, E.; Cao, H.; Xiao, J. Technological aspects and stability of polyphenols. In *Polyphenols: Properties, Recovery, and Applications*; Woodhead Publishing: Duxford, UK, 2018; pp. 295–323.
16. Jackman, R.L.; Yada, R.Y.; Tung, M.A.; Speers, R.A. Anthocyanins as Food Colorants—A Review. *J. Food Biochem.* **1987**, *11*, 201–247. [CrossRef]
17. Jiang, T.; Mao, Y.; Sui, L.; Yang, N.; Li, S.; Zhu, Z.; Wang, C.; Yin, S.; He, J.; He, Y. Degradation of Anthocyanins and Polymeric Color Formation during Heat Treatment of Purple Sweet Potato Extract at Different PH. *Food Chem.* **2019**, *274*, 460–470. [CrossRef]
18. Howard, L.R.; Prior, R.L.; Liyanage, R.; Lay, J.O. Processing and Storage Effect on Berry Polyphenols: Challenges and Implications for Bioactive Properties. *J. Agric. Food Chem.* **2012**, *60*, 6678–6693. [CrossRef]
19. Gonçalves, F.J.; Rocha, S.M.; Coimbra, M.A. Study of the Retention Capacity of Anthocyanins by Wine Polymeric Material. *Food Chem.* **2012**, *134*, 957–963. [CrossRef]
20. Mónica Giusti, M.; Wrolstad, R.E. Anthocyanins. In *Handbook of Food Analytical Chemistry*; John Wiley & Sons, Inc.: Hoboken, NJ, USA, 2005; Volume 2, pp. 5–69. ISBN 9780471709084.
21. Sun, B.; Ricardo-da-Silva, J.M.; Spranger, I. Critical Factors of Vanillin Assay for Catechins and Proanthocyanidins. *J. Agric. Food Chem.* **1998**, *46*, 4267–4274. [CrossRef]
22. Pękal, A.; Pyrzynska, K. Evaluation of Aluminium Complexation Reaction for Flavonoid Content Assay. *Food Anal. Methods* **2014**, *7*, 1776–1782. [CrossRef]
23. Ozgen, M.; Reese, R.N.; Tulio, A.Z.; Scheerens, J.C.; Miller, A.R. Modified 2,2-Azino-Bis-3-Ethylbenzothiazoline-6-Sulfonic Acid (ABTS) Method to Measure Antioxidant Capacity of Selected Small Fruits and Comparison to Ferric Reducing Antioxidant Power (FRAP) and 2,2′-Diphenyl-1-Picrylhydrazyl (DPPH) Methods. *J. Agric. Food Chem.* **2006**, *54*, 1151–1157. [CrossRef]

24. Kader, F.; Rovel, B.; Girardin, M.; Metche, M. Mechanism of Browning in Fresh Highbush Blueberry Fruit (*Vaccinium corymbosum* L.) Role of Blueberry Polyphenol Oxidase, Chlorogenic Acid and Anthocyanins. *J. Sci. Food Agric.* **1997**, *74*, 31–34. [CrossRef]
25. Kader, F.; Irmouli, M.; Nicolas, J.P.; Metche, M. Involvement of Blueberry Peroxidase in the Mechanisms of Anthocyanin Degradation in Blueberry Juice. *J. Food Sci.* **2002**, *67*, 910–915. [CrossRef]
26. Watanabe, T.; Yamamoto, A.; Nagai, S.; Terabe, S. Analysis of Elderberry Pigments in Commercial Food Samples by Micellar Electrokinetic Chromatography. *Anal. Sci.* **1998**, *14*, 839–844. [CrossRef]
27. Buckow, R.; Kastell, A.; Terefe, N.S.; Versteeg, C. Pressure and Temperature Effects on Degradation Kinetics and Storage Stability of Total Anthocyanins in Blueberry Juice. *J. Agric. Food Chem.* **2010**, *58*, 10076–10084. [CrossRef]
28. Brownmiller, C.; Howard, L.R.; Prior, R.L. Processing and Storage Effects on Monomeric Anthocyanins, Percent Polymeric Color, and Antioxidant Capacity of Processed Blueberry Products. *J. Food Sci.* **2008**, *73*, H72–H79. [CrossRef]
29. Hager, T.J.; Howard, L.R.; Prior, R.L. Processing and Storage Effects on Monomeric Anthocyanins, Percent Polymeric Color, and Antioxidant Capacity of Processed Blackberry Products. *J. Agric. Food Chem.* **2008**, *56*, 689–695. [CrossRef]
30. Hager, A.; Howard, L.R.; Prior, R.L.; Brownmiller, C. Processing and Storage Effects on Monomeric Anthocyanins, Percent Polymeric Color, and Antioxidant Capacity of Processed Black Raspberry Products. *J. Food Sci.* **2008**, *73*, H134–H140. [CrossRef]
31. Casati, C.B.; Baeza, R.; Sanchez, V.; Catalano, A.; López, P.; Zamora, M.C. Thermal Degradation Kinetics of Monomeric Anthocyanins, Colour Changes and Storage Effect in Elderberry Juices. *J. Berry Res.* **2015**, *5*, 29–39. [CrossRef]
32. Ferreira, D.; Guyot, S.; Marnet, N.; Delgadillo, I.; Renard, C.M.G.C.; Coimbra, M.A. Composition of Phenolic Compounds in a Portuguese Pear (*Pyrus communis* L. Var. *S. bartolomeu*) and Changes after Sun-Drying. *J. Agric. Food Chem.* **2002**, *50*, 4537–4544. [CrossRef]
33. Tsai, P.-J.; Huang, H.-P.; Huang, T.-C. Relationship between Anthocyanin Patterns and Antioxidant Capacity in Mulberry Wine during Storage. *J. Food Qual.* **2004**, *27*, 497–505. [CrossRef]
34. Zhao, C.L.; Chen, Z.J.; Bai, X.S.; Ding, C.; Long, T.J.; Wei, F.G.; Miao, K.R. Structure–Activity Relationships of Anthocyanidin Glycosylation. *Mol. Divers.* **2014**, *18*, 687–700. [CrossRef] [PubMed]
35. Martí, N.; Pérez-Vicente, A.; García-Viguera, C. Influence of Storage Temperature and Ascorbic Acid Addition on Pomegranate Juice. *J. Sci. Food Agric.* **2002**, *82*, 217–221. [CrossRef]
36. Brauch, J.E.; Kroner, M.; Schweiggert, R.M.; Carle, R. Studies into the Stability of 3-O-Glycosylated and 3,5-O-Diglycosylated Anthocyanins in Differently Purified Liquid and Dried Maqui (Aristotelia Chilensis (Mol.) Stuntz) Preparations during Storage and Thermal Treatment. *J. Agric. Food Chem.* **2015**, *63*, 8705–8714. [CrossRef] [PubMed]
37. Timberlake, C.F.; Bridle, P. Anthocyanins: Colour Augmentation with Catechin and Acetaldehyde. *J. Sci. Food Agric.* **1977**, *28*, 539–544. [CrossRef]
38. Kırca, A.; Özkan, M.; Cemeroğlu, B. Effects of Temperature, Solid Content and PH on the Stability of Black Carrot Anthocyanins. *Food Chem.* **2007**, *101*, 212–218. [CrossRef]

Article

Validation of a Commercial Loop-Mediated Isothermal Amplification (LAMP)-Based Kit for the Detection of *Salmonella* spp. According to ISO 16140:2016

Calogero Di Bella [1], Antonella Costa [2], Sonia Sciortino [2], Giuseppa Oliveri [2], Gaetano Cammilleri [3], Francesco Geraci [1], Daniela Lo Monaco [1], Davide Carpintieri [1], Giuseppe Lo Bue [1], Carmelo Bongiorno [4], Alessandro Altomare [1], Valentina Ciprì [1,*], Rosario Pitti [5], Carmine Lanzillo [5], Giuseppe Arcoleo [5] and Rosalinda Allegro [1]

1 Area Sorveglianza Epidemiologica, Istituto Zooprofilattico Sperimentale della Sicilia (IZSSi), Via G. Marinuzzi 3, 90129 Palermo, Italy; calogero.dibella@izssicilia.it (C.D.B.); francesco.geraci@izssicilia.it (F.G.); daniela.lomonaco@izssicilia.it (D.L.M.); da.car19@gmail.com (D.C.); lobuegiuseppe1981izs@gmail.com (G.L.B.); alessandro.altomare@izssicilia.it (A.A.); rosalinda.allegro@gmail.com (R.A.)
2 Area Microbiologia degli Alimenti, Istituto Zooprofilattico Sperimentale della Sicilia (IZSSi), Via G. Marinuzzi 3, 90129 Palermo, Italy; antonella.costa@izssicilia.it (A.C.); sonia.sciortino@izssicilia.it (S.S.); giuseppa.oliveri@izssicilia.it (G.O.)
3 Dipartimento Alimenti, Area Chimica e Tecnologie Alimentari, Istituto Zooprofilattico Sperimentale della Sicilia (IZSSi), Via G. Marinuzzi 3, 90129 Palermo, Italy; gaetano.cammilleri@izssicilia.it
4 Area Affari Generali Legali e Contenzioso, Istituto Zooprofilattico Sperimentale della Sicilia (IZSSi), Via G. Marinuzzi 3, 90129 Palermo, Italy; carmelo.bongiorno@izssicilia.it
5 Enbiotech s.r.l. Via Aquileia 34, 90144 Palermo, Italy; r.pitti@enbiotech.eu (R.P.); c.lanzillo@enbiotech.eu (C.L.); g.arcoleo@enbiotech.eu (G.A.)
* Correspondence: valentinacipri@gmail.com; Tel.: +39-3201-816-981

Citation: Di Bella, C.; Costa, A.; Sciortino, S.; Oliveri, G.; Cammilleri, G.; Geraci, F.; Lo Monaco, D.; Carpintieri, D.; Lo Bue, G.; Bongiorno, C.; et al. Validation of a Commercial Loop-Mediated Isothermal Amplification (LAMP)-Based Kit for the Detection of *Salmonella* spp. According to ISO 16140:2016. *Appl. Sci.* **2021**, *11*, 6669. https://doi.org/10.3390/app11156669

Academic Editors: Francesco Cappello and Magdalena Gorska-Ponikowska

Received: 23 June 2021
Accepted: 19 July 2021
Published: 21 July 2021

Featured Application: The validated LAMP kit provides an accurate method for the rapid detection of *Salmonella* spp., offering significant advantages over the traditional method, as it is characterised by a high sensitivity, easiness of use for laboratory testing, and a large reduction in the analysis time, making it a valuable asset to the food industry.

Abstract: The traditional cultural method (PCR and Real-Time PCR) for *Salmonella* spp. detection and identification is laborious and time-consuming. A qualitative LAMP method detecting *Salmonella* spp. was validated in compliance with ISO 16140:2016. The results show a relative accuracy, sensitivity, and specificity of 100% in comparison with the reference method ISO 6579-1:2017; the LOD50 was set as 0.4 CFU/g. Additionally, a field study was carried out comparing the LAMP kit, a commercially available Real-Time PCR kit (FoodProof *Salmonella*, Biotecon Diagnostics), and the reference cultural method. The *Salmonella* spp. LAMP kit was suitable for reliable detection of *Salmonella* spp., simplifying and reducing the extent and the steps of the analytical process. A total of 105 samples of raw poultry meat were screened for the presence of *Salmonella* spp. according to three methods: the LAMP kit *Salmonella* spp. (Enbiotech), the Real-Time PCR kit FoodProof *Salmonella* (Biotecon), and the reference cultural method. Using these three methods, only one sample out of the 105 (0.95%) tested was positive for *Salmonella* spp. This sample was further investigated using the reference method described in ISO 6579-3:2014, in order to characterise the *Salmonella* strain. Following this further biochemical identification and serological typing, the isolate was characterised as *Salmonella* Infantis.

Keywords: LAMP; *Salmonella*; ISO 16140:2016; food-borne; validation; specificity; sensitivity; accuracy; kit

1. Introduction

Salmonella is a highly relevant food-borne pathogen of large economic significance for both animals and humans. *Salmonella* outbreaks caused 94,530 human cases in the EU only in 2016, with the highest burden relating to the number of hospitalizations (1766) and deaths (10) [1]. *S.* enteritidis, in particular, accounted for 59% of all *Salmonella* infection cases originating in the EU. The principal reservoirs are the intestines of a wide variety of animals, resulting in the contamination of different foodstuffs, both of animal and plant origin [2,3]. Indeed, *Salmonella* is associated mainly with raw food, subject to faecal contamination, including poultry, raw meat, seafood, egg, and dairy products [4]. In addition, it was also found to be the most common bacterial pathogen responsible for produce outbreaks [5].

Therefore, in EU countries, surveillance of *Salmonella* infections in humans is compulsory, and also for food-producing animals and food thereof. In order to guarantee food security, the availability of reliable methods to identify this pathogenic bacterium is becoming increasingly relevant to the food industry, as well as for official controls.

The traditional cultural method [6] requires more than five days to determine a positive result, besides being laborious and time-consuming. For this reason, to use the traditional cultural method is not very suitable for high-throughput screening of large numbers of food samples for the presence of *Salmonella* cells [7,8]. Less laborious and faster alternative methods for pathogen detection in foods have been developed [8–12]. Among them, the method that combines loop-mediated isothermal amplification (LAMP) with bioluminescence detection stands out as a reliable, faster, and simpler approach than conventional culture methods. LAMP was developed by Notomi et al. in 2000 [9]. It is a method that amplifies DNA with high specificity, efficiency, and rapidity, utilizing a DNA polymerase enzyme with high strand displacement activity and two pairs of primers recognizing six independent sequences of a target gene under isothermal conditions [12]. Subsequently this method has been implemented by Nagamine et al. in 2002, incorporating forward loop primers that accelerate the LAMP reaction and reducing costs as a consequence [13]. Due to its high sensitivity and low cost, LAMP has been applied for pathogen detection screening of large numbers of food samples, and has successfully been used to detect many pathogens, including *Salmonella* spp. [14]. In recent years, some kits based on LAMP have been commercialised and their performance has been positively evaluated [8,12]. Hence alternative methods are catching on, as molecular methods are more rapid and have an interesting potential to be used for screening, revealing a preliminary result, even if ISO 6579-1:2017 [6] still remains necessary to isolate the microorganisms for further characterization. ISO 16140-2:2016 [15] defines the procedures for validation of alternative microbiological methods against the cultural method, measuring the concordance of the results for both methods.

A *Salmonella* LAMP (loop-mediated isothermal amplification) assay was validated in this study as a novel specific and cost-effective nucleic acid amplification method for bacterial detection and identification. This innovative method is characterised by six primers that specifically recognise eight different regions on the target gene [9].

In comparison to PCR and Real-Time PCR, LAMP has many advantages: reaction simplicity, as it can be performed by semi-skilled staff, even in a heating block without any thermal cycler; and detection sensitivity, displaying a 10–100-fold higher sensitivity than PCR [16]. In addition, LAMP shows a higher amplification efficiency and the enzyme commonly involved, Bst DNA polymerase, has shown not to be inhibited by the presence of anticoagulants, NaCl, hemin, and other PCR-interfering substances [17]. Due to its simplicity, the LAMP technique has initially been applied to diagnosis, but recently it has also been extended to genetically modified organisms and identification of meat and fish species in food products [18]. For these reasons, it represents an ideal candidate for point-of-care diagnosis and when rapid results are needed, such as in food industry, where *Salmonella* spp. positive samples need to be immediately blocked for public health and safety.

In this study, we evaluated the LAMP method through validating the kit by comparing it to different methods that are currently the most commonly used.

2. Materials and Methods

All the processed samples used for the method optimisation and validation came from a large-scale distribution, in order to reduce any bias from local food specialities and extend the range of the validation. Samples were chosen as positive samples for the validation of the method and for matrix effects evaluation. All the food samples came from Italian supermarkets.

2.1. DNA Extraction Genomic

Several samples were tested, such as heat-processed milk and dairy products, raw poultry and ready-to-cook poultry products, eggs and egg products (derivate), ready-to-eat and ready-to-reheat fishery products, as well as fresh produce and fruits.

The kit *Salmonella* spp. (Enbiotech) provides a rapid preliminary DNA extraction from food matrices. In the pre enrichment phase, 25 g of a sample was taken and homogenated with 225 mL of Buffered Peptone Water (BPW). After a 22 ± 2 h enrichment in Buffered Peptone Water (BPW) at 37 °C.

The DNA extraction was performed using a ready-to-use buffer contained in the Salmonella Screen Glow kit (Enbiotech). Then, 250 ± 50 mg of a sample was directly placed into 15 mL tubes containing 4 mL of the ready-to-use extraction buffer (Enbiotech) and then incubated for 40 ± 5 min at room temperature.

Genetic amplification using LAMP technology and real-time detection of the results using the dedicated device ICGENE mini (Enbiotech). The kit is ready-to-use and includes:

- DNA extraction buffer, through chemical lysis;
- Tubes strip containing lyophilised primers;
- Amplification master mix;
- Mineral oil;
- Positive control;
- Negative control;
- Sterile water.

2.2. LAMP Assays

The analytical and diagnostic assays to recognise *Salmonella* spp. DNA was performed using the Salmonella Screen Glow commercial kit (Enbiotech) with an ICGENE mini portable instrument (Enbiotech), consisting of a real-time fluorimeter, monitored and regulated by the the ICGENE application (Enbiotech), and downloadable on various smart devices. The Salmonella Screen Glow commercial kit includes ready-to-use reaction tubes (containing primers, fluorescent dye, etc.) to achieve a rapid amplification of the DNA template. The protocol to obtain the specific amplification of the target *Salmonella* spp. DNA was carried out in a mixture with a final volume of 55 µL, including 22 µL of the Salmonella Screen Glow LAMP mix (Enbiotech), 30 µL of mineral oil, and 3 µL of the extracted DNA samples. The mineral oil was added to the top of the reaction mixture to prevent evaporation. The amplification was optimised and performed at 65 °C for 35 min. Real-time monitoring of the fluorescence associated with the amplification was possible using the fluorimeter of the ICGENE portable instrument and the ICGENE application interface.

2.3. Validation Plan

The study was carried out at the Food Microbiology Laboratory of the Istituto Zooprofilattico Sperimentale of Sicily, Palermo (Italy), according to the validation process explained by ISO 16140-2:2016 [15]. According to ISO 16140, we demonstrated that the results obtained with the alternative method (LAMP kit) were comparable (at least equivalent) to the results obtained with the reference method. Following the validation protocol described by ISO 16140-2:2016 [15], we did a comparative study of the alternative method with the corresponding reference method, conducted by the Food Microbiology Laboratory of the Istituto Zooprofilattico Sperimentale of Sicily, Palermo (Italy). Validation determined the following parameters: relative limit of detection (RLOD), inclusivity, exclusivity and a method comparison study including relative accuracy (AC), relative specificity (SP), and sensitivity (SE). Examining the samples both with the alternative (LAMP kit) method and the reference method, the parameters of AC, SP, and SE were calculated as follows:

$$AC = (PA + NA)/N \times 100\% \tag{1}$$

$$SP = (NA)/N_{-} \times 100\% \tag{2}$$

$$SE = (PA)/N_{+} \times 100\% \tag{3}$$

where PA = positive agreement; NA = negative agreement; N = total number of samples; N_{-} = number of negative samples; and N_{+} = number of positive samples.

The RLOD was calculated as follows:

$$RLOD = LOD50 \text{ alternative method}/LOD50 \text{ reference method} \tag{4}$$

where LOD50 = the limit of detection (LOD) of 50% = the smallest amount of analyte that can be detected but not quantified with a 50% probability. Therefore, LOD50 is the level of detection for which 50% of tests give a positive result [19].

RLOD, SP, and SE were performed against the reference method ISO 6579-1:2017 [6] "Microbiology of the food chain-Horizontal method for the detection, enumeration and serotyping of *Salmonella*-Part 1: Detection of *Salmonella* spp."

2.4. Bacterial Strains

Bacterial strains were maintained on cryogenic beads at −20 °C; before use, the beads were placed in Columbia Blood Agar (CBA-Microbiol) plates and incubated at 37 °C for 18–24 h. Subsequently, the bacterial strains were placed in Tryptone Soya Agar (TSA-Microbiol) tubes, incubated at 37 °C for 18–24 h and maintained at 4 °C for two weeks. For RLOD and the comparative studies, a strain of *Salmonella* enteritidis ATCC 13,076 was used. For the inclusivity study, field strains cultures coming from food samples were used, which were identified and confirmed by ISO/TR 6579-3:2014 [20].

Inclusivity of the LAMP method was evaluated by testing 25 pure cultures of the target microorganisms, while exclusivity was determined by testing 30 pure cultures of species other than *Salmonella* spp. (Table 1).

A comparative study was performed comparing the LAMP kit and the reference method. In particular, this study allowed evaluating the relative accuracy, relative specificity, and sensitivity. Food samples were chosen based on the categories given in Table A1 of ISO 16140-2:2016 [15].

In particular, five categories were chosen among the most relevant:

- Heat-processed milk and dairy products;
- Raw poultry and ready-to-cook poultry products;
- Eggs and egg products (derivate);
- Ready-to-eat, ready-to-reheat fishery products;
- Fresh produce and fruits.

For each category, 60 samples were tested, made up of 3 specific typologies with 20 samples representative of each typology (3 typologies × 20 samples for each = 60 samples per category). Of tested samples per typology, 50% (i.e., 10) were negative and 50% were spiked and hence positive.

RLOD tests were run on the same five food matrices of the comparative study: heat-processed milk and dairy products; raw poultry and ready-to-cook poultry products; eggs and egg products (derivates); ready-to-eat, ready-to-reheat fishery products; and fresh produce and fruits.

Table 1. *Salmonella* spp. tested for inclusivity and species used for exclusivity testing.

Microorganism				
	Tested for Inclusivity		Species Used for Exclusivity Testing	Code
1	*S. Livingstone*	1	*Aeromonas hydrophila*	ATCC 35650
2	*S. Heron*	2	*Arcobacter butzleri*	NCTC 12481
3	*S. Corn*	3	*Bacillus cereus*	ATCC 11778
4	*S. Madelia*	4	*Bacillus cereus*	B25052
5	*S. Typhimurium (monophasic)*	5	*Bacillus subtilis*	BCS51
6	*S. Thompson*	6	*Campylobacter coli*	ATCC 33559
7	*S. Virchow*	7	*Campylobacter jejuni*	ATCC 33291
8	*S. London*	8	*Citrobacter freudii*	ATCC 8990
9	*S. Typhimurium*	9	*Clostridium bifermentans*	CBIF107
10	*S. Kissi*	10	*Clostridium perfringens*	ATCC 13124
11	*S. Blocklei*	11	*Escherichia coli*	ATCC 25922
12	*S. Toulon*	12	*Escherichia coli O157*	ATCC 35150
13	*S. Halle*	13	*Enterobacter cloacae*	Not available
14	*S. Abony*	14	*Enterococcus faecium*	EFC49
15	*S. Messina*	15	*Enterobacter sakazakii*	ATCC 29544
16	*S. Montevideo*	16	*Listeria innocua*	ATCC 33090
17	*S. Potsdam*	17	*Listeria ivanovii*	ATCC 19119
18	*S. Muenster*	18	*Listeria monocytogenes*	ATCC 7684
19	*S. Larochelle*	19	*Listeria seeligeri*	Not available
20	*S. Newport*	20	*Micrococcus luteus*	ATCC 9341
21	*S. Hadar*	21	*Psudomonas aeruginosa*	ATCC 10145
22	*S. Poona*	22	*Rhodococcus equi*	ATCC 6939
23	*S. Muenchen*	23	*Staphylococcus aureus*	ATCC 25923
24	*S. Derby*	24	*Staphylococcus aureus*	ATCC 38862
25	*S. Kottbus*	25	*Shigella sonnei*	ATCC 9290
		26	*Streptococccus agalactiae*	STRA41
		27	*Vibrio cholerae*	ATCC 1473A
		28	*Vibrio parahaemolyticus*	ATCC 17802
		29	*Vibrio vulnificus*	ATCC 27562
		30	*Yersinia enterocolitica*	ATCC 23715

From TSA, *Salmonella* enteritidis was inoculated in Xylose Lysine Deoxycholate agar (XLD-Microbiol) plates to obtain isolated colonies. XLD agar plates were incubated at 37 °C for 24 h.

In order to calculate the RLOD value, each of the five categories was spiked with the target microorganisms at three levels of contamination. In particular, they were 5 replicates of negative samples (0 CFU), 20 replicates of the lowest detection level (0.4 CFU/g), and 5 replicates of a higher contamination level (4 CFU/g), for a total of 30 contaminated samples for each food category. In total, 25 g of each the different food samples were inoculated with the corresponding level of contamination and the samples were then stabilized at room temperature or 4 °C, depending on food typology and its storage temperature. Subsequently, the reference and the alternative methods were performed.

3. Results

The method was optimised for the DNA extraction phase by testing in triplicate the initial weight of the samples at 250 mg. The extract was tested with three levels of contamination: 0, 0.4, and 4 CFU/g.

3.1. Validation

As for inclusivity and exclusivity, the results showed all samples were correctly recognised; as a matter of fact, all target microorganisms were identified, while the relevant range of other species tested did not interfere.

The relative accuracy, specificity, and sensitivity of each food category are reported in Table 2.

Table 2. Relative accuracy, relative sensitivity, and relative specificity of the alternative method (*Salmonella* spp.).

Category	PA	NA	PD	ND	N	AC	SE	SP
Heat-processed milk and dairy products	30	30	0	0	60	100%	100%	100%
Raw poultry and ready-to-cook poultry products	30	30	0	0	60	100%	100%	100%
Eggs and egg products (derivates)	30	30	0	0	60	100%	100%	100%
Fresh produce and fruits	30	30	0	0	60	100%	100%	100%
Ready-to-eat, ready-to-reheat fishery products	30	30	0	0	60	100%	100%	100%
Heat-processed milk and dairy products	30	30	0	0	60	100%	100%	100%
Raw poultry and ready-to-cook poultry products	30	30	0	0	60	100%	100%	100%
Eggs and egg products (derivates)	30	30	0	0	60	100%	100%	100%
Fresh produce and fruits	30	30	0	0	60	100%	100%	100%
Ready-to-eat, ready-to-reheat fishery products	30	30	0	0	60	100%	100%	100%
Heat-processed milk and dairy products	30	30	0	0	60	100%	100%	100%

The results showed 100% for all three parameters for all the food categories; hence, neither false-positive nor false-negative samples were detected, with this 100% performance consistent with the reference method.

The RLOD value obtained was 1 for all the categories; therefore, the same LOD was reached both for the LAMP kit and for the reference method, while the detection limit was set at 0.4 CFU/g for all food categories (Table 3). At this concentration, false-negative results were found with the LAMP method with the commercialised kit: a sample of the heat-processed milk and dairy products and three samples of raw poultry and ready-to-cook poultry products (Table 3). For the detection limit of 0.4 CFU/g, the alternative (LAMP kit) method proved to be less sensitive than the traditional ones.

Table 3. Data for RLOD calculation (N. tot = total number of samples; N. pos ref = number of positives with the reference method; N. pos kit = number of positives with the *Salmonella* spp.).

Category	Contamination Level	CFU/g	N. Tot	N. Pos Ref	N. Pos Kit
Heat-processed milk and dairy products	1	0	5	0	0
	2	0.4	20	20	19
	3	4	5	5	5
Raw poultry and ready-to-cook poultry products	1	0	5	0	0
	2	0.4	20	20	17
	3	4	5	5	5
Eggs and egg products (derivates)	1	0	5	0	0
	2	0.4	20	20	20
	3	4	5	5	5
Fresh produce and fruits	1	0	5	0	0
	2	0.4	20	20	20
	3	4	5	5	5
Ready-to-eat, ready-to-reheat fishery products	1	0	5	0	0
	2	0.4	20	20	20
	3	4	5	5	5

3.2. Field Study

A total of 105 samples of raw poultry meat were screened for the presence of *Salmonella* spp. according to three methods: the LAMP kit *Salmonella* spp. (Enbiotech), the Real-Time PCR kit FoodProof *Salmonella* (Biotecon), and the reference cultural method. Using these three methods, only one sample out of the 105 (0.95%) tested was positive for *Salmonella* spp. This sample was further investigated using the reference method described in ISO.

4. Discussion

In recent years, different diagnostic approaches have been developed for detecting various food pathogens, including innovative molecular methods.

Salmonella infections have been declining constantly since the implementation of EU control measures in poultry in 2007, although the data for 2016 showed a relevant increase of 11.5% in the number of cases compared to the previous year [1], underlining the need for continued risk management plans both at the state and at the food industry level.

In this context, the use of alternative methods, such as the "*Salmonella* spp." kit that can rapidly identify pathogenic bacteria, is of great relevance, provided they are validated against the standardized reference method as stated by ISO 16140-2:2016 [15].

The results obtained by the validated method in the comparative studies were equivalent to the microbiological reference method, hence providing a valid alternative to the cultural method. The relative sensitivity was found to be 100% for all the food typologies examined, confirming the absence of inhibition by different kinds of substrates. Moreover, the kit is characterised by a 100% specificity, as it does not amplify the other species tested, and it is inclusive of at least 25 serovars of *Salmonella* evaluated. At the detection limit of 0.4 CFU/g, the LAMP kit showed false-negative results for 4 out of 100 samples; but, even though there is a lower sensitivity than the traditional methods, it still has the advantage in terms of speed of execution and ease of use.

During this study, we also compared the LAMP kit with another commercially available diagnostic method: Real-Time PCR in raw poultry samples. The results indicate that all the methods were in good agreement, even if a limitation of this study is the scarcity of positive results (i.e., 1 sample) that could hinder a more deepened evaluation. No problems of PCR inhibition were found using the internal amplification control provided in the kit; hence, in negative samples, the absence of pathogenic microorganisms was effectively determined.

Another feature of this study is that out of the 105 poultry samples screened for the presence of *Salmonella* spp., only 1 positive sample occurred, with a prevalence of 0.95%. Samples were bought in different retail markets in order to have a more realistic representation of Sicilian poultry contamination. In fact, poultry flocks, particularly chicken, are frequently colonized with *Salmonella* without any detectable symptoms by horizontal and vertical transmission at the primary production level [21].

In European countries, the percentage of *Salmonella*-positive samples from fresh broiler meat is quite higher (4.85%) [22], even if, besides retail, also samples from slaughterhouse and processing plants are included.

Although our prevalence is rather low, this should be taken as an additional motivation for the continuous control of this pathogens, as an effective implementation of control measures could still decrease the prevalence, producing safer food. Constant monitoring is mandatory to avoid new difficulties, such as the increasing antibiotic resistance in *Salmonella* spp. that has become a severe issue for public health at a global level [23].

5. Conclusions

The data in this study support the suitability of the *Salmonella* spp. kit for commercial use on different food samples, including egg products and poultry meat, which are the foods most associated with salmonellosis.

Therefore, the validated LAMP kit provides an accurate method for the rapid detection of *Salmonella* spp., offering significant advantages over the traditional method, as it is characterised by a high sensitivity (up to 0.4 CFU/g), easiness of use for laboratory testing, and a large reduction in the analysis time (about 26 h to obtain definitive results), making it a valuable asset to the food industry. Despite the LAMP kit being less sensitive than the traditional methods, the great rapidity and ease of use suggest that the LAMP assay can be a valid alternative for routine examination in the food sector and for screening of large numbers of food samples.

In our study we isolated the S. Infantis strain from the positive sample. In the EU, an increased occurrence of various serotypes implicated in human infections, including S. Infantis, has been reported, related to poultry meat [24]. The increase in S. Infantis has been associated with the propagation of various clones of broiler origin in different European countries, including the dominant Hungarian clone [25].

Continuous monitoring to detect *Salmonella* along the food chain is of critical importance for public health, above all in the poultry meat industry, as poultry meat is one of the most consumed meats globally and thus one of the most traded meat products.

Author Contributions: Conceptualization, V.C. and C.D.B.; methodology, V.C., A.C., S.S., G.O., D.C., R.P., C.L. and G.A.; software, G.L.B.; validation, F.G.; formal analysis, R.A.; investigation, G.C. and A.A.; resources, C.D.B.; data curation, R.A.; writing—original draft preparation, V.C.; writing—review and editing, V.C. and D.L.M.; supervision, C.D.B.; project administration, C.B. All authors have read and agreed to the published version of the manuscript.

Funding: This research received no external funding.

Institutional Review Board Statement: Not applicable.

Informed Consent Statement: Not applicable.

Data Availability Statement: The raw data supporting the conclusions of this article will be made available by the authors, without undue reservation, to any qualified researcher.

Acknowledgments: We wish to thank our colleagues involved in this scientific collaboration at the Istituto Zooprofilattico Sperimentale della Sicilia (IZSSi).

Conflicts of Interest: The authors declare that the research was conducted in the absence of any commercial or financial relationships that could be construed as a potential conflict of interest. In the interest of transparency, it should be pointed out that authors R.P., C.L. and G.A. are employed by Enbiotech srl. They offered only a methodological background contribution but had no role in the experimentation; in the design of the study; in the collection, analyses, or interpretation of data; in the writing of the manuscript; or in the decision to publish the results.

References

1. European Food Safety Authority (EFSA). The European Union summary report on trends and sources of zoonoses, zoonotic agents and food-borne outbreaks in 2016. *EFSA J.* **2017**, *15*, 5077. [CrossRef]
2. Humphrey, T.; Jorgensen, F. Pathogens on meat and infection in animals—Establishing a relationship using Campylobacter and Salmonella as examples. *Meat Sci.* **2006**, *74*, 89–97. [CrossRef] [PubMed]
3. Obukhovska, O. The Natural Reservoirs of Salmonella Enteritidis in Populations of Wild Birds. *Online J. Public Health Inform.* **2013**, *5*, e171. [CrossRef]
4. Hugas, M.; Beloeil, P.A. Controlling Salmonella along the Food Chain in the European Union-Progress over the Last Ten Years. *Eurosurveillance* **2014**, *19*, 20804. Available online: http://www.eurosurveillance.org/ViewArticle.aspx?ArticleId=20804 (accessed on 1 April 2020). [CrossRef] [PubMed]
5. Callejon, R.M.; Rodriguez-Naranjo, M.I.; Ubeda, C.; Hornedo-Ortega, R.; Garcia-Parrilla, M.C.; Troncoso, A.M. Reported Foodborne Outbreaks Due to Fresh Produce in the United States and European Union: Trends and Causes. *Foodborne Pathog. Dis.* **2015**, *12*, 32–38. [CrossRef] [PubMed]
6. ISO 6579-1:2017; *Microbiology of the Food Chain—Horizontal Method for the Detection, Enumeration and Serotyping of Salmonella—Part 1: Detection of Salmonella spp.*; International Organisation for Standardisation: Geneva, Switzerland, 2017.
7. Liang, N.; Dong, J.; Luo, L.; Li, Y. Detection of viable Salmonella in lettuce by propidium monoazide real-time PCR. *J. Food Sci.* **2011**, *76*, M234–M237. [CrossRef] [PubMed]
8. Lim, H.S.; Zheng, Q.; Miks-Krajnik, M.; Turner, M.; Yuk, H.G. Evaluation of commercial kit based on loop-mediated isothermal amplification for rapid detection of low levels of uninjured and injured Salmonella on duck meat, bean sprouts, and fishballs in Singapore. *J. Food Prot.* **2015**, *78*, 1203–1207. [CrossRef] [PubMed]
9. Notomi, T.; Okayama, H.; Masubuchi, H.; Yonekawa, T.; Watanabe, K.; Amino, N.; Hase, T. Loop-mediated isothermal amplification of DNA. *Nucleic Acids Res.* **2000**, *28*, e63. [CrossRef] [PubMed]
10. Fratamico, P.M. Comparison of culture, polymerase chain reaction (PCR), TaqMan Salmonella, and Transia Card Salmonella assays for detection of Salmonella spp. in naturally-contaminated ground chicken, ground turkey, and ground beef. *Mol. Cell. Probes* **2003**, *17*, 215–221. [CrossRef]
11. Kokkinos, P.A.; Ziros, P.G.; Bellou, M.; Vantarakis, A. Loop-mediated isothermal amplification (LAMP) for the detection of Salmonella in food. *Food Anal. Methods* **2014**, *7*, 512–526. [CrossRef]

12. Wang, D.; Wang, Y.; Xiao, F.; Guo, W.; Zhang, Y.; Wang, A.; Liu, Y. A Comparison of In-House Real-Time LAMP Assays with a Commercial Assay for the Detection of Pathogenic Bacteria. *Molecules* **2015**, *20*, 9487–9495. [CrossRef] [PubMed]
13. Nagamine, K.; Hase, T.; Notomi, T. Accelerated reaction by loop-mediated isothermal amplification using loop primers. *Mol. Cell. Probes* **2002**, *16*, 223–229. [CrossRef] [PubMed]
14. Hara-Kudo, Y.; Yoshino, M.; Kojima, T.; Ikedo, M. Loop-mediated isothermal amplification for the rapid detection of Salmonella. *FEMS Microbiol. Lett.* **2005**, *253*, 155–161. [CrossRef] [PubMed]
15. ISO 16140-2:2016; *Microbiology of the Food Chain—Method Validation—Part 2: Protocol for the Validation of Alternative (Proprietary) Methods against a Reference Method*; International Organisation for Standardisation: Geneva, Switzerland, 2016.
16. Parida, M.; Posadas, G.; Inoue, S.; Hasebe, F.; Morita, K. Real-time reverse transcription loop-mediated isothermal amplification for rapid detection of West Nile virus. *J. Clin. Microbiol.* **2004**, *42*, 257–263. [CrossRef]
17. Francois, P.; Tangomo, M.; Hibbs, J.; Bonetti, E.J.; Boehme, C.C.; Notomi, T.; Perkins, M.D.; Schrenzel, J. Robustness of a loop-mediated isothermal amplification reaction for diagnostic applications. *FEMS Immunol. Med. Microbiol.* **2011**, *62*, 41–48. [CrossRef]
18. Tasrip, N.A.; Mokhtar, K.N.; Hanapi, U.K.; Abdul Manaf, A.Y.; Ali, M.E.; Cheah, Y.K.; Mustafa, S.; Mohd Desa, M.N. Loop mediated isothermal amplification; a review on its application and strategy in animal species authentication of meat based food products. *Int. Food Res. J.* **2019**, *26*, 1–10.
19. ISO 16140-1:2016; *Microbiology of the Food Chain—Method Validation—Part 1: Vocabulary*; International Organisation for Standardisation: Geneva, Switzerland, 2016.
20. ISO/TR 6579-3:2014; *Microbiology of the Food Chain—Horizontal Method for the Detection, Enumeration and Serotyping of Salmonella—Part 3: Guidelines for Serotyping of Salmonella spp.*; International Organisation for Standardisation: Geneva, Switzerland, 2014.
21. Cosby, D.E.; Cox, N.A.; Harrison, M.A.; Wilson, J.L.; Buhr, R.J.; Fedorka-Cray, P.J. Salmonella and antimicrobial resistance in broilers: A review. *J. Appl. Poultry Res.* **2015**, *24*, 408–426. [CrossRef]
22. European Food Safety Authority (EFSA); European Centre for Disease Prevention and Control (ECDC). The European Union summary report on trends and sources of zoonoses, zoonotic agents and food-borne outbreaks in 2017. *EFSA J.* **2018**, *16*, 262. [CrossRef]
23. Rašeta, M.; Mrdović, B.; Janković, V.; Bečkei, Z.; Lakićević, B.; Vidanović, D.; Polaček, V. Prevalence and antibiotic resistance of *Salmonella* spp. in meat products, meat preparations and minced meat. In Proceedings of the IOP Conference Series: Earth and Environmental Science, 59th International Meat Industry Conference MEATCON2017, Zlatibor, Serbia, 1–4 October 2017; Volume 85. [CrossRef]
24. Antunes, P.; Mourao, J.; Campos, J.; Peixe, L. Salmonellosis: The role of poultry meat. *Clin. Microbiol. Infect.* **2016**, *22*, 110–121. [CrossRef]
25. Nogrady, N.; Kiraly, M.; Davies, R.; Nagy, B. Multidrug resistant clones of Salmonella Infantis of broiler origin in Europe. *Int. J. Food Microbiol.* **2012**, *157*, 108–112. [CrossRef] [PubMed]

Article

Hsp60 Quantification in Human Gastric Mucosa Shows Differences between Pathologies with Various Degrees of Proliferation and Malignancy Grade

Alessandro Pitruzzella [1,2,3], Stefano Burgio [1], Pietro Lo Presti [1], Sabrina Ingrao [4], Alberto Fucarino [1], Fabio Bucchieri [1], Daniela Cabibi [4], Francesco Cappello [1,2], Everly Conway de Macario [5], Alberto J. L. Macario [2,5], Sabrina David [1,†] and Francesca Rappa [1,*,†]

1 Department of Biomedicine, Neurosciences and Advanced Diagnostics, Institute of Human Anatomy and Histology University of Palermo, 90127 Palermo, Italy; alessandro.pitruzzella@unipa.it (A.P.); stefano.burgio01@unipa.it (S.B.); pietro.lopresti@unipa.it (P.L.P.); alberto.fucarino@unipa.it (A.F.); fabio.bucchieri@unipa.it (F.B.); francesco.cappello@unipa.it (F.C.); sabrina.david@unipa.it (S.D.)
2 Euro-Mediterranean Institute of Science and Technology (IEMEST), via Michele Miraglia 20, 90139 Palermo, Italy; ajlmacario@som.umaryland.edu
3 University Consortium of Caltanissetta, 93100 Caltanissetta, Italy
4 Department of Sciences for the Promotion of Health and Mother and Child Care, Anatomic Pathology, University of Palermo, 90127 Palermo, Italy; sabrina.ingrao@unipa.it (S.I.); daniela.cabibi@unipa.it (D.C.)
5 Department of Microbiology and Immunology, School of Medicine, University of Maryland at Baltimore-Institute of Marine and Environmental Technology (IMET), Baltimore, MD 21202, USA; econwaydemacario@som.umaryland.edu
* Correspondence: francesca.rappa@unipa.it
† These authors contributed equally.

Citation: Pitruzzella, A.; Burgio, S.; Lo Presti, P.; Ingrao, S.; Fucarino, A.; Bucchieri, F.; Cabibi, D.; Cappello, F.; Conway de Macario, E.; Macario, A.J.L.; et al. Hsp60 Quantification in Human Gastric Mucosa Shows Differences between Pathologies with Various Degrees of Proliferation and Malignancy Grade. *Appl. Sci.* **2021**, *11*, 3582. https://doi.org/10.3390/app11083582

Academic Editor: Rosario Caltabiano

Received: 8 March 2021
Accepted: 13 April 2021
Published: 16 April 2021

Abstract: Background: Stomach diseases are an important sector of gastroenterology, including proliferative benign; premalignant; and malignant pathologies of the gastric mucosa, such as gastritis, hyperplastic polyps, metaplasia, dysplasia, and adenocarcinoma. There are data showing quantitative changes in chaperone system (CS) components in inflammatory pathologies and tumorigenesis, but their roles are poorly understood, and information pertaining to the stomach is scarce. Here, we report our findings on one CS component, the chaperone Hsp60, which we studied first considering its essential functions inside and outside mitochondria. **Methods:** We performed immunohistochemical experiments for Hsp60 in different samples of gastric mucosa. **Results:** The data obtained by quantitative analysis showed that the average percentages of Hsp60 were of 32.8 in normal mucosa; 33.5 in mild-to-moderate gastritis; 51.8 in severe gastritis; 58.5 in hyperplastic polyps; 67.0 in intestinal metaplasia; 89.4 in gastric dysplasia; and 92.5 in adenocarcinomas. Noteworthy were: (i) the difference between dysplasia and adenocarcinoma with the other pathologies; (ii) the progressive increase in Hsp60 from gastritis to hyperplastic polyp, gastric dysplasia, and gastric carcinoma; and (iii) the correlation of Hsp60 levels with histological patterns of cell proliferation and, especially, with tissue malignancy grades. **Conclusions:** This trend likely reflects the mounting need for cells for Hsp60 as they progress toward malignancy and is a useful indicator in differential diagnosis, as well as the call for research on the mechanisms underpinning the increase in Hsp60 and its possible roles in carcinogenesis.

Keywords: chaperone system; Hsp60; gastritis; gastric dysplasia; gastric carcinogenesis; intestinal metaplasia

1. Introduction

Quantitative variations of molecular chaperones in cells and tissues during carcinogenesis have been recognized for a long time. However, it is still unclear what these variations signify for the carcinogenic process. There is abundant information suggesting that tumors require one or more of the components of the chaperone system (CS), including Hsp60 discussed in this work, for growth and proliferation, epithelial-to-mesenchymal transition, dissemination, and anti-drug resistance [1–15]. Thus, it appears likely that the increase in the levels of chaperones observed in the cells of various types of cancers reflects the response of the cell to the needs of the carcinogenic process. However, due to the close correlation often observed between a progressive increase in the level of certain chaperones with advancing carcinogenesis, with metastasization, and with resistance to anti-cancer drugs, it appears possible that chaperones play a distinct etiologic–pathogenic role in the initiation of malignancies, or at least in their maintenance and progression.

Hsp60 is among the components of the CS studied during carcinogenesis in various tumors by several groups of investigators. For example, a steady increase in Hsp60 levels in tumor tissue paralleling tumor progression in uterine and colon cancers is among the earliest reported observations of the quantitative changes in a chaperone in relation to carcinogenesis [16]. Along the same lines, other reports pertain to other tumors, such as prostate [17–20], thyroid [21], and salivary gland [22] cancers. Here, we present data from a recent study of Hsp60 levels in gastric mucosa obtained with immunohistochemistry, comparing cancer with other gastric pathologies. These pathologies cause tissue disorganization and remodeling as exemplified by benign inflammatory, benign proliferation (e.g., hyperplastic polyps), metaplastic and dysplastic conditions, and malignant proliferations. Metaplasia is a para-physiological condition in which gastric glandular epithelium is replaced by another type of cell. Metaplasia occurs in response to chronic inflammation and is benign and typically reversible, but it can sometimes degenerate and evolve to dysplasia, a pre-cancerous condition [23]. Gastric adenocarcinoma is the most frequent gastric epithelial malignancy [24], and is aggressive and invasive. Gastric carcinogenesis is an orderly process that requires time for the tumor to become established and start its expansion. However, tumor progression can be triggered and accelerated by pre-cancerous conditions, such as gastric dysplasia, which facilitates cell proliferation, escape from pro-apoptotic stimuli, and genetic mutations [25–27]. In the study reported here, we performed an immunomorphological evaluation of the tissue levels of Hsp60 in specimens of gastric mucosa with inflammation (mild or moderate/severe gastritis), hyperplastic polyps, metaplasia, dysplasia, or cancer. The objective was to present an immunomorphological standard of Hsp60 quantitative variations that would be useful for the microscopic analysis of gastric pathologies aiming at differential diagnosis. Elucidation of the underlying molecular mechanisms is beyond the scope of the present work, but the data reported provide the basis for initiating a mechanistic–molecular dissection of the Hsp60 roles in various pathologies that are different from one another at the immunomorphological level.

2. Materials and Methods

2.1. Samples

Biopsy specimens of normal gastric tissue and samples with mild-to-moderate gastritis and hyperplastic polyp were obtained from the archives of the Biotechnology Laboratory, Euro-Mediterranean Institute of Sciences and Technologies (IEMEST). Gastric samples with severe gastritis, intestinal metaplasia, dysplasia, and carcinoma were collected from the archive of the Surgical Pathology laboratory, Department of Sciences for the Promotion of Health and Mother and Child Care, University of Palermo. Formalin fixed and paraffin embedded blocks of human gastric tissue were collected (10 cases for each group). The group of normal mucosa (NM) consisted of 5 males and 5 females with an average age of 51 ± 4 years. The group of specimens with mild-to-moderate gastritis (MMG) consisted of 3 males and 7 females, with an average age of 54 ± 4 years, with no signs of activity and a negative history of cancer or polyps. The group of hyperplastic polyp samples comprised 4 males and 6 females, with an average age of 56 ± 5 years, and a negative history of tumors and/or familial hereditary polyposis syndromes. Endoscopically, these polypoid formations had a diameter below 1 cm. The group of severe gastritis comprised 8 males and 2 females, with an average age of 52 ± 2 years. The group of gastric dysplasia comprised 7 males and 3 females, with an average age of 51 ± 2 years. All patients had a negative history of *Helicobacter pylori*. The grade of dysplasia was moderate to severe. The group of gastric carcinoma samples comprised 5 males and 5 females, with an average age of 58 ± 5 years. These samples were obtained from total surgical gastrectomy and presented the histopathological diagnosis of intestinal type adenocarcinoma (type I according to Lauren), with a moderate degree of differentiation (G2). Together with infiltrating neoplasm, all samples also showed areas of normal mucosa and areas of intestinal metaplasia. The pathological staging (pTNM) of the gastric carcinoma group was $T_2N_0M_X$ and $T_3N_0M_X$ for 6 and 4 cases, respectively.

2.2. Immunohistochemistry

Immunohistochemistry (IHC) reactions for Hsp60 were carried out on 5 μm thick tissue sections obtained from paraffin blocks with a cutting microtome. The IHC reactions were performed using the automated IHC system of the Biotechnology Laboratory of the Euro-Mediterranean Institute of Sciences and Technologies (IEMEST) (IntelliPath Flx, Biocare Medical, distributed by Bio-Optica, Milan, Italy). The primary antibody used was a rabbit polyclonal anti-human Hsp60 (Clone H300, Santa Cruz Biotechnology Inc., Santa Cruz, CA, USA, catalog no. Sc-13966, dilution 1:300). After the end of the immunostaining cycle, the slides were prepared for observation with coverslips, using a permanent mounting medium (Vecta Mount, Vector, H-5000). The observation of the sections was performed with an optical microscope (Microscope Axioscope 5/7 KMAT, Carl Zeiss, Oberkochen, Germany) connected to a digital camera (Microscopy Camera Axiocam 208 color, Carl Zeiss, Oberkochen, Germany). Two independent pathologists (F.C. and F.R.) examined the specimens on two separate occasions, blinded, i.e., the slides were unidentifiable by the pathologist performing the examination (the k values for each group are: NM k = 0.88; MMG k = 0.88; SG k = 1; HP k = 0.88; IM k = 1; GD k = 0.88 and GC k = 0.75) and performed a quantitative analysis to determine the percentage of cells positive for Hsp60. The evaluation of the percentage of immunopositivity was calculated in a high-power field (HPF) at 400× of magnification and repeated for 10 HPF. The average of the percentages of all immuno-quantifications performed in each case for each group described was considered as a conclusive result, and this value was used for the statistical investigation.

2.3. Statistical Analysis

The one-way analysis of variance (one-way ANOVA with Bonferroni post hoc multiple comparison) was applied to comparatively evaluate the results, using the GraphPad Prism 4.0 software (GraphPad Inc., San Diego, CA, USA). The data are presented as the arithmetic mean (AM) ± the standard deviation (SD), and the statistical significance limit was set at mboxemphp ≤ 0.05.

3. Results

The immunomorphological analysis was performed on epithelial cells and revealed that the immunolocalization of Hsp60 was uniformly widespread in the cytoplasm, but with some granular appearance at times. The average percentages of Hsp60-positive epithelial cells were 32.8 ± 6.9 in normal mucosa; 33.5 ± 15.28 in mild-to-moderate gastritis; 51.8 ± 14.3 in severe gastritis; 58.5 ± 17.4 in hyperplastic polyps; 67 ± 11.8 in intestinal metaplasia; 89.4 ± 3.4 in gastric dysplasia; and 92.5 ± 4.4 in gastric adenocarcinomas.

The statistical analysis revealed significant differences between some of the groups analyzed, showing a gradual increase in Hsp60-positive cell numbers from gastritis to hyperplastic polyp, gastric dysplasia, and gastric carcinoma (Figure 1).

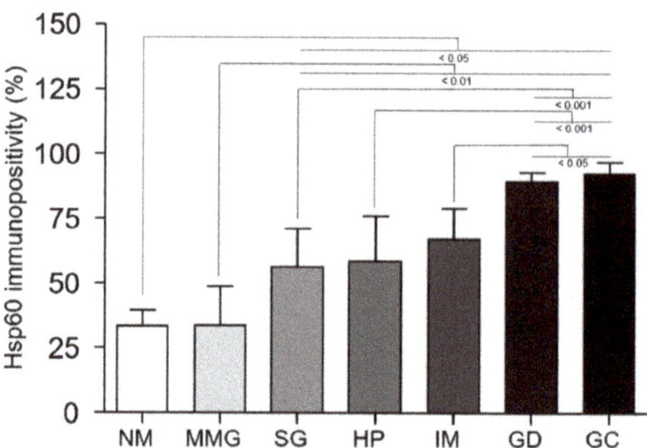

Figure 1. Histogram showing the percentage of Hsp60-positive epithelial cells in normal mucosa (NM), mild-to-moderate gastritis (MMG), severe gastritis (SG), hyperplastic polyp (HP), intestinal metaplasia (IM), gastric dysplasia (GD), and gastric adenocarcinoma (GC). Each group consisted of 10 cases. Data are presented as the arithmetic mean (AM) ± SD.

Gastric dysplasia was no different from gastric carcinoma. Representative immunohistochemical images are displayed in Figure 2.

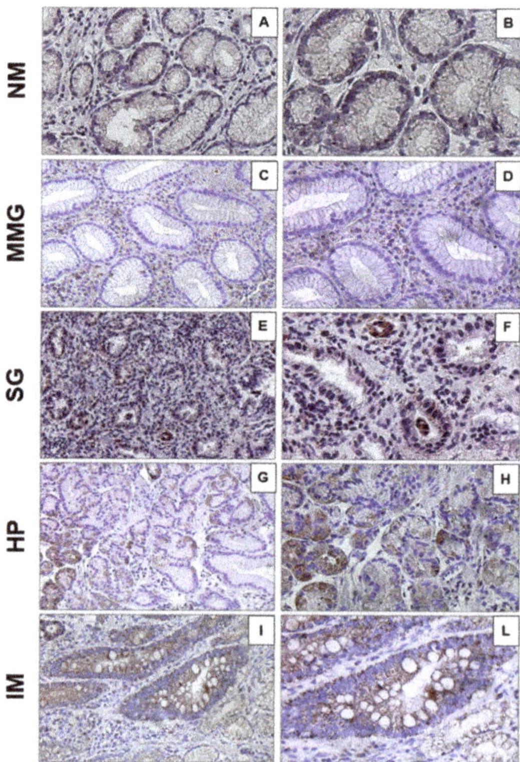

Figure 2. *Cont.*

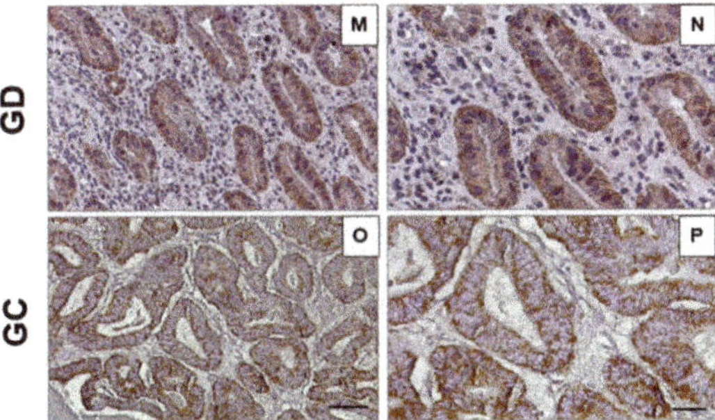

Figure 2. Representative images of immunohistochemical detection of Hsp60 in biopsies of human gastric tissue: (**A**,**B**) normal mucosa (NM), (**C**,**D**) mild-to-moderate gastritis (MMG), (**E**,**F**) severe gastritis (SG), (**G**,**H**) hyperplastic polyp (HP), (**I**,**L**) intestinal metaplasia (IM), (**M**,**N**) gastric dysplasia (GD), (**O**,**P**) gastric adenocarcinoma (GC). (**A**,**C**,**E**,**G**,**I**,**M**,**O**) Magnification 200×, scale bar 50 µm. (**B**,**D**,**F**,**H**,**L**,**N**,**P**) Magnification 400×, scale bar 20 µm.

Details of the normal mucosa, intestinal metaplasia, and gastric carcinoma with increasing numbers of Hsp60-positive cells (see percentages above) are shown in Figure 3.

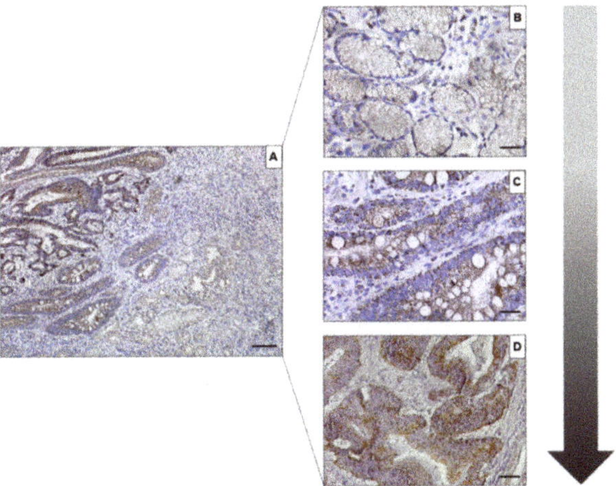

Figure 3. (**A**) Representative image of immunohistochemical detection of Hsp60 in gastric adenocarcinoma with associated areas of normal mucosa and intestinal metaplasia (magnification 100×, scale bar 100 µm). (**B**) Normal gastric mucosa (magnification 400×, scale bar 20 µm). (**C**) Intestinal metaplasia (magnification 400×, scale bar 20 µm). (**D**) Gastric adenocarcinoma (magnification 400×, scale bar 20 µm).

In some biopsies, there were areas of severe gastritis associated with areas of gastric dysplasia. The latter areas showed changes in nuclear and gland morphology typical of dysplasia with increased numbers of Hsp60-positive cells (Figure 4).

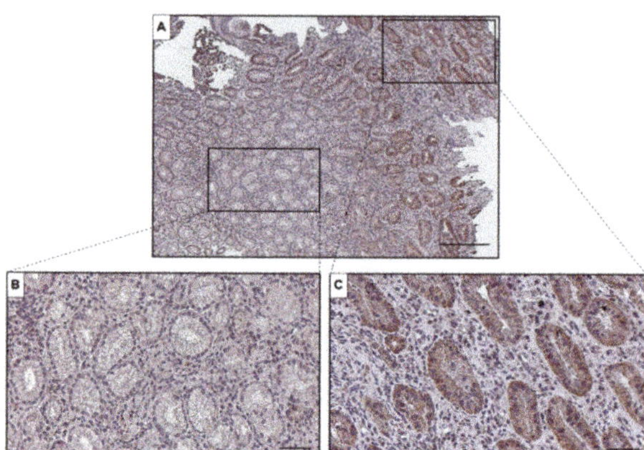

Figure 4. (**A**) Representative image of immunohistochemical detection of Hsp60 in gastric mucosa with areas of severe gastritis and areas with glandular dysplasia (magnification 50×, scale bar 200 μm). (**B**) Higher magnification of severe gastritis (magnification 200×, scale bar 50 μm). (**C**) Higher magnification of gastric dysplasia (magnification 200×, scale bar 50 μm).

4. Discussion

We found that the number of cells positive for Hsp60 and the positivity of the individual cells in gastric adenocarcinoma (GC) are high compared to the normal mucosa (NM). We also compared GC with other gastric pathologies, as follows: (1) mild-to-moderate gastritis, an inflammatory process without damage or destruction of the glandular structure; (2) severe gastritis, also an inflammatory process but associated with the injury and destruction of the glandular structure; (3) hyperplastic polyp, a paradigm of benign proliferation; (4) intestinal metaplasia, a benign reversible condition for the most part, characterized by morphological variations in the mucosa that probably reflect adaptive changes to a chronic inflammation; and (5) gastric dysplasia, a precancerous condition.

The highest levels of Hsp60 positivity were found in GC, but they were also high in gastric dysplasia (GD) in comparison with all the other pathologies studied, which reaffirms the close relationship between GC and GD, with the latter being a prelude to malignant transformation with the atypical alteration of epithelial cells. Additionally, it is of interest that (i) the Hsp60 positivity in hyperplastic polyps (HP) was higher than in normal mucosa (NM) but lower than in GC, even though HP and GC are both characterized by increased cell proliferation; (ii) the Hsp60 positivity was higher in the malignant version of these two examples of increased cell proliferation, HP and GC, but with the difference that GC is undifferentiated and disrupts the tissue structure and infiltrates neighboring spaces, properties not shown by polyps. This indicates that Hsp60 quantitative changes are strongly associated with malignancy (Figure 5).

Epithelial cell proliferation in polyps is characterized by an increase in mitotic replication cycles and a reduction in pro-apoptotic stimuli [28,29] also seen in carcinogenic proliferation, in which genetic mutations may occur [28,30]. Therefore, both pathologies share tissue remodeling events and factors, including an increase in Hsp60. Previous studies have found that Hsp60 increases in cases of alteration of the transmission of the apoptotic signal [31,32], and this may occur in the gastric polyps such as those included in this study, in which an altered cell turnover leads to an increase in proliferation. The involvement of Hsp60 in the carcinogenic process also includes its functional participation in many metabolic and biomolecular mechanisms of the cancer cells, encompassing interactions with various other molecules involved in programmed cell death, cell proliferation, and other pathways leading to malignant transformation [31].

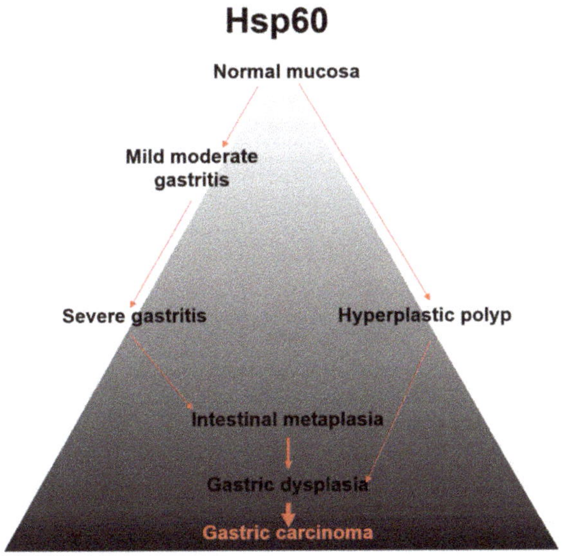

Figure 5. Schematic image visualizing the progressive increase in the number of Hsp60-positive epithelial cells in gastric mucosa as measured in this work. The increase is represented by the increment in darkness from top (lower proliferation) to bottom (higher proliferation), during the progression from normal mucosa (**top**) toward gastric carcinoma (**bottom**), passing through intermediate stages, such severe gastritis, hyperplastic polyp, intestinal metaplasia, and gastric dysplasia.

Other authors have also reported an increase in Hsp60 in the process of gastric carcinogenesis [11,33], ascribing to the chaperonin a fundamental role in the replication mechanisms and in cell survival [33]. Thus, our data are consistent with those from other investigators, showing that the Hsp60 increase is not only associated with the onset of GC but also correlates with the progression of the tumor mass and worsening of the clinical prognosis [11]. Regardless of the mechanisms by which Hsp60 increases in tumor cells, it is likely that the chaperonin aids them and should, therefore, be the target of anti-cancer treatments. The use of compounds to block Hsp60 activity, i.e., negative chaperonotherapy, has been proposed and is a promising field for investigation to develop means to defeat not only GC but also various other malignancies [34,35].

The Hsp60 positivity in mild-to-moderate gastritis is the same as that in normal mucosa, but in severe gastritis, it reaches higher levels, suggesting that in inflammatory processes the chaperonin is also involved to some extent. Here, the non-canonical functions of Hsp60, for example, stimulation of the production of pro-inflammatory cytokines [36–40], may be at play, an issue that deserves further investigation. Along the same lines, it has to be mentioned that infection with *H. pylori* can be associated with an increased risk of gastric neoplastic transformation [41]. Likewise, it has been observed that the *H. pylori* cytotoxin CagA assists in suppressing the heat shock response in gastric cancer cells, including Hsp60 expression [42]. In our study, it is relatively unlikely that *H. pylori* infection contributes to the results in any way, because the specimens we examined were derived from patients with a negative history of this infection.

5. Conclusions and Challenges for the Future

The increase in Hsp60 levels in the gastric mucosa during the progression from hyperplasia to carcinoma, passing through dysplasia, reflects the mounting need for cells for the CS, including Hsp60, to deal with their escalating metabolism and proliferation as they become malignant. It is possible that Hsp60 plays a distinct etiologic role in these processes, different from its canonical functions pertaining to the maintenance of protein homeostasis inside mitochondria. It is likely that the chaperonin helps the tumor to grow and metastasize and, therefore, a Hsp60 chaperonopathy by mistake or collaborationism is in operation: a normal chaperone (as far as it can be determined with current technology) functions to favor cancer rather than protect the organism against it. This is a key concept to consider while developing treatment strategies because it puts the chaperone at the center of the carcinogenic mechanism. Consequently, Hsp60 becomes a preferential target against which one must develop anti-cancer drugs. This concept also paves the way to investigating the mechanism underpinning the quantitative increase in Hsp60 in tumor cells. Is it the increased expression of the *hsp60* gene? Is it an increase in the life span (low degradation rate) and/or rate of translation of the Hsp60 mRNA? Is it the low degradation rate of the protein Hsp60? What is the role of Hsp60-related miRNAs in the regulation of Hsp60 levels in cancer cells? Finding answers to at least one of these questions will greatly help in choosing approaches to develop anti-cancer drugs targeting Hsp60 as an inside collaborator with the "enemy." Another key issue to keep in mind is that Hsp60 could play other roles unrelated to the maintenance of protein homeostasis inside mitochondria that could favor tumor growth and dissemination, as suggested by the redistribution of the chaperonin outside the mitochondria in tumor cells, and even outside these cells. These extramitochondrial Hsp60 molecules could play non-canonical roles, some of which could favor the tumor. Whether these Hsp60 extramitochondrial and extracellular Hsp60 molecules are normal or abnormal is unknown, but they could bear post-translational modifications, enabling them to play roles that favor tumor growth and dissemination, and resistance to stressors and anti-cancer compounds. The above considerations indicate that research focusing on Hsp60 in gastric tumors could provide information suitable not just for the possible uses of the chaperonin and its quantitative variations as a biomarker in diagnosis and patient monitoring but also for developing therapeutic drugs targeting the chaperonin.

Author Contributions: Conceptualization: F.C. and F.R.; Methodology: A.P. and P.L.P.; Validation: D.C.; Investigation: F.R. and P.L.P.; Resources: A.P., S.I., A.F. and S.D.; Data Curation: F.C. and F.R.; Writing—Original Draft Preparation: F.R. and S.B.; Writing—Review and Editing, F.B., A.J.L.M. and E.C.d.M.; Supervision: A.J.L.M. and E.C.d.M. All authors have read and agreed to the published version of the manuscript.

Funding: The work was partially supported by the Euro-Mediterranean Institute of Science and Technology (IEMEST) and the University of Palermo. A.J.L.M. and E.C.d.M. were partially supported by IEMEST and IMET. This is IMET contribution number is IMET 21-006.

Institutional Review Board Statement: This study was ethically approved by the institutional review board of the Euro-Mediterranean Institute of Science and Technology of Palermo, Palermo, Italy (PIC 01/2018).

Informed Consent Statement: This study is an observational study in which all patients involved gave informed consent before performing the gastroscopy / biopsy or surgery. The same biological material (tissue) used for histopathological diagnosis was used for this work.

Acknowledgments: A.J.L.M and E.C.d.M. were partially supported by IMET and IEMEST. This is IMET contribution number IMET 21-006.

Conflicts of Interest: The authors declare that they have no conflicts of interest.

References

1. Luo, B.; Lee, A.S. The critical roles of endoplasmic reticulum chaperones and unfolded protein response in tumorigenesis and anticancer therapies. *Oncogene* **2013**, *32*, 805–818. [CrossRef] [PubMed]
2. Lipinski, K.A.; Britschgi, C.; Schrader, K.; Christinat, Y.; Frischknecht, L.; Krek, W. Colorectal cancer cells display chaperone dependency for the unconventional prefoldin URI1. *Oncotarget* **2016**, *7*, 29635–29647. [CrossRef] [PubMed]
3. Chatterjee, S.; Burns, T.F. Targeting Heat Shock Proteins in Cancer: A Promising Therapeutic Approach. *Int. J. Mol. Sci.* **2017**, *18*, 1978. [CrossRef] [PubMed]

4. Shi, C.; Yang, X.; Bu, X.; Hou, N.; Chen, P. Alpha B-crystallin promotes the invasion and metastasis of colorectal cancer via epithelial-mesenchymal transition. *Biochem. Biophys. Res. Commun.* **2017**, *489*, 369–374. [CrossRef]
5. Guo, J.; Li, X.; Zhang, W.; Chen, Y.; Zhu, S.; Chen, L.; Xu, R.; Lv, Y.; Wu, D.; Guo, M.; et al. HSP60-regulated Mitochondrial Proteostasis and Protein Translation Promote Tumor Growth of Ovarian Cancer. *Sci. Rep.* **2019**, *9*, 12628. [CrossRef]
6. Tian, Y.; Wang, C.; Chen, S.; Liu, J.; Fu, Y.; Luo, Y. Extracellular Hsp90α and clusterin synergistically promote breast cancer epithelial-to-mesenchymal transition and metastasis via LRP1. *J. Cell Sci.* **2019**, *132*. [CrossRef]
7. Uretmen Kagiali, Z.C.; Sanal, E.; Karayel, Ö.; Polat, A.N.; Saatci, Ö.; Ersan, P.G.; Trappe, K.; Renard, B.Y.; Önder, T.T.; Tuncbag, N.; et al. Systems-level Analysis Reveals Multiple Modulators of Epithelial-mesenchymal Transition and Identifies DNAJB4 and CD81 as Novel Metastasis Inducers in Breast Cancer. *Mol. Cell Proteom.* **2019**, *18*, 1756–1771. [CrossRef]
8. Xiong, G.; Chen, J.; Zhang, G.; Wang, S.; Kawasaki, K.; Zhu, J.; Zhang, Y.; Nagata, K.; Li, Z.; Zhou, B.P.; et al. Hsp47 promotes cancer metastasis by enhancing collagen-dependent cancer cell-platelet interaction. *Proc. Natl. Acad. Sci. USA* **2020**, *117*, 3748–3758. [CrossRef]
9. Zeng, G.; Wang, J.; Huang, Y.; Lian, Y.; Chen, D.; Wei, H.; Lin, C.; Huang, Y. Overexpressing CCT6A Contributes To Cancer Cell Growth By Affecting The G1-To-S Phase Transition And Predicts A Negative Prognosis In Hepatocellular Carcinoma. *OncoTargets Ther.* **2019**, *12*, 10427–10439. [CrossRef]
10. Harper, A.K.; Fletcher, N.M.; Fan, R.; Morris, R.T.; Saed, G.M. Heat Shock Protein 60 (HSP60) Serves as a Potential Target for the Sensitization of Chemoresistant Ovarian Cancer Cells. *Reprod. Sci.* **2020**, *27*, 1030–1036. [CrossRef]
11. Li, X.S.; Xu, Q.; Fu, X.Y.; Luo, W.S. Heat shock protein 60 overexpression is associated with the progression and prognosis in gastric cancer. *PLoS ONE* **2014**, *9*, e107507. [CrossRef]
12. Qu, H.; Zhu, F.; Dong, H.; Hu, X.; Han, M. Upregulation of CCT-3 Induces Breast Cancer Cell Proliferation Through miR-223 Competition and Wnt/β-Catenin Signaling Pathway Activation. *Front. Oncol.* **2020**, *10*. [CrossRef]
13. Showalter, A.E.; Martini, A.C.; Nierenberg, D.; Hosang, K.; Fahmi, N.A.; Gopalan, P.; Khaled, A.S.; Zhang, W.; Khaled, A.R. Investigating Chaperonin-Containing TCP-1 subunit 2 as an essential component of the chaperonin complex for tumorigenesis. *Sci. Rep.* **2020**, *10*, 798. [CrossRef]
14. Tang, Y.; Yang, Y.; Luo, J.; Liu, S.; Zhan, Y.; Zang, H.; Zheng, H.; Zhang, Y.; Feng, J.; Fan, S.; et al. Overexpression of HSP10 correlates with HSP60 and Mcl-1 levels and predicts poor prognosis in non-small cell lung cancer patients. *Cancer Biomark.* **2020**, 1–10. [CrossRef]
15. Xiong, H.; Xiao, H.; Luo, C.; Chen, L.; Liu, X.; Hu, Z.; Zou, S.; Guan, J.; Yang, D.; Wang, K. GRP78 activates the Wnt/HOXB9 pathway to promote invasion and metastasis of hepatocellular carcinoma by chaperoning LRP6. *Exp. Cell Res.* **2019**, *383*, 111493. [CrossRef]
16. Cappello, F.; Bellafiore, M.; Palma, A.; David, S.; Marcianò, V.; Bartolotta, T.; Sciumè, C.; Modica, G.; Farina, F.; Zummo, G.; et al. 60KDa chaperonin (HSP60) is over-expressed during colorectal carcinogenesis. *Eur. J. Histochem.* **2003**, *47*, 105–110. [CrossRef]
17. Cappello, F.; Bellafiore, M.; Palma, A.; Marciano, V.; Martorana, G.; Belfiore, P.; Martorana, A.; Farina, F.; Zummo, G.; Bucchieri, F. Expression of 60-kD Heat shock protein increases during carcinogenesis in the uterine exocervix. *Pathobiology* **2002**, *70*, 83–88. [CrossRef]
18. Johansson, B.; Pourian, M.R.; Chuan, Y.-C.; Byman, I.; Bergh, A.; Pang, S.-T.; Norstedt, G.; Bergman, T.; Pousette, A. Proteomic comparison of prostate cancer cell lines LNCaP-FGC and LNCaP-r reveals heatshock protein 60 as a marker for prostate malignancy. *Prostate* **2006**, *66*, 1235–1244. [CrossRef]
19. Glaessgen, A.; Jonmarker, S.; Lindberg, A.; Nilsson, B.; Lewensohn, R.; Ekman, P.; Valdman, A.; Egevad, L. Heat shock proteins 27, 60 and 70 as prognostic markers of prostate cancer. *APMIS* **2008**, *116*, 888–895. [CrossRef]
20. Castilla, C.; Congregado, B.; Conde, J.M.; Medina, R.; Torrubia, F.J.; Japón, M.A.; Sáez, C. Immunohistochemical expression of Hsp60 correlates with tumor progression and hormone resistance in prostate cancer. *Urology* **2010**, *76*, 1017.e1–1017.e6. [CrossRef]
21. Pitruzzella, A.; Paladino, L.; Vitale, A.M.; Martorana, S.; Cipolla, C.; Graceffa, G.; Cabibi, D.; David, S.; Fucarino, A.; Bucchieri, F.; et al. Quantitative immunomorphological analysis of heat shock proteins in thyroid follicular adenoma and carcinoma tissues reveals their potential for differential diagnosis and points to a role in carcinogenesis. *Appl. Sci.* **2019**, *9*, 4324. [CrossRef]
22. Basset, C.A.; Cappello, F.; Rappa, F.; Lentini, V.L.; Jurjus, A.R.; Conway de Macario, E.; Macario, A.J.L.; Leone, A. Molecular chaperones in tumors of salivary glands. *J. Mol. Histol.* **2020**, *51*, 109–115. [CrossRef]
23. Giroux, V.; Rustgi, A.K. Metaplasia: Tissue injury adaptation and a precursor to the dysplasia-cancer sequence. *Nat. Rev. Cancer* **2017**, *17*, 594–604. [CrossRef]
24. Karimi, P.; Islami, F.; Anandasabapathy, S.; Freedman, N.D.; Kamangar, F. Gastric cancer: Descriptive epidemiology, risk factors, screening, and prevention. *Cancer Epidemiol. Biomark. Prev.* **2014**, *23*, 700–713. [CrossRef]
25. Tan, P.; Yeoh, K.G. Genetics and Molecular Pathogenesis of Gastric Adenocarcinoma. *Gastroenterology* **2015**, *149*, 1153–1162.e3. [CrossRef]
26. Tomasello, G.; Rodolico, V.; Zerilli, M.; Martorana, A.; Bucchieri, F.; Pitruzzella, A.; Marino Gammazza, A.; David, S.; Rappa, F.; Zummo, G.; et al. Changes in immunohistochemical levels and subcellular localization after therapy and correlation and colocalization with CD68 suggest a pathogenetic role of Hsp60 in ulcerative colitis. *Appl. Immunohistochem. Mol. Morphol.* **2011**, *19*, 552–561. [CrossRef]
27. Yakirevich, E.; Resnick, M.B. Pathology of gastric cancer and its precursor lesions. *Gastroenterol. Clin. N. Am.* **2013**, *42*, 261–284. [CrossRef]

28. Bosari, S.; Moneghini, L.; Graziani, D.; Lee, A.K.; Murray, J.J.; Coggi, G.; Viale, G. bcl-2 oncoprotein in colorectal hyperplastic polyps, adenomas, and adenocarcinomas. *Human Pathol.* **1995**, *26*, 534–540. [CrossRef]
29. Flohil, C.C.; Janssen, P.A.; Bosman, F.T. Expression of Bcl-2 protein in hyperplastic polyps, adenomas, and carcinomas of the colon. *J. Pathol.* **1996**, *178*, 393–397. [CrossRef]
30. Phelps, R.A.; Chidester, S.; Dehghanizadeh, S.; Phelps, J.; Sandoval, I.T.; Rai, K.; Broadbent, T.; Sarkar, S.; Burt, R.W.; Jones, D.A. A two-step model for colon adenoma initiation and progression caused by APC loss. *Cell* **2009**, *137*, 623–634. [CrossRef] [PubMed]
31. Rappa, F.; Farina, F.; Zummo, G.; David, S.; Campanella, C.; Carini, F.; Tomasello, G.; Damiani, P.; Cappello, F.; Conway de Macario, E.; et al. HSP-molecular chaperones in cancer biogenesis and tumor therapy: An overview. *Anticancer Res.* **2012**, *32*, 5139–5150. [PubMed]
32. Takayama, S.; Reed, J.C.; Homma, S. Heat-shock proteins as regulators of apoptosis. *Oncogene* **2003**, *22*, 9041–9047. [CrossRef]
33. Lianos, G.D.; Alexiou, G.A.; Mangano, A.; Mangano, A.; Rausei, S.; Boni, L.; Dionigi, G.; Roukos, D.H. The role of heat shock proteins in cancer. *Cancer Lett.* **2015**, *360*, 114–118. [CrossRef]
34. Macario, A.J.L.; Conway de Macario, E. Chaperonopathies and chaperonotherapy. *FEBS Lett.* **2007**, *581*, 3681–3688. [CrossRef]
35. Meng, Q.; Li, B.X.; Xiao, X. Toward Developing Chemical Modulators of Hsp60 as Potential Therapeutics. *Front. Mol. Biosci.* **2018**, *5*, 35. [CrossRef]
36. Habich, C.; Burkart, V. Heat shock protein 60: Regulatory role on innate immune cells. *Cell. Mol. Life Sci.* **2007**, *64*, 742–751. [CrossRef]
37. Tsan, M.F.; Gao, B. Heat shock protein and innate immunity. *Cell. Mol. Immunol.* **2004**, *1*, 274–279.
38. Ohashi, K.; Burkart, V.; Flohé, S.; Kolb, H. Cutting edge: Heat shock protein 60 is a putative endogenous ligand of the toll-like receptor-4 complex. *J. Immunol.* **2000**, *164*, 558–561. [CrossRef]
39. Sangiorgi, C.; Vallese, D.; Gnemmi, I.; Bucchieri, F.; Balbi, B.; Brun, P.; Leone, A.; Giordano, A.; Conway de Macario, E.; Macario, A.J.L.; et al. HSP60 activity on human bronchial epithelial cells. *Int. J. Immunopathol. Pharmacol.* **2017**, *30*, 333–340. [CrossRef]
40. Swaroop, S.; Mahadevan, A.; Shankar, S.K.; Adlakha, Y.K.; Basu, A. HSP60 critically regulates endogenous IL-1β production in activated microglia by stimulating NLRP3 inflammasome pathway. *J. Neuroinflamm.* **2018**, *15*, 177. [CrossRef]
41. Tanaka, A.; Kamada, T.; Yokota, K.; Shiotani, A.; Hata, J.; Oguma, K.; Haruma, K. Helicobacter pylori heat shock protein 60 antibodies are associated with gastric cancer. *Pathol. Res. Pract.* **2009**, *205*, 690–694. [CrossRef] [PubMed]
42. Lang, B.J.; Gorrell, R.J.; Tafreshi, M.; Hatakeyama, M.; Kwok, T.; Price, J.T. The Helicobacter pylori cytotoxin CagA is essential for suppressing host heat shock protein expression. *Cell Stress Chaperones* **2016**, *21*, 523–533. [CrossRef] [PubMed]

applied sciences

MDPI

Article

Gold-Modified Micellar Composites as Colorimetric Probes for the Determination of Low Molecular Weight Thiols in Biological Fluids Using Consumer Electronic Devices

Elli A. Akrivi [1,2], Athanasios G. Vlessidis [3], Dimosthenis L. Giokas [3,*] and Nikolaos Kourkoumelis [1,*]

1 Department of Medical Physics, School of Health Sciences, University of Ioannina, 45110 Ioannina, Greece; elliakrivi@gmail.com
2 Neurology Clinic, University Hospital of Ioannina, 45110 Ioannina, Greece
3 Department of Chemistry, School of Natural Sciences, University of Ioannina, 45110 Ioannina, Greece; avlessid@uoi.gr
* Correspondence: dgiokas@uoi.gr (D.L.G.); nkourkou@uoi.gr (N.K.)

Abstract: This work describes a new, low-cost and simple-to-use method for the determination of free biothiols in biological fluids. The developed method utilizes the interaction of biothiols with gold ions, previously anchored on micellar assemblies through electrostatic interactions with the hydrophilic headgroup of cationic surfactant micelles. Specifically, the reaction of $AuCl_4^-$ with the cationic surfactant cetyltrimethyl ammonium bromide (CTAB) produces an intense orange coloration, due to the ligand substitution reaction of the Br^- for Cl^- anions, followed by the coordination of the $AuBr_4^-$ anions on the micelle surface through electrostatic interactions. When biothiols are added to the solution, they complex with the gold ions and disrupt the $AuBr_4^-$–CTAB complex, quenching the initial coloration and inducing a decrease in the light absorbance of the solution. Biothiols are assessed by monitoring their color quenching in an RGB color model, using a flatbed scanner operating in transmittance mode as an inexpensive microtiter plate photometer. The method was applied to determine the biothiol content in urine and blood plasma samples, with satisfactory recoveries (i.e., >67.3–123% using external calibration and 103.8–115% using standard addition calibration) and good reproducibility (RSD < 8.4%, $n = 3$).

Keywords: biothiols; blood plasma; gold nanoparticles; colorimetric probes

Citation: Akrivi, E.A.; Vlessidis, A.G.; Giokas, D.L.; Kourkoumelis, N. Gold-Modified Micellar Composites as Colorimetric Probes for the Determination of Low Molecular Weight Thiols in Biological Fluids Using Consumer Electronic Devices. *Appl. Sci.* **2021**, *11*, 2705. https://doi.org/10.3390/app11062705

Academic Editor: Francesco Cappello

Received: 23 February 2021
Accepted: 16 March 2021
Published: 17 March 2021

Publisher's Note: MDPI stays neutral with regard to jurisdictional claims in published maps and institutional affiliations.

1. Introduction

More than 100 biothiols, which play a significant role in many biological processes and metabolic pathways (e.g., protein synthesis, antioxidant defense system, cell metabolism, etc.), have been identified in the human organism [1,2]. Of these biothiols, glutathione (GSH), cysteine (Cys) and homocysteine (Hcy) are the most abundant, with concentration levels that vary significantly in biological fluids. Under normal (non-pathological) conditions, GSH is present at concentrations as high as 3 mM in whole blood and less than 5 μM in blood plasma and urine. Cys, on the other hand, typically ranges from 135 to 300 μM in blood plasma and 20–80 μM in urine while Hcy concentrations are lower than 15 μM in both plasma and urine. Abnormal levels of biothiols have been related to clinical disorders and diseases, such as liver damage, Alzheimer's disease, osteoporosis and cardiovascular diseases, among others [3–6].

Due to the important biological implications of biothiols, there has been a great interest in their determination in biological fluids. To that end, several analytical methods and techniques have been developed: HPLC, LC-MS, UV-Vis spectrometry, fluorescence, chemiluminescence, electrochemical techniques, etc. [7,8]. Over the past decade, nanomaterials have also been a popular scaffold for the development of biothiol probes and chemosensors [9]. By exploiting the high affinity of the sulfhydryl group for metal nanoparticles, a large variety of optical methods have been proposed. These methods are based on the generation or quenching of an optical event when biothiols bind on the surface of nanomaterials through a variety of mechanisms, such as host–guest interactions of biothiols with receptor molecules on the nanomaterials surface, the replacement of functional molecules from the nanomaterials surface and the disruption of inter-nanoparticle bonds [9]. The main advantage of nanomaterial probes is their high sensitivity as a result of the strong distance- and size-dependent optical properties of metal-based nanomaterials, especially those made of noble metals. However, the use of nanomaterials also faces several shortcomings: although synthesis may be relatively easy in a research laboratory, the reproducible large-scale synthesis and functionalization of nanomaterials for biomedical applications is a resource-demanding task that requires substantial nanotechnology skills and advanced equipment for their proper characterization. Additionally, due to a high surface area, which imbues nanomaterials with high reactivity, and surface functionalization, non-specific interactions with the matrix components may be observed. Furthermore, many of these methods require laboratory facilities, expensive scientific equipment (e.g., plate spectrophotometers or spectrofluorometers) and trained operators for their application, which are probably unavailable at the point-of-need and remote healthcare units.

An alternative approach to overcome the challenges associated with the use of nanomaterials is to use thiols as size and growth regulators of nanomaterials. In these methods, thiols firstly react with the precursor metal ions. Upon addition of a reducing agent, the produced nanomaterials exhibit a different size, morphology or aggregation state compared to those obtained in the absence of thiols. The difference in the colorimetric and spectral response in the presence and absence of biothiols is therefore used to determine their concentration. Based on this effect, several analytical methods using gold and silver nanoparticles have been developed, also from our group [10–13]. The main challenge in these methods, however, is that the kinetics of the reactions evolve over time and thus requiring relatively strict timing to obtain reproducible results [13,14].

Motivated by the latter methods, in this work we describe a simpler approach that does not rely on the formation of metallic nanomaterials but exploits the direct interaction of thiols with gold anions incorporated on the surface of cationic surfactant micelles. Gold anions form colored ion-pair complexes with the positively charged headgroup of cationic surfactant micelles through electrostatic interactions. These complexes have been widely used as a template for the controlled synthesis of noble metal nanoparticles [15–17] by chemical reduction of the Au–CTAB complex with appropriate reducing agents. However, the use of Au–CTAB complexes as colorimetric probes for analytical purposes has not been reported yet. In that sense, we exploit the intense colorimetric response generated from of Au halides associated on the surface of CTAB cationic micelles, which is quenched upon the addition of thiols due to the complexation of Au ions and the formation of colorless Au–thiolate complexes. In this manner, a concentration-dependent colorimetric response was obtained that was quantified with a simple flatbed scanner operated as an inexpensive microtiter plate photometer and was used to determine biothiols in real biological fluids (urine and blood plasma).

2. Materials and Methods

2.1. Reagents and Chemicals

L-cysteine (Cys), glycine (Gly), histidine (His), valine (Val), lysine (Lys), cystine (Cys-Cys), arginine (Arg), DL-homocysteine (HCy), glutamine (Glu), uric acid, magnesium chloride hexahydrate, sodium chloride, sodium sulfate, sodium hydrogen bicarbonate, sodium acetate, disodium hydrogen phosphate, potassium chloride, calcium chloride dehydrate, D(+)-glucose, and glacial acetic acid were obtained from Sigma-Aldrich (Steinheim, Germany). L-Glutathione (reduced), hydrogen tetrachloroaurate trihydrate (min. 99.9%), urea (>99%), trisodium citrate dihydrate, bovine serum albumin (crystalline, 98%) and tris(2-carboxyethyl)phosphinehydrochloride (TCEP, 95%, 0.5 M) were obtained from Alfa Aesar (Karlsruhe, Germany). Nanosep® centrifugal vials with modified polyethersulfone membranes of 3 kDa molecular cut-off size were obtained from Pall Corp. (New York,

NY, USA). Lastly, 96-well microtiter plates (Nuclon 400 μL) with a clear flat surface were purchased from Thermo Fischer Scientific (Waltham, MA, USA).

2.2. Equipment and Instrumentation

A single-beam spectrophotometer (Jenway 6405 UV/Visible, Staffordshire, UK) with matched quartz cells of 1 cm path length was used to obtain the UV-vis spectra of the solutions. The IR spectra were acquired with a Perkin Elmer Spectrum Two^{TM} attenuated total reflectance—infrared (ATR-IR) spectrometer. A flatbed scanner (PerfectionV550 Photo, Epson Corp., Suwa, Japan) operating in transmittance mode was used to obtain photometric measurements by placing the microtiter plate containing the samples between the imaging surface and the transparency unit of the scanner. In this manner, the sample solution was aligned between the white LED light source with the CCD strip detector establishing an optical path length [14,18]. During scanning all automatic correction functions embedded in the software (Easy Photo Scan, v.1.00.08, Epson Corp.) were disabled to ensure that the photometric data were not manipulated. Images of the microtiter plate were saved as Joint Photographic Experts Group (JPEG) files at a resolution of 300 dpi and the color intensity was determined as the mean gray intensity and the intensity of the color in the RGB color system using Image J [19].

2.3. Experimental Procedure

The determination of biothiols was performed by adding 0.25 mL of the sodium acetate/acetic acid buffer (0.1 M, pH 6), 0.125 mL CTAB (0.16 M) and 0.05 mL $AuCl_4^-$ (0.01 M) to 1.575 mL of the sample or standard solution. The mixture was incubated at room temperature for 40 min and an aliquot of 250 μL was transferred to a 96-well microtiter plate. The color intensity of the light transmitted through the sample was recorded as a colored image of the sample using a flatbed scanner operating in transmittance mode. The analytical signal was calculated as the difference in the mean grey area color intensity of the blank and the samples.

2.4. Samples

Matrix effects and interferences were evaluated using artificial and simulated body fluids (urine and blood plasma). Detailed information for the composition of each solution is given in the Supplementary Materials.

A few mL of whole blood was donated from a group member voluntarily after informed consent was obtained. The blood sample was taken in a designated room by a trained nurse of the university hospital and was not cultured or examined in any other way than the experimental protocol presented here. After collection, red blood cells were separated from the plasma using centrifugation at 800× *g* for 10 min. The collected blood plasma (1000 μL) was first treated with 1.0 mM TCEP for 30 min and interim vortex mixing to reduce the oxidized biothiols. Then, plasma proteins were removed by centrifugation at 12,000× *g* for 20 min at room temperature through centrifugal filters (MWCO = 3 kDa). This procedure was repeated twice to ensure complete removal of proteins. The collected (supernatant) liquid was diluted with distilled water (approximately 20-fold) and 1.575 mL of the sample was used for the determination of total biothiols according to the experimental procedure described above using the method of standard additions.

Urine was also provided by a group member after informed consent was obtained. Within 30 min of collection, the urine sample was centrifuged at 6000 rpm for 20 min to remove the insoluble materials. The clarified liquid was collected and treated with TCEP as above to reduce the oxidized biothiols. Then, proteins were removed with centrifugal filters (MWCO = 3 kDa) at 12,000× *g* for 20 min at room temperature. The collected liquid was diluted 20-fold with distilled water and 1.575 mL of the sample was used for the determination of total biothiols according to the experimental procedure described above using the method of standard additions.

To evaluate the accuracy of the method, known concentrations of cysteine were spiked in urine and blood plasma after protein precipitation in order to avoid potential losses during sample pre-treatment (which was based on a standard procedure and was out of the scope of this study).

3. Results and Discussion

3.1. Mechanism of Biothiol Detection

The sensing mechanism of the biothiols was investigated by studying the color and spectral transitions of the $AuCl_4^-$ and Au–CTAB complex solutions in the presence and absence of cysteine, which is the most representative species of biothiols in biological fluids. The UV-Vis spectra of CTAB showed no absorption band while the $AuCl_4^-$ solutions exhibit two absorption bands at 220 and 290 nm due to the ligand-to-metal charge transfer (LMCT) band and the ligand field (LF) band of $AuCl_4^-$, respectively [15]. When Cys was added into the solution, it formed a metal complex with the Au ions and the absorption of the Au–SH complex solution increased and red-shifted to 235 and 300 nm, respectively. On both occasions the colors were barely conceivable by the bare eye and the scanning imaging device (Figure 1b).

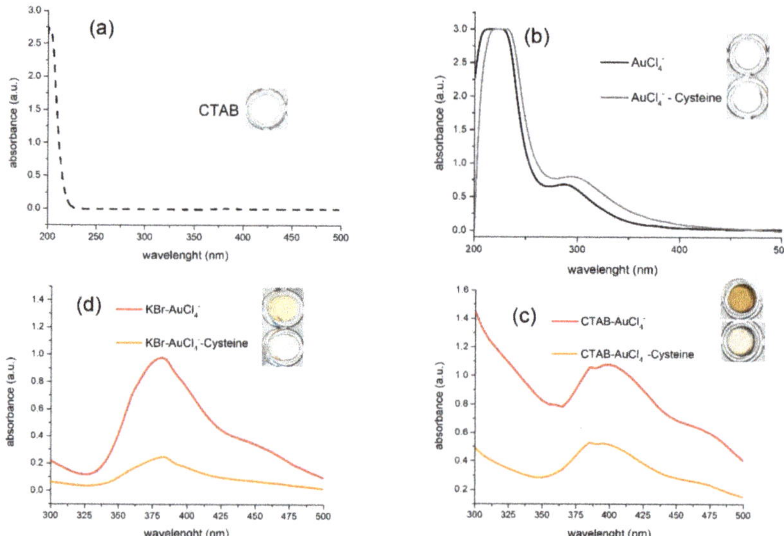

Figure 1. Color and UV-Vis absorbance spectra of (**a**) the CTAB solution (10 mM); (**b**) 0.25 mM $AuCl_4^-$ solutions in the absence and presence of 50 µM Cys; (**c**) 0.25 mM $AuCl_4^-$ and 10 mM KBr solutions in the absence and presence of 50 µM Cys; (**d**) 0.25 mM $AuCl_4^-$ and 10 mM CTAB solutions in the absence and presence of 50 µM Cys.

When CTAB was mixed with $AuCl_4^-$ ions, the color of the solution turned orange and the absorption red-shifted (as compared to Au solutions) to 260 and 405 nm (Figure 1c). These absorption bands can be explained on the basis of two phenomena: The first, which explains the appearance of an orange coloration in the Au–CTAB solution, is the formation of $AuBr_4^-$, due to the fast, multi-step, ligand substitution reaction of Cl^- from Br^- for anions, released from the dissociation of CTAB [15,17]. In an aqueous solution, Au species may be distributed among various hydroxyl-containing gold complexes as a function of the pH, in the general form of $(AuCl_x(OH)_y)^-$ (where x + y = 4). As the Br^- anions released from the dissociation of CTAB are in large excess compared to $AuCl_4^-$, we assume that all gold chloride complexes are transformed to gold bromide complexes so that the Au species will actually be present in various bromide/hydroxyl complexes in the general form of $(AuBr_x(OH)_y)^-$ (x + y = 4).

The UV-vis spectrum of the $(AuBr_x(OH)_y)^-$ species, however, show a peak at 385 nm (Figure 1d), while the spectrum of the Au–CTAB complex is red-shifted to 405 nm. In addition, the color of the Au–CTAB solution is darker than that of the Au–KBr solution at equimolar concentration levels, possibly due to the high local concentration of $(AuBr_x(OH)_y)^-$ species on the micelle surface. This observation is a strong indication that the $(AuBr_x(OH)_y)^-$ anions are not free in the solution but incorporated into the Stern layer of the CTAB micelles, forming an ion-pair complex with the positive head groups of the CTAB. The formation of an orange precipitate in the Au–CTAB mixture after a few hours also supports the above reaction combined with the fact that no precipitate was formed in the Au–KBr solutions even after prolonged incubation (1 week).

In the presence of Cys, the UV-Vis spectra of the Au–CTAB-cysteine solution blue-shifts to 395–400 nm, the absorbance intensity decreases, and the color of the solution turns yellow. This colorimetric and spectral change can be attributed to the formation of Au–SH complexes that disrupt the $(AuBr_x(OH)_y)^-$–CTA^+ complexes and release gold ions from the micelle surface. Other phenomena, such as the hydrolysis of Au, shall not have any effect since all experiments were performed in an acetate/acetic acid buffer (0.1 M, pH 6). In addition, no new peaks at 500–540 nm were observed, which excludes the possibility of Au reduction neither by CTAB nor by thiols and the formation of gold nanoparticles, (Figure S1). The sensing mechanism is graphically demonstrated in Figure 2.

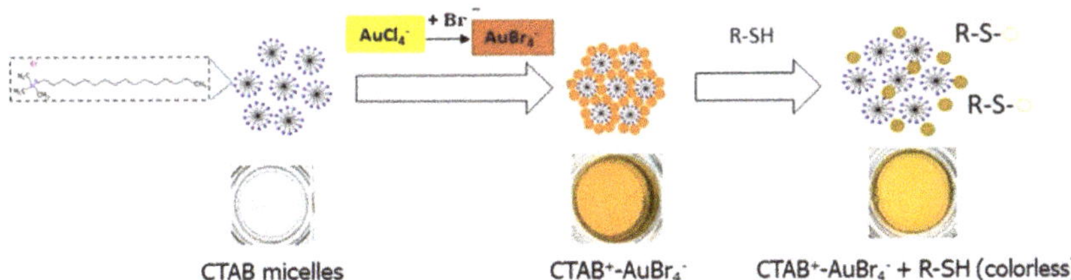

Figure 2. Graphical representation of the potential mechanism of biothiol sensing using gold-coated CTAB micellar assemblies.

3.2. Effect of Gold Ion Concentration

According to previous reports, $AuCl_4^-$ reacts quantitatively with CTAB at a 1:1 ratio to form water-insoluble precipitates [20]. As the concentration of CTAB increases above the critical micellar concentration (cmc) (i.e., the Au/CTAB ratio decreases), the Au–CTAB complex solubilizes in the CTAB micelles, reaching its maximum solubility at Au/CTAB = 1/60 [20]. Therefore, the effect of the Au ion concentration was investigated at an excess amount of CTAB (i.e., 75 mM) to ensure that the Au/CTAB ratio is lower than 1/60. This value also prevents the formation of insoluble precipitates of the Au–CTAB complex. According to the results depicted in Figure 3a, the maximum analytical signal was obtained at $AuCl_4^-$ concentration of 0.25 mM, which was used as the optimum value. Based on the retail prices of gold chloride salts (which is the most expensive reagent used in this method), this concentration corresponds to a cost of <1.2 cent per sample (i.e., less than 1.2 € per 100 samples), which is affordable even in resource-limited settings. Lower Au concentrations produce solutions with faint colors and as a result the colorimetric changes induced by the addition of Cys are minor. On the contrary, when the Au ions are in excess, the added Cys (50 μM) cannot induce any significant color changes because there is an excess amount of Au ions complexed with the CTAB micelles. On both occasions, the net analytical signal, which is defined as the difference in the color intensity of the blank and the sample solutions, decreases.

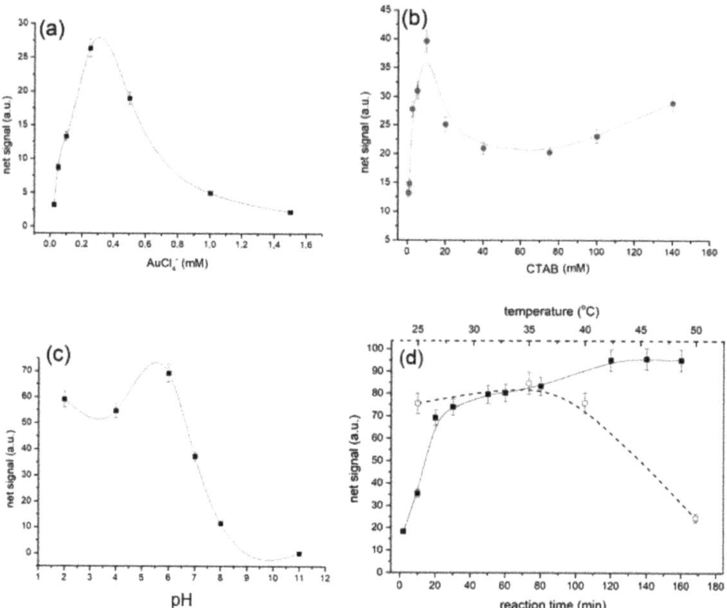

Figure 3. Optimization of the experimental conditions: (**a**) $AuCl_4^-$ concentration (other conditions: 75 mM CTAB, 50 μM Cys, no buffer, room temperature, reaction time 30 min); (**b**) CTAB concentration (other conditions: 0.25 mM $AuCl_4^-$, 50 μM Cys, no buffer, room temperature, reaction time 30 min); (**c**) pH (other conditions: 0.25 mM $AuCl_4^-$, 10 mM CTAB, 50 μM Cys, no buffer, room temperature, reaction time 30 min); (**d**) reaction time and temperature (other conditions: 0.25 mM $AuCl_4^-$, 10 mM CTAB, 50 μM Cys, sodium acetate/acetic acid buffer pH 6).

3.3. Selection of Cationic Surfactant Concentration

The effect of CTAB concentration was investigated over a wide concentration range, from 0.5 mM (i.e., below the cmc of $\approx$ 0.9 mM) to 140 mM. The signal gradually increases up to 10 mM and decreases at higher concentrations (Figure 3b). This observation agrees with the solubility of the Au–CTAB complex, which reaches its maximum value at 1/60 [20]. At CTAB concentrations lower than 10 mM, the Au/CTAB ratio is higher than 1/60; therefore, the complex has limited solubility and forms precipitates that scavenge the incident light and produce intense signals (darker images) as less light is transmitted through the sample. At CTAB concentrations higher than 20 mM, the Au/CTAB ratio is lower than 1/60, hence the Au–CTAB complexes are solubilized, and more light is transmitted through the solution producing brighter images. When cysteine is added into the solutions, it complexes Au, releasing it from the Au–CTAB complex. In this manner, the Au/CTAB ratio decreases (as the Au concentration on the Au–CTAB complex also decreases), thus increasing the solubility of the Au–CTAB complex. As a result, the signal of the blank is lower than that of the sample due to the increased solubility of the Au–CTAB complex. Therefore, the net analytical signal response, which is calculated as the difference between the signals of the blank and the sample, reaches its maximum value. The above discussion also explains why at CTAB concentrations lower than 10 mM, the (net) analytical signals are higher than those observed at CTAB > 20 mM (except for the cases where the CTAB concentration is lower than its cmc) due to the solubilizing action of Cys.

3.4. Optimization of the Working pH

The influence of pH in the detection of cysteine was investigated by varying the pH from 2 to 11 using dilute HCl and NaOH solutions. According to the results in Figure 3c, the highest signals were obtained at acidic pH values and specifically at pH 6, while at alkaline conditions the signal decreased. Since CTAB is a non-pH responsive surfactant [21], we reasoned that the pH-dependent response may be related to the hydrolysis of Au salts as well as to the ionization state of cysteine and its interaction with CTAB. Au forms anionic complexes with halogens and hydroxyl anions over a wide pH range [22]; therefore, the formation of Au–CTAB complexes is feasible over a wide pH range. In addition, the formation of ion-pair $AuCl_4^-$–CTAB complexes precedes the addition of cysteine. Therefore, hydrolysis of Au species at different pH values should not play any role since Au is not free in the solution but as a complex with CTAB. Based on these phenomena, we argue that the effect of pH is mainly related to the ionization state of cysteine.

The maximum signal obtained at a pH of 6 is close to the isoelectric point of cysteine, suggesting that electrically neutral biothiols favor the formation of the S–Au bond. At pH $\geq$ 7, the signal decreases and completely diminishes at pH > 10. At these conditions, cysteine is negatively charged since the amine group is protonated at pKa_2 = 8.3 while the thiol group is deprotonated at the pka_3 = 10.8 [23]. The reason for the reduction in the analytical signal may lie in the electrostatic attraction of cysteine to CTAB micelles. CTAB is present in a very large excess compared to Cys (CTAB:Cys > 200); therefore, the unprotonated COO- groups may interact with the positively charged amine headgroup of the CTAB micelles, and possibly reduce the mobility and reactivity of Cys and thus its ability to form Au–S complexes. The inhibitory effect of electrostatic interactions between the free thiols and charged molecules on the surface of the AuNP assemblies has been reported to affect the formation of the S–Au bonds, and is more favorable when the thiol groups were electrically neutral [24]. Based on these observations, the pH of the solution was regulated at pH = 6, where the predominant Au anions are $(AuCl_3(OH))^-$ and $(AuCl_2(OH)_2)^-$, using sodium acetate/acetic acid buffer.

3.5. Optimization of the Reaction Kinetics

The reaction kinetics were optimized by varying (i) the reaction time from 2–120 min, and (ii) the temperature towards warm conditions, from room temperature to 50 °C. The graph of Figure 3d shows that, at ambient conditions, the difference in the signal intensity between the blank and the sample increases rapidly during the first 30 min. At higher incubation times, the reactions slow down and only a small increase of the net signal is obtained (~14%) for incubation times up to 80 min. A significant gain in the net signal intensity (~30%), on the other hand, can be accomplished at longer incubation times (>120 min). These observations are in agreement with earlier observations regarding the kinetics of the formation of Au–CTAB complexes [25]. However, the analysis was performed in a microtiter plate (including calibration); therefore, there is no need to strictly monitor the reaction kinetics since all samples were analyzed simultaneously.

The signal response with increasing temperature shows that the signal decreases with increasing temperature. The highest signal is observed between 25 and 35 °C, which is higher than the Krafft temperature of CTAB (about 25 °C) [26]. Below the Krafft temperature, CTAB forms a bilayer-structured, hydrated solid, but above the Krafft temperature cylindrical micelles are formed, which increases the local concentration of Au anions per surface area. At higher temperatures, the signal decreases significantly, which can be attributed to the structural perturbations of CTAB micelles, because CTAB micelles are temperature sensitive [27]. In fact, at temperatures higher than 50 °C, micelles may destabilize and undergo structural and morphological changes [28,29]. Based on these observations, the optimum kinetic conditions of the assay were decided at 40 min of incubation time and ambient temperature.

3.6. Interferences and Selectivity

The selectivity of the assay was investigated by comparing the net signal intensity obtained from Cys and binary mixtures of Cys with other common biomolecules and inorganic electrolytes typically found in biological fluids at concentrations equal or higher than their physiological levels [30]. Although the formation of complexes between amino acids and $AuCl_4^-$ ions is well documented [31,32], the bar plots in Figure 4 show that the recovery of cysteine in the presence of amino acids is higher than 95%, suggesting the lack of interference from amino acids, including the cationic amino acids arginine and lysine. We believe that the complex formation between $AuBr_4^-$ with the amine group of CTAB, prior to sample addition, deters the complexation of free amino acids with Au anions. The only exception to this observation is cystine, which can be tolerated up to equimolar concentration levels to that of cysteine. The interference of cystine stems from the ability of Au(III) ions to oxidatively cleave the disulfide bonds of cystine, forming the respective sulfonic acid derivatives [33,34]. This interference, however, poses no threat to the analysis because a reducing agent is commonly used before the analysis of thiols in order to reduce the disulfide bonds (oxidized thiols) to the more reactive sulfhydryl moieties (reduced thiols).

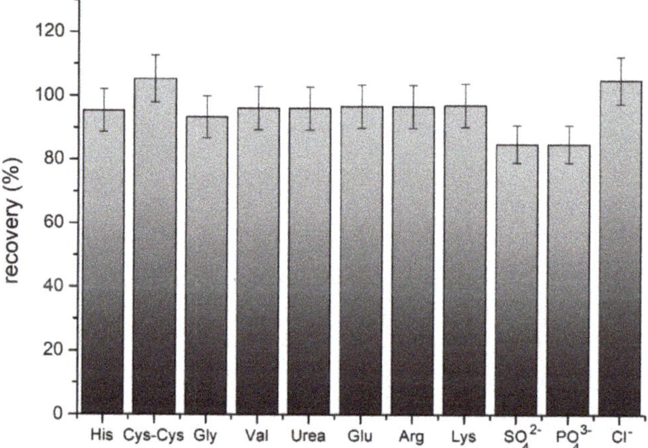

Figure 4. Investigation of interferences from common biomolecules andinorganic electrolytes at physiologically relevant concentration levels. The control sample contains 200 μM Cys while all other solutions contain 200 μM Cys and the potentially interfering biomolecule or ion as follows: histidine, glycine, valine, glutamine: 4.0 mM; cystine: 50 μM; urea: 2.0 mM; glucose: 5.0 mM; Na_2SO_4: 12.0 mM; Na_3PO_4: 12.0 mM; NaCl: 150.0 mM). All solutions were diluted 4-fold before analysis. Error bars represent the standard error calculated for triplicate samples.

Regarding inorganic electrolytes, they were found to influence the analytical signal, possibly due to the effect of counter-anions on the properties (e.g., micellization, Krafft temperature, surface tension, aggregation number, etc.) and structure of cationic micelles [26,35]. Biological fluids contain a variety of inorganic electrolytes; therefore, we investigated the influence of ionic strength and inorganic electrolytes using artificial body fluids (for details see the experimental section). A calibration plot was then constructed in each matrix in order to study the effect of counter anions at various concentration levels of biothiols. Figure S2 shows that the artificial urine solution (AUS) had the lowest effect as compared to artificial blood plasma (ABP) while NaCl caused a significant suppression of the analytical signal. Therefore, dilution of the samples is necessary to reduce the relative abundance of inorganic electrolytes and thus minimize their effect on the properties of CTAB micelles. Technical aspects regarding sample pre-treatment are discussed further below in the analysis of biological samples.

3.7. Analytical Figures of Merit

The analytical merits of the method in the determination of different biothiols were evaluated by preparing standard solutions of Cys, Hcy and GSH at different concentration levels and constructing calibration curves of the analytical signal (i.e., mean color intensity in each well) vs. the concentration of Cys, Hcy or GSH. For all the examined biothiols, the calibration functions correlating the analytical signal to the concentration of each biothiol was linear up to 100 μM. The quantification limit (10σ) of the assay was 5 μM for Cys and GSH and 3.5 μM for Hcy. The working concentration range and the quantification limit are lower than the physiological concentration levels of Cys in blood plasma (150–300 μM), which enables method application with minimal sample volume requirements since dilution is necessary to bring the Cys concentrations within the working concentration range of the method.

The repeatability, calculated as the relative standard deviation of seven replicate measurements, was below 10% (for solutions containing 50 μM cysteine). The calibration functions, linearity, precision and detection limits (DLs) of the assays are summarized in Table 1.

Table 1. Analytical figures of merit of the assay [a].

Biothiol	Cysteine (Cys)	Homocysteine (Hcy)	Glutathione (GSH)
Linear range Regression function Quantification limit [b]	5–100 μM $y = 1.3x + 5.3$, $R^2 = 0.97$ 5.0 μM	5–100 μM $y = 0.73x + 2.9$, $R^2 = 0.99$ 3.5 μM	5–100 μM $Y = x + 8.6$, $R^2 = 0.98$ 5.0 μM
Scanner image (transmittance mode)			

[a] The concentrations that were used to calculate the calibration curves were 0, 5, 20, 30, 50, 70 and 100 μM for Cys and Glu and 0, 20, 30, 50, 70 and 100 μM for Hcy. The inset graphs show a digitally cropped image (to facilitate the aspect view) of the sensing zones (micro titter plates). The first plate is the control (blank) sample. A circular area occupying 80% of the total well surface was used to acquire the analytical signal, which was recorded as the mean grey area in the RGB color system. [b] The quantification limit was determined as 10 times the signal to noise ratio.

Overall, the analytical merits of the assay, as expressed by the working range and the detection limits, are comparable to other published assays [9] and well below the concentration of total biothiols in biological fluids. Nevertheless, other methods that use sophisticated nanomaterial probes or reagents and advanced detectors have achieved even lower detection limits (Table 2).

Table 2. Comparison of the method with recent methods for the determination of biothiols.

Detection Technique	Material or Sensing Element	Real Samples	Linear Range (μM) [a]	LOD (μM) [a]	Recoveries (%)	Reference
Colorimetric (paper-based)	Asp-AuNPs	Blood plasma	99.9–998.7	1.0	99.2–101.1	[36]
Turbidimetry/Colorimetry	AgCl/AgNPs	Blood plasma	10–100	8.1	92–97	[37]
Fluorescence	VS-CDs	Spiked blood plasma [b]	5–200	0.3	98.6–111.5	[38]
Colorimetry/Fluorescence	RhB–MnO2 NFs	Spiked blood plasma [b]	0–15	0.14	89.3–116.3%	[39]
Colorimetric	IrO2/rGO nanocomposites	5 wt% BSA, Blood serum	0.1–50	0.04	99.2–122	[40]
Visual/Colorimetric	$AuCl_4^-$/AuNP seeds/AuNPs	Whole blood, blood plasma	3–300	1.0	88.7–114	[14]
Visual/Colorimetric	$AuCl_4^-$ CTAB	Urine, Blood plasma	5–100	5		This method

Asp-AuNPs: Aspartic-acid-modified gold nanoparticles (AuNPs); VS-CDs: Vinyl sulfone—carbon dots; RhB-MnO2 NFs: Rhodamine B–manganese dioxide nanoflakes; rGO: reduced graphene oxide. [a] Linear range and LODs may vary for individual biothiols. [b] Biothiols were determined after spiking of the known concentrations in the pretreated serum.

3.8. Analysis of Biological Samples

The method was applied to simulated body fluid in order to assess potential matrix effects and identify the appropriate sample pre-treatment conditions. Using simulated blood plasma (for details, see the Experimental section), we investigated the reduction of oxidized thiols (i.e., cystine), as the major biothiol species in biofluids, and the conditions for precipitating proteins, which carry a significant amount of disulfide bonds and hence may produce false-positive results.

The reduction of biothiols with TCEP was optimized through trial-and-error tests in simulated blood plasma (SBP) solutions containing 250 μM of cystine, 250 μM of cysteine and no cysteine as control. We found that the TCEP concentration in the final solution that is used for analysis should be <0.05 mM. At higher concentrations, a positive interference was observed because an excess of residual (unreacted) TCEP could reduce the gold ions to metal gold. Therefore, the reduction of cystine to cysteine was performed with 1.0 mM of TCEP as reducing agent for 30 min before protein removal. The final sample was then diluted as appropriate to adjust the concentration of TCEP in the final extract below the tolerance limit.

Despite the lack of interference observed in the selectivity study, the application of the method to the simulated body fluids showed recoveries that varied from 67 to 76% in simulated urine and 94 to 123% in simulated blood plasma (Table 3).

Table 3. Application of the method to the analysis of real samples.

Sample	Measured (μM)	Spiked (μM)	Found (μM) [a]	Recovery (%) [b]	RSD (%, n = 3)
Simulated urine [c]	0	150.0	101.0	67.3	8.4
	0	250.0	190.5	76.2	6.3
Simulated blood plasma [c]	0	150.0	142.0	94.6	5.9
	0	250.0	307.5	123.0	7.0
Urine [d]	108	100.0	212.1	103.8	6.3
Blood plasma [d]	201	100.0	331.1	115.0	5.8

[a] Values refer to undiluted samples. [b] Recoveries were calculated based on IUPAC recommendations according to the formulae: Recovery (%) = (total − spiked) × 100/found. [c] Concentrations were determined by external calibration using standard solutions of Cys. [d] Concentrations were determined using the method of standard additions.

Student's t-test analysis in the recovery values revealed the presence of matrix effects in the urine samples but not in the blood plasma (i.e., the calculated t-value in the simulated urine and simulated blood plasma was higher and lower than the critical t-value at the p = 0.05 probability level, respectively). Since no AuNP formation was observed (i.e., no absorbance peak at 500–540 nm appeared), this interference was not caused by the matrix components present in the real samples that could reduce gold ions. Therefore, to mitigate potential matrix effects, the method of standard additions was also examined for analysis of biological samples. The spiking levels were 2, 4 and 10 times higher than the expected concentration of the analyte, which ensured good linearity and a zero intercept (i.e., the confidence internal of the intercept b $\pm$ t_{sb}, passed through zero) [41]. The recovery of Cys in the biological samples ranged from 103.8 to 115%, which offered an improvement compared to the direct application of external calibration (the calculated t-value was lower than t-critical at p = 0.05).

4. Conclusions

In this work, we have shown that the colorimetric reaction product of Au anions with the cationic surfactant CTAB can be quenched by thiols proportionally to the thiol concentration. Based on this observation, low molecular weight biothiols were determined in simulated and physiological biological fluids. The method afforded satisfactory recoveries, low detection limits in relation to biothiol levels in biofluids, and good reproducibility. The assay is simple to perform since the end-user just mixes the necessary reagents with the sample and uses a flatbed scanner as a simple and inexpensive microtiter photometric detector. Importantly, all reagents are commercially available at low cost and are stable under normal conditions. These features render the method suitable for applications in low-resource settings, such as decentralized or remote healthcare units (e.g., rural), and for in-clinic analysis (e.g., nurse's bench) where there is access to basic equipment and infrastructure capabilities. The experimental procedure can be further simplified by using the appropriate filters to remove red blood cells and plasma proteins, alleviating the need for centrifugation. In this manner, the method can be applied even in non-laboratory conditions, i.e., where there is a lack benchtop instrumentation (such as centrifuges) and trained operators.

Supplementary Materials: The following are available online at https://www.mdpi.com/2076-3417/11/6/2705/s1, Figure S1: Absorbance spectra of Au-CTAB complex (black line) and Au-CTAB in the presence of 50 μM of glutathione. No peaks above 500 nm are observed suggesting that gold has not been reduced to its respective gold nanoparticle species under the optimum experimental conditions (0.25 mM $AuCl4^-$, 10 mM CTAB, 50 μM GSH, sodium acetate/acetic acid buffer pH 6, 15 min incubation time at room temperature); Figure S2: Response of the colorimetric assay in various artificial biofluids. The linear curves are the result of linear regression while error bar represent the standard error calculated for triplicate samples. AUS: Artificial urine solution, ABP: artificial blood plasma, DW: distilled water.

Author Contributions: Conceptualization, D.L.G.; methodology, E.A.A. and D.L.G.; formal analysis, D.L.G. and N.K.; investigation, E.A.A.; data curation, A.G.V.; visualization, E.A.A. and A.G.V.; writing—original draft preparation, D.L.G.; writing—review and editing, N.K. All authors have read and agreed to the published version of the manuscript.

Funding: This research received no external funding.

Institutional Review Board Statement: Not applicable.

Informed Consent Statement: Informed consent was obtained from all subjects involved in the study.

Data Availability Statement: Data are available from the corresponding authors upon reasonable request.

Conflicts of Interest: The authors declare no conflict of interest.

References

1. Turell, L.; Radi, R.; Alvarez, B. The thiol pool in human plasma: The central contribution of albumin to redox processes. *Free. Radic. Biol. Med.* **2013**, *65*, 244–253. [CrossRef]
2. Valko, M.; Leibfritz, D.; Moncol, J.; Cronin, M.T.D.; Mazur, M.; Telser, J. Free radicals and antioxidants in normal physiological functions and human disease. *Int. J. Biochem. Cell Biol.* **2007**, *39*, 44–84. [CrossRef]
3. Van Meurs, J.B.; Dhonukshe-Rutten, R.A.; De Groot, L.C.; Hofman, A.; Witteman, J.C.; Van Leeuwen, J.P.; Breteler, M.M.; Lips, P.; Pols, H.A.; Uitterlinden, A.G.; et al. Homocysteine Levels and the Risk of Osteoporotic Fracture. *N. Engl. J. Med.* **2004**, *350*, 2033–2041. [CrossRef] [PubMed]
4. Rehman, T.; Shabbir, M.A.; Inam-Ur-Raheem, M.; Manzoor, M.F.; Ahmad, N.; Liu, Z.; Ahmad, M.H.; Siddeeg, A.; Abid, M.; Aadil, R.M. Cysteine and homocysteine as biomarker of various diseases. *Food Sci. Nutr.* **2020**, *8*, 4696–4707. [CrossRef]
5. Teskey, G.; Abrahem, R.; Cao, R.; Gyurjian, K.; Islamoglu, H.; Lucero, M.; Martinez, A.; Paredes, E.; Salaiz, O.; Robinson, B.; et al. Glutathione as a Marker for Human Disease. In *Advances in Clinical Chemistry*; Elsevier: Amsterdam, The Netherlands, 2018; Volume 87, pp. 141–159. ISBN 978-0-12-815203-4.
6. Prendecki, M.; Florczak-Wyspianska, J.; Kowalska, M.; Ilkowski, J.; Grzelak, T.; Bialas, K.; Wiszniewska, M.; Kozubski, W.; Dorszewska, J. Biothiols and oxidative stress markers and polymorphisms of TOMM40 and APOC1 genes in Alzheimer's disease patients. *Oncotarget* **2018**, *9*, 35207–35225. [CrossRef]
7. Isokawa, M.; Kanamori, T.; Funatsu, T.; Tsunoda, M. Analytical methods involving separation techniques for determination of low-molecular-weight biothiols in human plasma and blood. *J. Chromatogr. B* **2014**, *964*, 103–115. [CrossRef]
8. Chen, X.; Zhou, Y.; Peng, X.; Yoon, J. Fluorescent and colorimetric probes for detection of thiols. *Chem. Soc. Rev.* **2010**, *39*, 2120–2135. [CrossRef] [PubMed]
9. Tsogas, G.Z.; Kappi, F.A.; Vlessidis, A.G.; Giokas, D.L. Recent Advances in Nanomaterial Probes for Optical Biothiol Sensing: A Review. *Anal. Lett.* **2018**, *51*, 443–468. [CrossRef]
10. Coronado-Puchau, M.; Saa, L.; Grzelczak, M.; Pavlov, V.; Liz-Marzán, L.M. Enzymatic modulation of gold nanorod growth and application to nerve gas detection. *Nano Today* **2013**, *8*, 461–468. [CrossRef]

11. Shen, L.-M.; Chen, Q.; Sun, Z.-Y.; Chen, X.-W.; Wang, J.-H. Assay of Biothiols by Regulating the Growth of Silver Nanoparticles with C-Dots as Reducing Agent. *Anal. Chem.* **2014**, *86*, 5002–5008. [CrossRef]
12. Jung, Y.L.; Park, J.H.; Kim, M.I.; Park, H.G. Label-free colorimetric detection of biological thiols based on target-triggered inhibition of photoinduced formation of AuNPs. *Nanotechnology* **2016**, *27*, 055501. [CrossRef]
13. Kostara, A.; Tsogas, G.Z.; Vlessidis, A.G.; Giokas, D.L. Generic Assay of Sulfur-Containing Compounds Based on Kinetics Inhibition of Gold Nanoparticle Photochemical Growth. *ACS Omega* **2018**, *3*, 16831–16838. [CrossRef]
14. Akrivi, E.; Kappi, F.; Gouma, V.; Vlessidis, A.G.; Giokas, D.L.; Kourkoumelis, N. Biothiol modulated growth and aggregation of gold nanoparticles and their determination in biological fluids using digital photometry. *Spectrochim. Acta Part A Mol. Biomol. Spectrosc.* **2021**, *249*, 119337. [CrossRef] [PubMed]
15. Torigoe, K.; Esumi, K. Preparation of colloidal gold by photoreduction of tetracyanoaurate(1-)-cationic surfactant complexes. *Langmuir* **1992**, *8*, 59–63. [CrossRef]
16. Nikoobakht, B.; El-Sayed, M.A. Preparation and Growth Mechanism of Gold Nanorods (NRs) Using Seed-Mediated Growth Method. *Chem. Mater.* **2003**, *15*, 1957–1962. [CrossRef]
17. Bai, T.; Tan, Y.; Zou, J.; Nie, M.; Guo, Z.; Lu, X.; Gu, N. AuBr2—Engaged Galvanic Replacement for Citrate-Capped Au–Ag Alloy Nanostructures and Their Solution-Based Surface-Enhanced Raman Scattering Activity. *J. Phys. Chem. C* **2015**, *119*, 28597–28604. [CrossRef]
18. Christodouleas, D.C.; Nemiroski, A.; Kumar, A.A.; Whitesides, G.M. Broadly Available Imaging Devices Enable High-Quality Low-Cost Photometry. *Anal. Chem.* **2015**, *87*, 9170–9178. [CrossRef] [PubMed]
19. Eliceiri, K. ImageJ2: ImageJ for the next Generation of Scientific Image Data. *BMC Bioinform.* **2017**, *18*, 1–26.
20. Perez-Juste, J.; Liz-Marzan, L.M.; Carnie, S.; Chan, D.Y.C.; Mulvaney, P. Electric-Field-Directed Growth of Gold Nanorods in Aqueous Surfactant Solutions. *Adv. Funct. Mater.* **2004**, *14*, 571–579. [CrossRef]
21. Patel, V.; Dharaiya, N.; Ray, D.; Aswal, V.K.; Bahadur, P. pH controlled size/shape in CTAB micelles with solubilized polar additives: A viscometry, scattering and spectral evaluation. *Colloids Surf. A Physicochem. Eng. Asp.* **2014**, *455*, 67–75. [CrossRef]
22. Pacławski, K.; Fitzner, K. Kinetics of gold(III) chloride complex reduction using sulfur(IV). *Met. Mater. Trans. A* **2004**, *35*, 1071–1085. [CrossRef]
23. O'Neil, M. *The Merck Index—An Encyclopedia of Chemicals, Drugs, and Biologicals*; Royal Society of Chemistry: London, UK, 2013.
24. Ma, Q.; Fang, X.; Zhang, J.; Zhu, L.; Rao, X.; Lu, Q.; Sun, Z.; Yu, H.; Zhang, Q. Discrimination of cysteamine from mercapto amino acids through isoelectric point-mediated surface ligand exchange of β-cyclodextrin-modified gold nanoparticles. *J. Mater. Chem. B* **2020**, *8*, 4039–4045. [CrossRef]
25. Khan, M.N.; Alam Khan, T.; Al-Thabaiti, S.A.; Khan, Z. Spectrophotometric evidence to the formation of AuCl4–CTA complex and synthesis of gold nano-flowers with tailored surface textures. *Spectrochim. Acta Part A Mol. Biomol. Spectrosc.* **2015**, *149*, 889–897. [CrossRef]
26. Roy, J.C.; Islam, N.; Aktaruzzaman, G. The Effect of NaCl on the Krafft Temperature and Related Behavior of Cetyltrimethylammonium Bromide in Aqueous Solution. *J. Surfactants Deterg.* **2013**, *17*, 231–242. [CrossRef]
27. Becker, R.; Liedberg, B.; Käll, P.-O. CTAB promoted synthesis of Au nanorods–Temperature effects and stability considerations. *J. Colloid Interface Sci.* **2010**, *343*, 25–30. [CrossRef] [PubMed]
28. Jana, N.R. Gram-Scale Synthesis of Soluble, Near-Monodisperse Gold Nanorods and Other Anisotropic Nanoparticles. *Small* **2005**, *1*, 875–882. [CrossRef]
29. Zijlstra, P.; Bullen, C.; Chon, J.W.M.; Gu, M. High-Temperature Seedless Synthesis of Gold Nanorods. *J. Phys. Chem. B* **2006**, *110*, 19315–19318. [CrossRef]
30. De, M.; Rana, S.; Akpinar, H.; Miranda, O.R.; Arvizo, R.R.; Bunz, U.H.F.; Rotello, V.M. Sensing of proteins in human serum using conjugates of nanoparticles and green fluorescent protein. *Nat. Chem.* **2009**, *1*, 461–465. [CrossRef]
31. Csapó, E.; Ungor, D.; Kele, Z.; Baranyai, P.; Deák, A.; Juhász, Á.; Janovák, L.; Dékány, I. Influence of pH and aurate/amino acid ratios on the tuneable optical features of gold nanoparticles and nanoclusters. *Colloids Surf. A Physicochem. Eng. Asp.* **2017**, *532*, 601–608. [CrossRef]
32. Zou, J.; Guo, Z.; Parkinson, J.A.; Chen, Y.; Sadler, P.J. Gold(III)-induced oxidation of glycine. *Chem. Commun.* **1999**, 1359–1360. [CrossRef]
33. Shaw, C.F.; Cancro, M.P.; Witkiewicz, P.L.; Eldridge, J.E. Gold(III) oxidation of disulfides in aqueous solution. *Inorg. Chem.* **1980**, *19*, 3198–3201. [CrossRef]
34. Franskhaw, P.; Awitkiewic, C. Oxidative Cleavage of Peptide and Protein Disulphide Bonds by Gold(II1): A Mechanism for Gold Toxicity. *J. Chem. Soc. Chem. Commun.* **1981**, *21*, 1111–1114.
35. Oelschlaeger, C.; Suwita, P.; Willenbacher, N. Effect of Counterion Binding Efficiency on Structure and Dynamics of Wormlike Micelles. *Langmuir* **2010**, *26*, 7045–7053. [CrossRef] [PubMed]
36. Liu, C.; Miao, Y.; Zhang, X.; Zhang, S.; Zhao, X. Colorimetric determination of cysteine by a paper-based assay system using aspartic acid modified gold nanoparticles. *Microchim. Acta* **2020**, *187*, 362. [CrossRef] [PubMed]
37. Kappi, F.A.; Papadopoulos, G.A.; Tsogas, G.Z.; Giokas, D.L. Low-cost colorimetric assay of biothiols based on the photochemical reduction of silver halides and consumer electronic imaging devices. *Talanta* **2017**, *172*, 15–22. [CrossRef]

38. Ortiz-Gomez, I.; Ortega-Muñoz, M.; Marín-Sánchez, A.; De Orbe-Payá, I.; Hernandez-Mateo, F.; Capitan-Vallvey, L.F.; Santoyo-Gonzalez, F.; Salinas-Castillo, A. A vinyl sulfone clicked carbon dot-engineered microfluidic paper-based analytical device for fluorometric determination of biothiols. *Microchim. Acta* **2020**, *187*, 1–11. [CrossRef] [PubMed]
39. Xue, H.; Yu, M.; He, K.; Liu, Y.; Cao, Y.; Shui, Y.; Li, J.; Farooq, M.; Wang, L. A novel colorimetric and fluorometric probe for biothiols based on MnO2 NFs-Rhodamine B system. *Anal. Chim. Acta* **2020**, *1127*, 39–48. [CrossRef]
40. Tang, J.; Kong, B.; Wang, Y.; Xu, M.; Wang, Y.; Wu, H.; Zheng, G. Photoelectrochemical Detection of Glutathione by IrO2–Hemin–TiO2Nanowire Arrays. *Nano Lett.* **2013**, *13*, 5350–5354. [CrossRef]
41. Ellison, S.L.R.; Thompson, M. Standard additions: Myth and reality. *Analyst* **2008**, *133*, 992–997. [CrossRef]

applied sciences

MDPI

Review

Therapeutic Applications of Halloysite

Mohammadmahdi Mobaraki [1], Sonali Karnik [2], Yue Li [3] and David K. Mills [3,4,*]

1 Biomaterials Group, Department of Biomedical Engineering, Amirkabir University of Technology, Tehran 15916-34311, Iran; m.mahdimobaraki70@gmail.com

2 Department of Mechanical and Energy Engineering, IUPUI, Indianapolis, IN 46202, USA; karnik.sonu01@gmail.com

3 Center for Biomedical Engineering & Rehabilitation Science, Louisiana Tech University, Ruston, LA 71272, USA; yli021@latech.edu

4 School of Biological Sciences, Louisiana Tech University, Ruston, LA 71272, USA

* Correspondence: dkmills@latech.edu; Tel.: +1-318-257-2640

Abstract: In recent years, nanomaterials have attracted significant research interest for applications in biomedicine. Many kinds of engineered nanomaterials, such as lipid nanoparticles, polymeric nanoparticles, porous nanomaterials, silica, and clay nanoparticles, have been investigated for use in drug delivery systems, regenerative medicine, and scaffolds for tissue engineering. Some of the most attractive nanoparticles for biomedical applications are nanoclays. According to their mineralogical composition, approximately 30 different nanoclays exist, and the more commonly used clays are bentonite, halloysite, kaolinite, laponite, and montmorillonite. For millennia, clay minerals have been extensively investigated for use in antidiarrhea solutions, anti-inflammatory agents, blood purification, reducing infections, and healing of stomach ulcers. This widespread use is due to their high porosity, surface properties, large surface area, excellent biocompatibility, the potential for sustained drug release, thermal and chemical stability. We begin this review by discussing the major nanoclay types and their application in biomedicine, focusing on current research areas for halloysite in biomedicine. Finally, recent trends and future directions in HNT research for biomedical application are explored.

Keywords: drug delivery; halloysite; nanoclay; regenerative medicine; tissue engineering; wound healing

Citation: Mobaraki, M.; Karnik, S.; Li, Y.; Mills, D.K. Therapeutic Applications of Halloysite. *Appl. Sci.* **2022**, *12*, 87. https://doi.org/10.3390/app12010087

Academic Editors: Francesco Cappello and Magdalena Gorska-Ponikowska

Received: 27 July 2021

Accepted: 9 December 2021

Published: 22 December 2021

1. Introduction

Nanoclays are inexpensive materials that constitute sedimentary rocks and derived soils and are classified into natural and synthetic clays [1–3]. They have at least one dimension in the order of 1–100 nm [4–6], have a high aspect ratio, a thickness of less than one nanometer, and a surface area in the range of 700 m squared per gram [6,7]. Nanoclay nanoparticles are mineral silicates with layered structural units that can form complex clay crystallites by stacking these layers [4]. The principal clay types include kaolinite, laponite, montmorillonite, and halloysite (Figure 1, Table 1) [1]. Nanoclays are abundant, mined at low cost, and do not threaten the environment. Accordingly, nanoclays have been studied and developed for aerospace, biomedical, commercial, and industrial applications [2]. Nanoclays, in general, have no mutagenic effect on the body [4], are cyto- and biocompatible [6], and are environmentally friendly [6,7].

Nanoclays are typically used as additives for polymeric materials, and their addition results in significant improvements in mechanical and thermal resistance and overall durability [8,9]. The market for nanoclays is in automotive, biomaterials, biomedicine, cosmetics, flame retardant materials, paints, pigments and dyes, packaging, and textiles [9,10]. For example, bentonite is used for exterior waterproofing treatment, an agent for removing impurities in oil, and as an absorbent and carrier for fertilizers or pesticides. Kaolinite is widely used in producing ceramics and porcelain. As with many nanoclays, it is also used as a bulk filler for paint, rubber, and plastics.

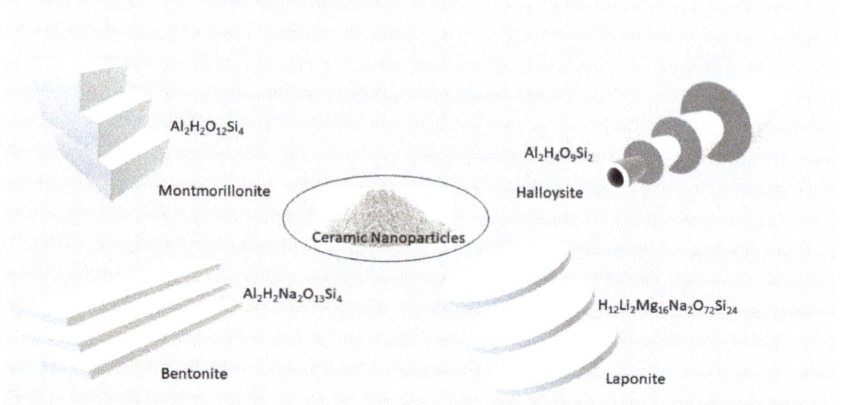

Figure 1. Ceramic nanoparticle types and their morphologies.

Table 1. Major types of clay nanoparticles.

Clay Type	General Formula	Structure	Therapeutic Applications
Bentonite	$Al_2H_2Na_2O_{13}Si_4$	Layered	Absorption, detoxification, drug delivery, filler
Halloysite	$Al_2Si_2O_5(OH)_4$	Nanotube	Absorption, bioink additive bioremediation, detoxification, drug delivery, filler, tissue engineering
Kaolinite	$Al_2H_4O_9Si_2$	Layered	Absorption, drug delivery, filler, tissue engineering
Laponite	$H_{12}Li_2Mg_{16}Na_2O_{72}Si_{24}$	Discoidal	Bioink additive, bioimaging, drug delivery, tissue engineering
Montmorillonite	$Al_2H_2O_{12}Si_4$	Multi-layered	Antimicrobial, drug delivery, filler, tissue engineering

Due to their wide availability, relatively low cost, and relatively low and environmental impact, nanoclays have been widely used in preparing polymer matrix–nanoclay biomedical composites. These include bone repair, cancer therapy, drug delivery, tissue engineering, wound healing, and 3D printing [3,6,8,10]. Laponite, halloysite, and montmorillonite are among the most widely used clays. Halloysite nanotubes (HNTs) and montmorillonite have been used for drug delivery, gene delivery, cancer therapeutics, tissue engineering, and wound healing applications. Halloysite has been intensively studied because of such properties as high mechanical strength, high porosity, thermal resistance, and sustained drug release capability. Because of these properties, increased research is being focused on using HNTs as drug delivery systems and in bioactive bandages, tissue engineering scaffolds, and regenerative medicine. After a brief overview of the major nanoclays, this review will focus on the therapeutic applications of HNTs.

2. Nanoclay Types

2.1. Montmorillonite

Montmorillonite is an abundant phyllosilicate clay material composed of layered silica sheets. Each layer consists of two sheets, octahedral and tetrahedral sheets [11]. The octahedral sheet is aluminum and magnesium bonded with six oxygen and a hydroxyl group. The tetrahedral sheet is composed of linked silicon-oxygen tetrahedral and bonds with octahedra. It can be modified chemically to form nanocomposites. The loosely packed silicates layers can let water infiltrate the sheets, and the clay can swell. It also has excellent cation exchange capacity, producing nanocomposites from the naturally occurring montmorillonite [11,12]. Montmorillonite is the main constituent of bentonite clay.

As a nanoclay, montmorillonite can have particle length and breadth between 1.5 μm to 1/10th of a micron. The pore diameter is small compared to the length of the particle (~1 nm). Refs. [11,12] Because of its property to hold water and hydrophilic molecules, montmorillonite nanocomposites have been used as a filler material to make bioactive scaffolds. As montmorillonite also enhances the mechanical properties of the scaffold materials, it has been used as an additive to hydrogel or polymer scaffolds [11–13]. Montmorillonite has been combined with different scaffold materials such as chitosan, methyl methacrylate, gelatin, starch, and polycaprolactone for tissue engineering applications [14–16]. Montmorillonite nanoclay composite scaffolds have been studied for their applications in bone tissue engineering [15,16], controlled drug delivery [17], and wound healing [18,19].

2.2. Bentonite

Bentonite is an aluminum and phyllosilicate clay formed by weathering volcanic ash by water. Montmorillonite is a significant component of bentonite in addition to feldspar and quartz [11]. Being hydrophilic makes bentonite a very absorbent clay. Exfoliation of sodium and potassium salts from bentonite might result in plates 1 nm in thickness [11]. Bentonite being absorptive would make a suitable wound dressing additive. As a nanoclay in combination with scaffold materials, Bentonite has been studied for hemostatic effect in wound healing [20–23]. Bentonite or any other nanoclay material is a cost-efficient alternative to biological hemostatic agents such as fibrin, as biological hemostatic agents cost more in production and purification [24,25]. Since Bentonite has montmorillonite as its primary component, it has also been explored to manufacture scaffold materials used for skeletal tissue engineering applications. It would improve the mechanical properties of the soft scaffolding materials such as hydrogels [25–27]. Montmorillonite layers and sheets in bentonite make it a suitable nanoclay for drug delivery [28,29]. Its use in sustained and targeted drug release for targeted chemotherapy has also been investigated [30–32].

2.3. Laponite

Laponite®, synthetic clay nanoparticles (25–30 nm diameter, 1 nm thickness), closely resembles the natural clay mineral hectorite in both structure and composition [33,34]. Laponite is a 1:2 layered clay nanoparticle. One central octahedral magnesia sheet is sandwiched between two tetrahedral silica layered sheets. It is widely used in conserving stone, metals, organic materials, ceramics, and paintings. As a bulk filler and reinforcement agent, it is employed in agrochemical, cosmetics, mining, petroleum, and pharmaceutical industries. Laponite possesses an anisotropic nanometric shape and has different charge distribution [33]. Laponite, as a biomedical material, has been applied in drug delivery [34] and tissue engineering [35,36].

3. Halloysite Structure and Applications

3.1. HNT Structure

Halloysite nanotubes (HNTs) are naturally occurring aluminosilicate nanoparticles empirical formula $Al_2Si_2O_5(OH)_4$) with a chemical composition similar to kaolinite, dickite, or nacrite [37–39] (Figure 2). However, unlike kaolinite, dickite, and nacrite, the unit layers in halloysite are separated by a monolayer of water molecules [38,39]. As a result, a hydrated halloysite has a basal (d001) spacing of 10 Å, approximately 3 Å larger than kaolinite. Halloysite-(10 Å) can readily and irreversibly dehydrate to give the corresponding halloysite-(7 Å) form when halloysite-(10 Å) is heated to 90–150 °C. HNTs can be found in China, France, Belgium, New Zealand, America, and Brazil [40]. The chemical composition for halloysite-(7 Å) and halloysite-(10 Å) is $Al_2Si_2O_5(OH)_4 \cdot nH_2O$ where n = 0 and 2, respectively [41–43]. If n is 2, the HNTs are hydrated, and if n is 0, the HNTs are dehydrated [10,11]. Therefore, AIPEA Nomenclature Committee recommended terms halloysite-(10 Å) for the hydrated mineral and halloysite-(7 Å) for the dehydrated form.

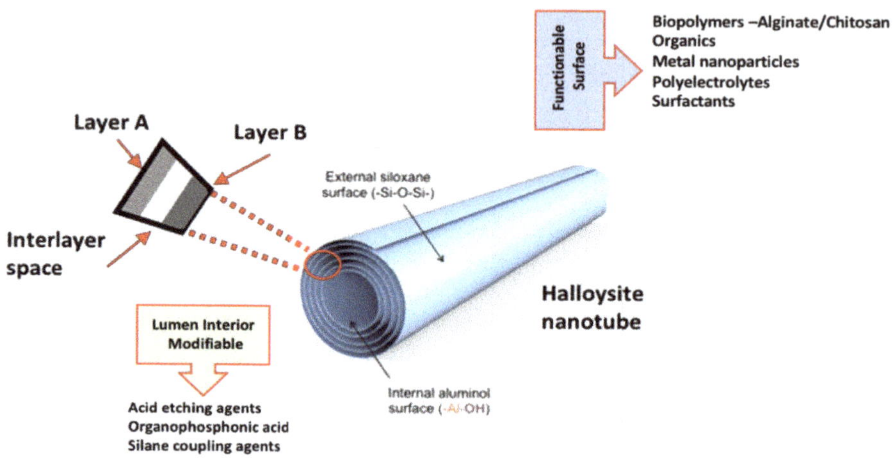

Figure 2. The structure of halloysite and its potential for surface modification.

The layers of aluminum and silicate with alternating positive and negatively charged layers in halloysite make them suitable adsorbents for cations and anions [41]. Apart from the conventional use of halloysite to manufacture porcelain and in petrochemical applications, halloysite has been explored as a carrier material for drug delivery [37,38], tissue engineering [41,42], and wound healing applications [43]. HNTs can adopt a variety of morphologies. For example, an elongated tubule short tubular, spheroidal and platy nanoparticle shapes have been reported [44]. Spheroidal halloysite occurs widely, and it is common to find pseudo-spherical or spheroidal particles in weathered volcanic ashes and pumices [38,39,45].

The hollow tubular form in the submicrometer range is most commonly used. These tubules may be extended and thin, short and stubby, or emerging from other tubes [10]. The halloysite tubules' size varies from 500–1000 nm in length with an outer diameter of 10–50 nm and an inner diameter measuring 5–20 nm depending on the deposit [43–45]. The neighboring alumina and silica layers, their hydration layers, create a packing disorder that induces curvature and the layers roll up, forming multilayer tubes. The HNT external surface comprises O-Si-O bonds with terminal hydroxyl groups [37,38]. The inner lumen comprises O-Al-O bonds, terminating in hydroxyl groups [46,47]. At pH 8.5 and below, these inner hydroxyl groups are mostly protonated, resulting in a positively charged inner lumen.

A wide range of active agents, including antibiotics, cancer drugs, marine biocides, and biological molecules, can be entrapped within the inner lumen and void spaces within the aluminosilicate shells [47–50]. HNTs nanotubes are non-cytotoxic on several cell types (up to concentrations of 0.1 mg/mL), including chondrocytes, dermal fibroblasts, osteoblasts, and stem cells on halloysite nanofilms or within HNT-hydrogel composites [48–54]. Examination of halloysite with in-vitro assays showed cells proliferated and maintained their cellular phenotype. Recent biocompatibility studies have shown that HNTs do not provoke a cytotoxic or host immune response [51,52]. As halloysite nanotubes have been shown to exhibit high biocompatibility levels and very low cytotoxicity, it represents an ideal candidate for new drug delivery and polymer systems.

3.2. HNTs in Cancer Therapeutics

HNTs have been studied for multiple applications. Currently, modified surface HNTs are being researched as an efficient delivery system for cancer drugs. For example, Chitosan oligosaccharide modified HNTs (HNTs-g-COS) demonstrated the ability to enhance the therapeutic efficacy of the anticancer drug doxorubicin (DOX) [53]. In vitro, DOX loaded HNTs-g-COS released in cell lysate in a controlled manner and increased the apoptosis effects of MCF-7 cells in flow cytometry results [53]. In vivo, the tumor inhibition ratio of DOX loaded HNTs-g-COS was two times higher than free DOX and no apparent systemic toxicity in DOX loaded HNTs-g-COS groups [53].

Synthesized chitosan grafted HNTs (HNTs-g-CS) also showed great potential as nanovehicles for anticancer drug delivery in cancer therapy [54]. The research found that HNTs-g-CS had a significantly enhanced curcumin loading capability and good serum stability. However, the curcumin-loaded HNTs-g-CS show specific toxicity to various cancer cell lines, including HepG2, MCF-7, SV-HUC-1, EJ, Caski, and HeLa demonstrate an inhibition concentration of IC50 at 5.3–192 mM as assessed by cytotoxicity studies [54]. In addition, this nanocomposite has a too high anticancer activity in EJ cells compared to the other cancer cell lines [54].

Folate-conjugated HNTs can be an efficient drug carrier for targeted breast cancer therapy via intravenous injection [55]. HNT conjugated with polyethylene glycol and folate (HNTs-PEG-FA) is designed as a targeted drug delivery system [55]. Doxorubicin (DOX) loaded HNTs-PEG-FA shows significant inhibition of proliferation and induction of death in MCF-7 cells with a positive folate receptor [55]. DOX-loaded HNTs-PEG-FA leads to more mitochondrial damage and apoptosis than the same dose of DOX [55]. In contrast to DOX, DOX-loaded HNTs-PEG-FA effectively reduces heart toxicity and inhibits substantial tumor growth with higher cleaved caspase-3 protein levels in tumor tissue of 4T1-bearing mice [55]. DOX-loaded HNTs-PEG-FA reveals more DOX in tumor tissue than in other normal tissues, including the heart, spleen, lung, and kidney.

3.3. HNTs in Drug Delivery

HNTs have been used as a drug delivery carrier for many clinically meaningful drugs [56,57]. HNT can be loaded with different drugs, including anticancer drugs, antibiotics, analgesics, antihypertension, anti-inflammatory drugs, and therapeutic nucleic acids [57]. HNTs have also been used for the controlled release of antibiotics, including tetracycline, ofloxacin, norfloxacin, amoxicillin, and ciprofloxacin [57]. Amoxicillin (AMX) loaded HNT is incorporated into a polylactic acid-glycolic acid copolymer (PLGA) solution, which is electrospun with water-soluble chitosan nanofibers in two different syringes simultaneously, thereby making a composite material [58]. Compared to loading the drug directly into the polymer matrix, HNT extends the release time of AMX and reduces the initial burst release [58].

Analgesic drugs and anti-inflammatories, such as ibuprofen (IBU), diclofenac sodium, and aspirin, have low water solubility and bioavailability [59]. Therefore, developing an efficient drug delivery system by encapsulating drugs in a nanoparticle system to enhance their bioavailability is urgently needed [59]. 3-aminopropyltriethoxysilane (APTES) functionalized surface HNT as a carrier for IBU could promote IBU loading [60]. By restricting the APTES oligomerization in the lumen, free lumen space was preserved, resulting in a 25.4% greater loading rate than that in unmodified halloysite. In order to sustain a more significant release of IBU, an ideal hydrophobic sustained-release drug delivery system was designed [60]. The HNT lumen (EHNT) was enlarged, and hydrophobic modification of the external surface by organosilane (OS) was done prior to loading IBU [60]. The OS composite of EHNT demonstrated a sustained-release performance for IBU (100 h) [60].

Halloysite has been used in other drug delivery systems such as anti-hypertension and gene therapeutic agent-delivery systems. Polydopamine was used to cap HNT for a controlled drug release [61]. After dispersion in a sodium alginate matrix and crosslinking via Fe^{3+}, HNTs were used to deliver diltiazem hydrochloride, widely used in high blood pressure therapy [61]. In gene therapeutic agent-delivery systems, HNTs were surface-modified with γ-aminopropyltriethoxysilane and assembled with antisense oligodeoxynucleotides (ASODNs) [62]. These functional HNT complexes showed improved intracellular delivery efficiency and inhibited the tumor growth activity of ASODNs [62].

3.4. HNTs in Tissue Engineering

Halloysite has a variety of applications in the field of tissue engineering. They are used in bone implants, dental fillings, and tissue scaffolds [63]. HNTs mixed with bone cement and used as a drug carrier and release system are promising applications. The research found HNTs loaded with the antibiotic gentamicin sulfate with a concentration of 5–8 wt% in the cement (PMMA) provide sustained release up to 300–400 h [40]. This PMMA/halloysite/gentamicin composite tensile strength does not deteriorate compared with pure cement, and its adhesion to bone is significantly increased [53]. HNTs resin-dentin bond is similar to halloysite-PMMA bone cement [64]. HNTs and functionalized HNTs improved mechanical properties significantly [64–66]. Silver nanoparticle immobilized HNT (HNT/Ag) fillers significantly improved mechanical properties [67]. This filler also showed a significant antibacterial activity observed on *S. mutans* [67]. Karnik et al., 2015 were the first to show that a nanoenhanced hydrogel could significantly enhance the biological activity of bone progenitor cells and achieve a sustained release of BMP-2 for over a week [49].

Currently, hydrogel scaffolds are being applied to transplant cells and engineer nearly every tissue in the body, including cartilage, bone, and smooth muscle [68]. Compared to pure alginate scaffolds, alginate/halloysite nanotube (HNTs) composite scaffolds significantly enhance compressive strength and compressive modulus in dry and wet states [69]. Furthermore, HNTs increased the scaffold density, decreased the swelling ratio in water, and improved alginate's thermal stability [69]. In addition, the alginate/HNT composite scaffolds have better cytocompatibility [69]. Chitosan–halloysite nanotubes (HNTs) nanocomposite (NC) scaffolds have similar results as alginate HNTs composite scaffolds [70]. Compared to the pure chitosan scaffold, the NC scaffolds exhibited significantly improved compressive strength, compressive modulus, and thermal stability [70]. Furthermore, the chitosan–HNTs nanocomposites were cytocompatible even when the HNTs load was 80% [70].

3.5. HNTs in Wound Healing

The absorptive capacity of HNTs has been used in several wound healing applications. An HNT/chitosan oligosaccharide nanocomposite was tested for its healing capacity in a mouse model [71]. The nanocomposite allowed enhanced skin reepithelization and reorganization compared to controls and the. Results suggested it has potential as a medical device for wound healing. Chitosan has been combined with HNTs in many would bandage and healing applications. Li et al. (2014) showed that the HNT/chitosan sponges significantly increased wound closure ratio compared with pure chitosan. HNT addition also aided in re-epithelialization and collagen deposition [72].

HNTs and other nanoclays such as montmorillonite used as scaffolds also possess the capability of improving the wound healing response. For example, Sandri et al. (2020) produced electrospun scaffolds incorporating these nanoclays, and their results showed enhanced fibroblast cell attachment and proliferation with very little to no proinflammatory activity [73]. A final example of potential wound healing applications using HNTs as a key contributor is the study by Wali et al. (2019) [74]. This study loaded electrospun cellulose ether-PVA nanofiber mats with HNTs and gentamicin sulfate. As a result, the mats offered sustained gentamicin release and advanced wound healing in an animal model [74]. The above studies have demonstrated that HNTs, especially chitosan-HNTs nanocomposites, have significant potential for burns, chronic wounds, and diabetic foot ulcers.

4. Recent Trends in HNT Research

4.1. HNTs as a Drug Carrier

As mentioned previously, HNTs have been used for various biomedical areas because of unique properties such as hemostatic properties. However, one of the most critical applications of HNT is drug delivery systems, and it is used as a drug carrier. For instance, in 2020, Cheng et al. [75] used HNTs to fabricate smart hydrogels with H_2O_2—responsive release properties. The researchers first prepared drug loaded HNTs and then, they characterized the composite hydrogels by FTIR, TGA, XPS, XRD, and TEM. The smart hydrogel was prepared with polyvinyl alcohol and released the drugs under a pathological concentration ($[H_2O_2]$ = 200 μm) because of degradation of the hydrogel in the presence of H_2O_2 (Figure 3). Core-shell gel-based chitosan fabricated by a dropping method and HNTs because of hollow cavity used as a carrier [76]. In this study, Lisuzzo et al. [76] utilized alginate as a coating layer (shell). In the hybrid gel beads, HNTs showed proper dispersion within the beads, and because of the core-shell structure of gel, the release of the drug was controllable.

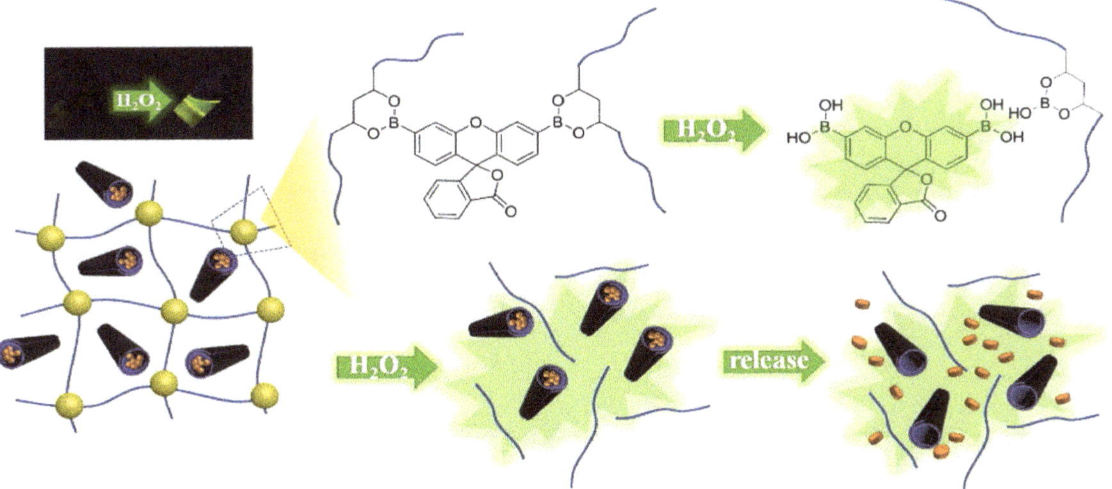

Figure 3. Schematic of the mechanism of release when hydrogel under pathological concentration. Reprinted by permission of the publisher.

4.2. HNTs in Coatings and Films

Akrami-Hasan-Kohal et al. [77] used the solution casting process to fabricate nanocomposite films with HNTs. In this study, the various concentrations of HNTs were loaded into the nanocomposites films. As expected, mechanical properties and water absorption were improved, and the films were cytocompatible. Xie et al. [78] used HNT for preparing nanocomposite films based on chitosan for biomedical applications. The film exhibited decent mechanical properties, and dosages of HNT had a direct effect on this property because of creating a three-dimensional network.

Additionally, water-resistance and the nanocomposite films' thermal stability were improved meaningfully. In 2019, Devi et al. [79] ternary nanocomposite films were fabricated using solution casting. This film used chitosan, starch, and HNTs to improve mechanical properties, water absorption capacity, water solubility, and water vapor transmission rate. Furthermore, these hydrophilic films were non-hemolytic and impermeable to bacteria. Mohebali et al. [80] prepared elastomeric nanocomposites based on HNT surface modification using APTES and then a polyvinyl alcohol (PVA) coating applied through the layer-by-layer assembly. As a result, the nanocomposite film exhibited antibacterial properties and had a synergistic anti-infection effect (Figure 4). Furthermore, drug release was sustained and controllable.

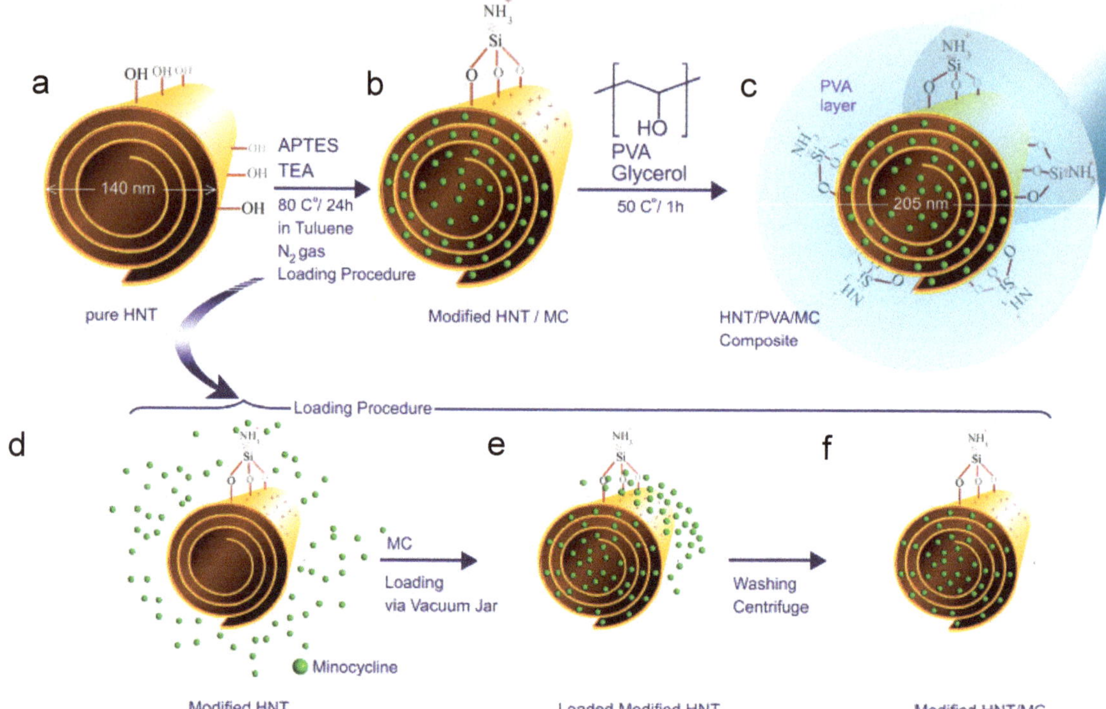

Figure 4. SEM image of pure HNT (**a**), surface of HNT/PVA (**b**), cross section of HNT/PVA (**c**), surface of HNT/PVA/MC, Schematic of drug loading, surface modification, and synthesis of HNT/PVA nanocomposites (**d**–**f**). Reprinted by permission of the publisher.

4.3. HNTs and 3D Printing

HNTs and 3D printing have also supported tissue regeneration of bone defects. For example, Weisman et al. [81] used PLA and gentamicin-doped HNTs to print a range of medical devices. For example, Tappa et al. [82] used doped HNTs and PLA to 3D print customized bioactive and absorbable surgical screws, pins, and bone plates for localized drug delivery.

HNTs and 3D printing have also been used to support tissue regeneration of bone defects. Illustration, Wu et al. [83] in 2019, used HNTs for coating on 3D printed scaffolds to guide human mesenchymal stem cells (hMSCs) orientation. Polylactic acid scaffolds consisting of different patterns were printed and functionalized by a polydopamine. Using field emission scanning electron microscopy (FE-SEM), the researcher found HNTs successfully coated on various patterns. Because of HNT addition, the scaffolds' surface roughness and hydrophilicity properties improved, and scaffolds could induce cell orientation. Moreover, the adhesion and proliferation of cells increased because of this supportive cell coating (Figure 5).

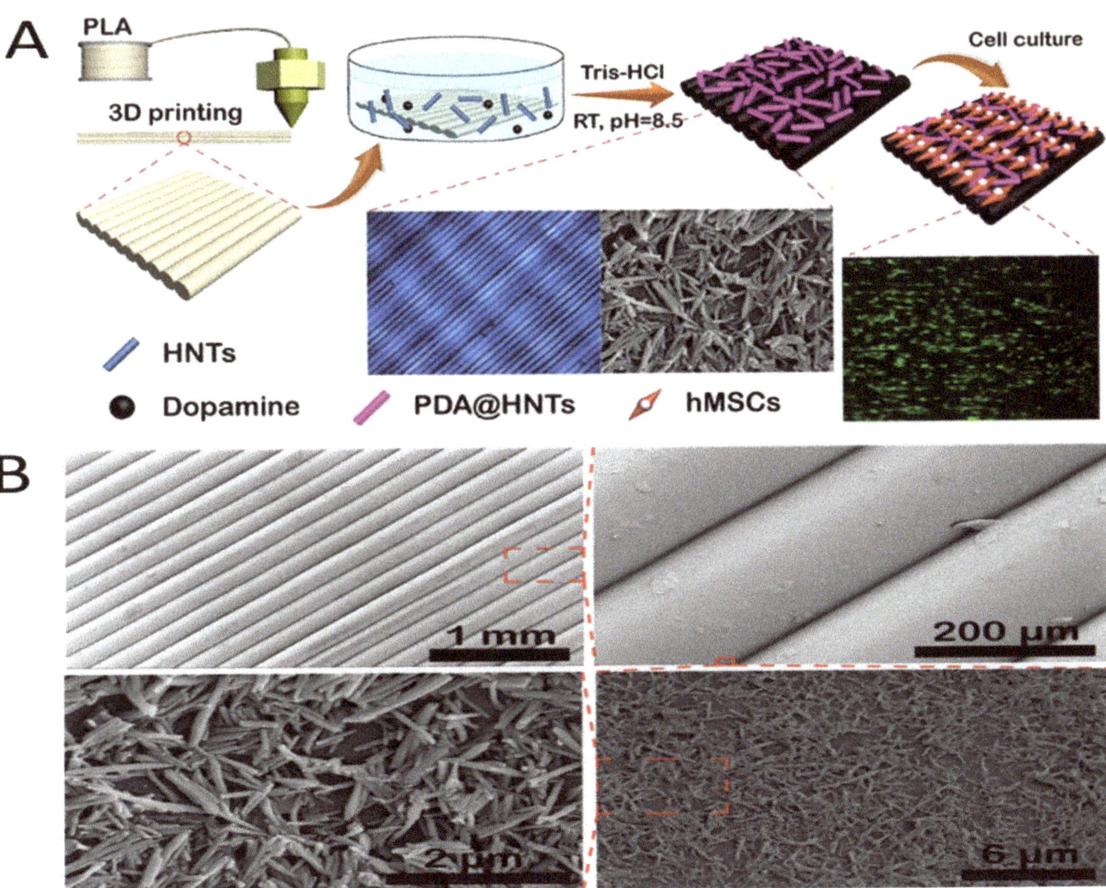

Figure 5. (**A**) Graphic abstract of the HNTs coated PLA printed scaffolds with different pattern for HNT patterns guiding cell orientation. (**B**) FE-SEM images of HNTs coated printed pattern surfaces with various magnification. Reprinted by permission of the publisher.

4.4. Metalized HNT Applications

Early in human history, metal and metal salts were used for various applications. For example, Greek and Roman cultures used silver vessels to preserve liquids [84]. Colloidal silver was used extensively in medical treatment, and copper-based medicines were used in treating eczema, lupus, anemia, chorea, syphilis, and zinc salts were used in wound treatment. Understanding how metallic nanostructures interact with microorganisms is a rapidly growing area of inquiry in the biomedical field will offer significant advantages in diagnostic and therapeutic applications (Table 2). Nanoparticles offer unique physical properties that have associated benefits for drug delivery. For example, some metal nanoparticles have bactericidal effects due to their high surface-to-volume ratio and small size, allowing them to interact with bacterial cell membranes rather than release metal ions into the solution. Copper, silver, and zinc nanoparticles, in particular, are known for their excellent antimicrobial properties and have additional beneficial functionalities [84,85].

Table 2. Potential applications of metalized HNTs.

Antimicrobial	Sustained Release	Targeted Drug Delivery	Tissue Engineering	Wound Healing
antibiotics	antimicrobials	brain	bone	anti-infection
antifungals	anti-cancer agents	breast cancer	guided nerve regeneration	hemostasis
anti-biofouling	dyes	colon cancer	cartilage	healing
anti-viral	growth factors	osteosarcoma	skin	revascularization

Surface functionalization of HNTs with different components, including metals, antibiotics, and bioactive compounds, is of increased importance for their potential use in biomedical devices, antimicrobial surface coatings, drug delivery systems, radiation absorbent composites, elastomer composites, electronic components, and as industrial catalysts (Figure 6) [86]. The outer surface of HNTs can be used for adsorption of metal NPs thus preventing their aggregation, reducing toxicity, and providing more sustained antimicrobial/viral action. Boyer and. Mills (2016) used a simple method (US Patent #9,981,074 B1) for the fabrication of HNT- supported metal nanoparticles [87]. Metals nanoparticles, such as silver, copper, gold, and other metal nanoparticles can be directly deposited onto the surfaces of HNTs. Existing fabrication methods include lengthy multi-step processes that include metal-salts or organic compounds, reducing agents, high temperatures, and expensive equipment to achieve metallization of the halloysite surface [88].

A strategy for controlled weight deposition of positively charged metal ions on negatively charged HNTs dispersed in an aqueous medium where the extent of metallization can be controlled through changes in voltage, solvent medium, time, and other electrolytic parameters. Additionally, this process does not require the use of any toxic chemicals, expensive reagents, or lengthy pre-processing steps [88].

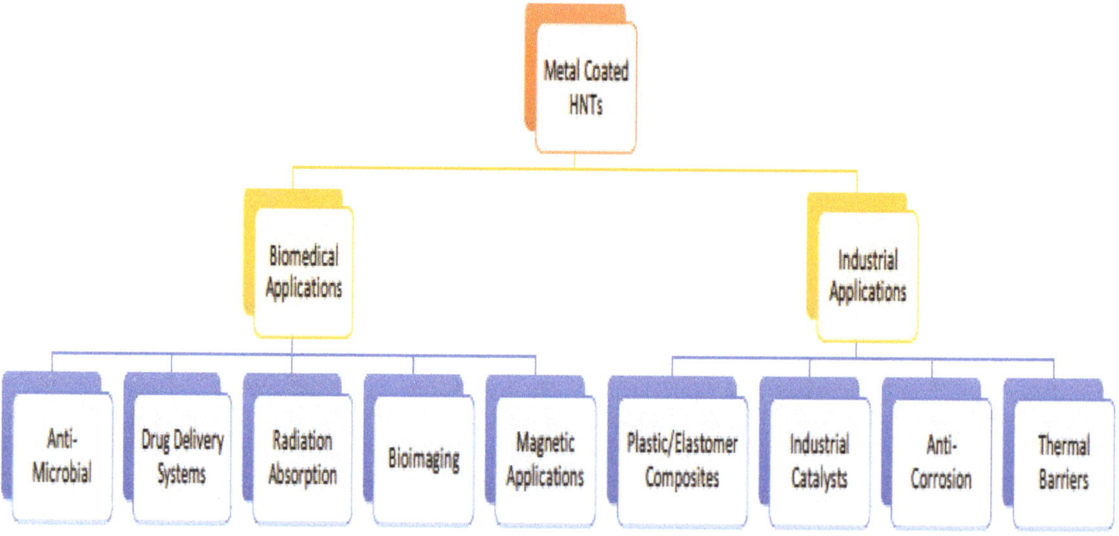

Figure 6. Biomedical and industrial applications of metallized HNTs.

This rapid and low-cost method was used in various applications, from 3D printing to surface coats. PLA 3D printed scaffolds incorporating metal coated HNTs for bone tissue regeneration have recently been published. Luo et al. [89] used a surface modification method to reduce the hydrophobicity of PLA. This study used fetal bovine serum and

NaOH to coat PLA scaffolds containing Zn-coated HNTs (Figure 7). The scaffolds possessed high mechanical strength and showed an osteoinductive potential. A strategy for controlled weight deposition of positively charged metal ions on negatively charged.

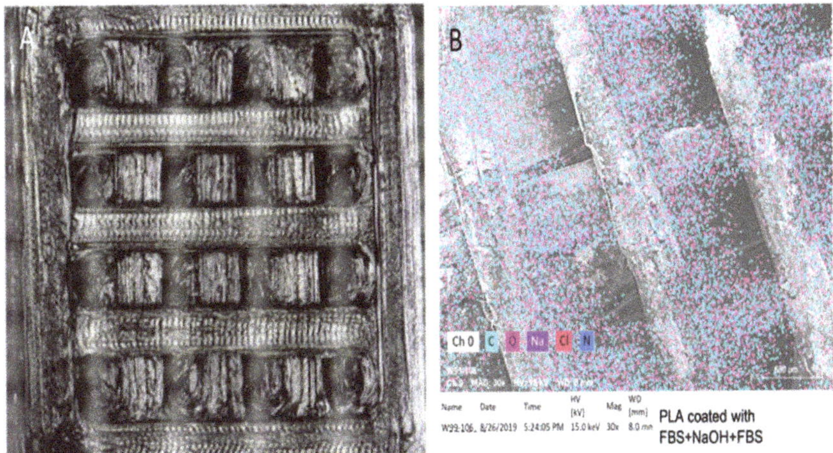

Figure 7. (**A**) Optical and laser combined image of 3D printed PLA square. (**B**) EDS elemental analysis for coated PLA square [87].

HNTs dispersed in an aqueous medium where metallization can be controlled using changes in voltage, solvent medium, time, and other electrolytic parameters without using any toxic chemicals, expensive reagents, or lengthy pre-processing steps [88].

This rapid and low-cost method was used in various applications, from 3D printing to surface coats. PLA 3D printed scaffolds incorporating metal coated HNTs for bone tissue regeneration have recently been published. Luo et al. [87] used a surface modification method to reduce the hydrophobicity of PLA. This study used fetal bovine serum and NaOH to coat PLA scaffolds containing Zn-coated HNTs (Figure 6). The scaffolds possessed high mechanical strength and showed an osteoinductive potential.

Furthermore, the external coating of antibiotics preserved the osteogenic properties but also significantly reduced bacterial growth. [87] A similar study used 3D printed PLA scaffolds modified using an alkali treatment to increase hydrophilicity and the surface-functionalized using a suspension of Zinc/HNTs-Ag-Chitosan Oligosaccharide Lactate [88].

5. Future Directions in HNT Research

Nanoclays are promising drug delivery carriers and additive or bioactive agents for tissue regenerative medicine and engineering. The reason is their biocompatibility, low toxicity, ease of surface modification, material enhancement properties, cost, and capability of encapsulating drugs. The application of clay nanomaterials allows the nanoparticle-based on size, solubility, surface charge, cationic exchange capacity, dispersibility, and drug release rate. Bioactive agents or drugs can be incorporated into the layered spaces or nanopores of montmorillonite, kaolinite, halloysite nanotube through various reactions. Current research on nanoclays explores novel drug/clay or drug/clay/polymer composites as sustained drug delivery systems. The objective is to deliver a 'focal and local' drug dosage to target sites with low toxicity. Nanoclay modification methods can modify surface properties for added functionalities or targeted drug delivery.

The future of HNTs in biomedicine looks promising. The US FDA regards them as generally regarded as safe (GRAS). HNTs have been used for drug delivery, gene delivery, cancer therapeutics, tissue engineering, and wound healing applications and are one of the most promising nanoclays. HNTs are biocompatible, possess high mechanical strength, high porosity, offer sustained drug release profiles, and their surfaces can be easily modified. As a result, HNTs have been used in numerous novel applications ranging from stem cell encapsulation and tissue engineering to intracellular and extracellular drug systems.

HNTs added to a polymer and fabricated as a scaffold can be fabricated through blow spinning, electrospinning, 3D printing, and bioprinting with appropriate pore size, mechanical strength and loaded with bioinstructive molecules.

During the last decade, increased research has been focused on using HNTs for non-medical applications, including adsorbents, chemical and corrosion resistance, curation of archeological and cultural materials, electromagnetic protection, and water purification. Here, surface modification of the HNT surface enables customized and tailored solutions for numerous biotechnological, aerospace, environmental, and military needs.

While HNTs have demonstrated promising potential in drug delivery, cancer therapy, gene therapy, and tissue engineering applications, many challenges are still ahead before their widespread clinical application. As an anti-cancer nanoparticle, a significant issue is the exact mechanism of cell death necrosis or apoptosis. They are entering inside the cells by this route, though several studies have examined cell uptake. In addition, while many studies suggest that HNTs are cyto- and biocompatible, there is no clear understanding of the mechanism(s) of interaction with living cells, organs, and organisms. Therefore, the biological responses to HNTs must be thoroughly investigated through additional in-vitro and in-vivo research leading to a complete understanding of HNT impact on cellular pathways, further toxicity profiling, and how HNTs are stored within the body or eliminated. Lacking this knowledge limits their usage for advanced drug delivery, bone regeneration, or therapeutic medical applications.

Author Contributions: All authors contributed equally to writing and editing of the manuscript. The conceptual framework for the manuscript was conceived by D.K.M. All authors have read and agreed to the published version of the manuscript.

Funding: This research was funded by NASA EPSCoR Rapid Response Research Opportunity. 21-EPSCoR-R3-0009 entitled Nano-based Ceramic-metal Composites to Support Planetary Agrosystems.

Institutional Review Board Statement: No humans or animals were used in this study.

Informed Consent Statement: Not applicable.

Data Availability Statement: Not applicable.

Conflicts of Interest: The authors declare no conflict of interest.

References

1. Savic, I.; Stojiljkovic, S.; Savic, I.; Gajic, D. Industrial application of clays and clay minerals. In *Clays and Clay Minerals: Geological Origin, Mechanical Properties and Industrial Applications*; Wesley, L.R., Ed.; Nova Science Publishers: New York, NY, USA, 2014; pp. 379–402.
2. Guo, F.; Aryana, S.; Han, Y.; Jiao, Y. A review of the synthesis and applications of polymer–nanoclay composites. *Appl. Sci.* **2018**, *8*, 1696. [CrossRef]
3. Sandri, G.; Bonferoni, M.C.; Rossi, S.; Ferrari, F.; Aguzzi, C.; Viseras, C.; Caramella, C. Clay minerals for tissue regeneration, repair, and engineering. In *Wound Healing Biomaterials*; Ågren, M.S., Ed.; Elsevier: Amsterdam, The Netherlands, 2016; pp. 385–402. ISBN 9781782424567.

4. Suresh, R.; Borkar, N.S.; Sawant, V.A.; Shende, V.S.; Dimble, S.K. Nanoclay drug delivery system. *Inter. J. Pharma. Sci. Nanotech.* **2010**, *3*, 901–905.
5. Kotal, M.; Bhowmick, A.K. Polymer nanocomposites from modified clays: Recent advances and challenges. *Prog. Polym. Sci.* **2015**, *51*, 127–187. [CrossRef]
6. Sánchez-Fernández, A.; Peña-Parás, L.; Vidaltamayo, R.; Cué-Sampedro, R.; Mendoza-Martínez, A.; Zomosa-Signoret, V.C.; Rivas-Estilla, A.M.; Riojas, P. Synthesization, characterization, and in vitro evaluation of cytotoxicity of biomaterials based on halloysite nanotubes. *Materials* **2014**, *7*, 7770–7780. [CrossRef] [PubMed]
7. Zhang, Y.; Long, M.; Huang, P.; Yang, H.; Chang, S.; Hu, Y.; Tang, A.; Mao, L. Emerging integrated nanoclay-facilitated drug delivery system for papillary thyroid cancer therapy. *Sci. Rep.* **2016**, *6*, 33335. [CrossRef]
8. Jawaid, M.; Qaiss, A.K.; Bouhfid, R. *Nanoclay Reinforced Polymer Composites: Nanocomposites and Bionanocomposites*; Springer: Singapore, 2016.
9. Lau, L.; Gu, C.; Hui, D. A critical review on nanotube and nanotube/nanoclay related polymer composite materials. *Compos. Part B Eng.* **2006**, *37*, 425–436. [CrossRef]
10. Uddin, F. *Montmorillonite: An Introduction to Properties and Utilization*; IntechOpen: London, UK, 2018.
11. Ali, A. Effect of incorporation of montmorillonite on Xylan/Chitosan conjugate scaffold. *Colloids Surf. Part B* **2019**, *180*, 75–82. [CrossRef]
12. Cui, Z.-K.; Kim, S.; Baljon, J.J.; Wu, B.M.; Aghaloo, T.; Lee, M. Microporous methacrylated glycol chitosan-montmorillonite nanocomposite hydrogel for bone tissue engineering. *Nat. Commun.* **2019**, *10*, 3523. [CrossRef] [PubMed]
13. Haroun, A.A.; Gamal-Eldeen, A.; Harding, D.R.K. Preparation, characterization and in vitro biological study of biomimetic three-dimensional gelatin–montmorillonite/cellulose scaffold for tissue engineering. *J. Mater. Sci. Mater. Med.* **2009**, *20*, 2527–2540. [CrossRef] [PubMed]
14. Mauro, N.; Chiellini, F.; Bartoli, C.; Gazzarri, M.; Laus, M.; Antonioli, D.; Griffiths, P.; Manfredi, A.; Ranucci, E.; Ferruti, P. RGD-mimic polyamidoamine–montmorillonite composites with tunable stiffness as scaffolds for bone tissue-engineering applications. *J. Tissue Eng. Regen. Med.* **2017**, *1*, 2164–2175. [CrossRef]
15. Jamshidi, M. Nanoclay reinforced starch-polycaprolactone scaffolds for bone tissue Engineering. *J. Tissues Mater.* **2019**, *2*, 55–63.
16. Chen, M. Fabrication and characterization of a rapid prototyped tissue engineering scaffold with embedded multicomponent matrix for controlled drug release. *Int. J. Nanomed.* **2012**, *2012*, 4285–4297. [CrossRef]
17. Noori, S. Nanoclay enhanced the mechanical properties of poly(vinylalcohol)/chitosan/montmorillonite nanocomposite hydrogel as wound dressing. *Procedia Mater. Sci.* **2015**, *1*, 52–156.
18. Garcia-Villen, F. Montmorillonite-norfloxacin nanocomposite intended for healing of infected wounds. *Int. J. Nanomed.* **2019**, *14*, 5051–5060. [CrossRef]
19. Emami-Razavi, S.H. Effect of bentonite on skin wound healing: Experimental study in the rat model. *Acta Med. Iran. Vol.* **2016**, *44*, 235–240.
20. Alavi, M.; Totonchi, A.; Okhovat, M.A.; Motazedian, M.; Rezaei, P.; Atefi, M. The effect of a new impregnated gauze containing bentonite and halloysite minerals on blood coagulation and wound healing. *Blood Coagul. Fbrinolysis* **2014**, *25*, 856–859. [CrossRef]
21. Khoshmohabat, H.; Dalfardi, B.; Dehghanian, A.; Rasouli, H.R.; Mortazavi, S.M.J.; Paydar, S. The effect of CoolClot hemostatic agent on skin wound healing in rats. *J. Surg. Res.* **2015**, *200*, 732–737. [CrossRef] [PubMed]
22. Wilson, R. Kaolin and halloysite deposits of China. *Clay Miner.* **2004**, *39*, 1–15. [CrossRef]
23. Liu, M.; Shen, Y.; Daib, L.; Zhou, C. The improvement of hemostatic and wound healing property of chitosan by halloysite nanotubes. *RSC Adv.* **2014**, *4*, 23540–23553. [CrossRef]
24. Fang, F.; Xu, Y.; Wang, Z.; Zhou, W.; Yan, L.; Fan, F.; Liu, H. 3D porous chitin sponge with high absorbency, rapid shape recovery, and excellent antibacterial activities for noncompressible wound. *Chem. Eng. J.* **2020**, *388*, 124169. [CrossRef]
25. Devi, N. Preparation and characterization of chitosan-bentonite nanocomposite films for wound healing application. *Int. J. Biol. Macromol.* **2017**, *104*, 1897–1904. [CrossRef] [PubMed]
26. Adams, L.A. Bentonite clay and waterglass porous monoliths Via the sol-gel process. *J. Met. Mat. Min.* **2011**, *21*, 1–6.
27. Essien, E.R.; Adams, L.A.; Shaibu, R.O.; Oki, A. Sol-gel bioceramic material from bentonite clay. *J. Biomed. Sci. Eng.* **2013**, *6*, 258–264. [CrossRef]
28. Shahbuddin, N.S. Alumina Foam (AF) Fabrication optimization and SBF immersion studies for AF, hydroxyapatite (HA) coated AF (HACAF) and HA-bentonite coated AF (HABCAF) Bone Tissue Scaffolds. *Procedia Chem.* **2016**, *19*, 884–890. [CrossRef]
29. Adams, L.A.; Essien, E.R.; Kaufmann, E.E. Mechanical and bioactivity assessment of wollastonite/PVA composite synthesized from bentonite clay. *Ceramica* **2019**, *65*, 246–251. [CrossRef]
30. Bajaj, H.; Kevadiya, B.; Joshi, G.; Patel, H.; Abdi, S. Montmorillonite-alginate composites as a drug delivery system: Intercalation and In vitro release of diclofenac sodium. *Indian J. Pharm. Sci.* **2010**, *72*, 732–737. [CrossRef]
31. Park, J.-H.; Shin, H.-J.; Kim, M.H.; Kim, J.-S.; Kang, N.; Lee, J.-Y.; Kim, K.-T.; Lee, J.I.; Kim, D.-D. Application of montmorillonite in bentonite as a pharmaceutical excipient in drug delivery systems. *J. Pharm. Investig.* **2016**, *46*, 363–375. [CrossRef]
32. Hosseini, F.; Hosseini, F.; Jafari, S.M.; Taheri, A. Bentonite nanoclay-based drug-delivery systems for treating melanoma. *Clay Miner.* **2018**, *53*, 53–63. [CrossRef]

33. Aguzzi, C.; Cerezo, P.; Viseras, C.; Caramella, C. Use of clays as drug delivery systems: Possibilities and limitations. *Appl. Clay Sci.* **2007**, *36*, 22–36. [CrossRef]
34. Thomas, H.; Alves, C.S.; Rodrigues, J. Laponite®: A key nanoplatform for biomedical applications? *Nanomed. Nanotechnol. Biol. Med.* **2018**, *14*, 2407–2420. [CrossRef]
35. Das, S.S.; Neelam; Hussain, K.; Singh, S.; Hussain, A.; Faruk, A.; Tebyetekerwa, M. Laponite-based nanomaterials for biomedical applications: A Review. *Curr. Pharm. Des.* **2019**, *25*, 424–443. [CrossRef] [PubMed]
36. Wang, C.; Wang, S.; Li, K.; Ju, Y.; Li, J.; Zhang, Y.; Li, J.; Liu, X.; Shi, X.; Zhao, Q. Preparation of laponite bioceramics for potential Bbne tissue engineering applications. *PLoS ONE* **2014**, *9*, e99585. [CrossRef]
37. Levis, S.; Deasy, P. Characterisation of halloysite for use as a microtubular drug delivery system. *Int. J. Pharm.* **2002**, *243*, 125–134. [CrossRef]
38. Jousseni, E.; Petit, S.; Churchman, J.; Theng, B.; Delvaux, D.R.B. Halloysite clay minerals: A review. *Clay Min.* **2005**, *40*, 383–426. [CrossRef]
39. Noro, H. Hexagonal platy halloysite in an altered tuff bed, Komaki City, Aichi Prefecture, Central Japan. *Clay Miner.* **1986**, *21*, 401. [CrossRef]
40. Wei, W.; Abdllayev, E.; Goeders, A.; Hollister, A.; Lvov, L.; Mills, D.K. Clay nanotube/poly(methyl methacrylate) bone cement composite with sustained antibiotic release. *Macromol. Mat. Eng.* **2012**, *297*, 645–653. [CrossRef]
41. De Silva, R.T.; Dissanayake, R.K.; Mantilaka, M.M.M.G.P.G.; Wijesinghe, W.P.S.L.; Kaleel, S.S.; Premachandra, T.N.; Weerasinghe, L.; Amaratunga, G.A.J.; de Silva, K.M.N. Drug-loaded halloysite nanotube-reinforced electrospun alginate-based nanofibrous scaffolds with sustained antimicrobial protection. *ACS Appl. Mater. Interfaces* **2018**, *10*, 33913–33922. [CrossRef]
42. Karnik, S.; Mills, D.K. Clay nanotubes as growth factor delivery vehicle for bone tissue engineering. *J. Nanomed. Nanotechnol.* **2013**, *4*, 102.
43. Satish, S.; Tharmavaram, M.; Rawtani, D. Halloysite nanotubes as a nature's boon for biomedical applications. *Nanobiomedicine* **2019**, *6*, 1849543519863625. [CrossRef]
44. Pasbakhsh, P.; De Silva, R.; Vahedi, V.; Churchman, G.J. Halloysite nanotubes: Prospects and challenges of their use as additives and carriers—A focused review. *Clay Miner.* **2016**, *51*, 479–487. [CrossRef]
45. Lvov, Y.; Abdullayev, E. Functional polymer–clay nanotube composites with sustained release of chemical agents. *Prog. Polym. Sci.* **2013**, *38*, 1690–1719. [CrossRef]
46. Mousavi, S.M.; Hashemi, S.A.; Salahi, S.; Hosseini, M.; Ali Mohammad Amani, A.M.; Aziz Babapoor, A. Development of clay nanoparticles toward bio and medical applications. In *Chapter 8 in Current Topics in the Utilization of Clay in Industrial and Medical Applications*; InTech Open: Zagreb, Croatia, 2018.
47. Tharmavaram, M.; Gaurav, P.; Deepak, R. Surface modified halloysite nanotubes: A flexible interface for biological, environmental and catalytic applications. *Adv. Colloid Interface Sci.* **2018**, *261*, 82–101. [CrossRef]
48. Patel, S.; Jammalamadaka, U.; Sun, L.; Tappa, K.; Mills, D.K. Sustained release of antibacterial agents from doped halloysite nanotubes. *Bioengineering* **2015**, *3*, 1. [CrossRef] [PubMed]
49. Karnik, S.; Hines, K.; Mills, D.K. Nanoenhanced hydrogel system with sustained release capabilities. *J. Biomed. Mater. Res. Part A* **2014**, *103*, 2416–2426. [CrossRef]
50. Romero-Trigueros, C.; Parra, M.; Bayona, J.M.; Nortes, P.; Alarcón, J.J.; Nicolás, E. Effect of deficit irrigation and reclaimed water on yield and quality of grapefruits at harvest and postharvest. *LWT Food Sci. Technol.* **2017**, *85*, 405–411. [CrossRef]
51. Fakhrullina, G.I.; Akhatova, F.S.; Lvov, Y.M.; Fakhrullin, R.F. Toxicity of halloysite clay nanotubes in vivo: A Caenorhabditis elegans study. *Environ. Sci. Nano* **2014**, *2*, 54–59. [CrossRef]
52. Kryuchkova, M.; Danilushkina, A.; Lvov, Y.; Fakhrullin, R. Evaluation of toxicity of nanoclays and graphene oxide in vivo: A Paramecium caudatum study. *Environ. Sci. Nano* **2016**, *3*, 442–452. [CrossRef]
53. Yang, J.; Wu, Y.; Shen, Y.; Zhou, C.; Li, Y.-F.; He, R.-R.; Liu, M. Enhanced Therapeutic Efficacy of Doxorubicin for Breast Cancer Using Chitosan Oligosaccharide-Modified Halloysite Nanotubes. *ACS Appl. Mater. Interfaces* **2016**, *8*, 26578–26590. [CrossRef] [PubMed]
54. Liu, M.; Chang, Y.; Yang, J.; You, Y.; He, R.; Chen, T.; Zhou, C. Functionalized halloysite nanotube by chitosan grafting for drug delivery of curcumin to achieve enhanced anticancer efficacy. *J. Mater. Chem. B* **2016**, *4*, 2253–2263. [CrossRef]
55. Wu, Y.-P.; Yang, J.; Gao, H.-Y.; Shen, Y.; Jiang, L.; Zhou, C.; Li, Y.-F.; He, R.-R.; Liu, M. Folate-Conjugated Halloysite Nanotubes, an Efficient Drug Carrier, Deliver Doxorubicin for Targeted Therapy of Breast Cancer. *ACS Appl. Nano Mater.* **2018**, *1*, 595–608. [CrossRef]
56. Lvov, Y.M.; Devilliers, M.M.; Fakhrullin, R.F. The application of halloysite tubule nanoclay in drug delivery. *Expert Opin. Drug Deliv.* **2016**, *13*, 977–986. [CrossRef] [PubMed]
57. Fizir, M.; Dramou, P.; Dahiru, N.S.; Ruya, W.; Huang, T.; He, H. Halloysite nanotubes in analytical sciences and in drug delivery: A review. *Microchim. Acta* **2018**, *185*, 389. [CrossRef] [PubMed]
58. Tohidi, S.; Ghaee, A.; Barzin, J. Preparation and characterization of poly (lactic-co-glycolic acid)/chitosan electrospun membrane containing amoxicillin-loaded halloysite nanoclay. *Poly. Adv. Techn.* **2016**, *27*, 1020–1028. [CrossRef]
59. Li, H.; Zhu, X.; Zhou, H.; Zhong, S. Functionalization of halloysite nanotubes by enlargement and hydrophobicity for sustained release of analgesic. *Colloids Surf. Physicochem. Eng. Asp.* **2015**, *487*, 154–161. [CrossRef]

60. Tan, D.; Yuan, P.; Annabi-Bergaya, F.; Yu, H.; Liu, D.; Liu, H.; He, H. Natural halloysite nanotubes as mesoporous carriers for the loading of ibuprofen. *Microporous Mesoporous Mater.* **2013**, *179*, 89–98. [CrossRef]
61. Ganguly, S.; Das, T.K.; Mondal, S.; Das, N.C. Synthesis of polydopamine-coated halloysite nanotube-based hydrogel for controlled release of a calcium channel blocker. *RSC Adv.* **2016**, *6*, 105350–105362. [CrossRef]
62. Shi, Y.-F.; Tian, Z.; Zhang, Y.; Shen, H.-B.; Jia, N.-Q. Functionalized halloysite nanotube-based carrier for intracellular delivery of antisense oligonucleotides. *Nanoscale Res. Lett.* **2011**, *6*, 608. [CrossRef]
63. Santos, A.C.; Ferreira, C.; Veiga, F.; Ribeiro, A.; Panchal, A.; Lvov, Y.; Agarwal, A. Halloysite clay nanotubes for life sciences applications: From drug encapsulation to bioscaffold. *Adv. Colloid Interface Sci.* **2018**, *257*, 58–70. [CrossRef]
64. Bottino, M.C.; Batarseh, G.; Palasuk, J.; Alkatheeri, M.S.; Windsor, L.J.; Platt, J.A. Nanotube-modified dentin adhesive—Physicochemical and dentin bonding characterizations. *Dent. Mater.* **2013**, *29*, 1158–1165. [CrossRef]
65. Chen, Q.; Zhao, Y.; Wu, W.; Xu, T.; Fong, H. Fabrication and evaluation of Bis-GMA/TEGDMA dental resins/composites containing halloysite nanotubes. *Dent. Mater.* **2012**, *28*, 1071–1079. [CrossRef]
66. Feitosa, S.A.; Münchow, E.A.; Al-Zain, A.O.; Kamocki, K.; Platt, J.A.; Bottino, M.C. Synthesis and characterization of novel halloysite-incorporated adhesive resins. *J. Dent.* **2015**, *43*, 1316–1322. [CrossRef]
67. Barot, T.; Rawtani, D.; Kulkarni, P. Physicochemical and biological assessment of silver nanoparticles immobilized halloysite nanotubes-based resin composite for dental applications. *Heliyon* **2020**, *6*, e03601. [CrossRef] [PubMed]
68. Drury, J.L.; Mooney, D.J. Hydrogels for tissue engineering: Scaffold design variables and applications. *Biomaterials* **2003**, *24*, 4337–4351. [CrossRef]
69. Fakhrullin, R.F.; Lvov, Y.M. Halloysite clay nanotubes for tissue engineering. *Nanomedicine* **2016**, *11*, 2243–2246. [CrossRef] [PubMed]
70. Kumar, A.; Han, S.S. Enhanced mechanical, biomineralization, and cellular response of nanocomposite hydrogels by bioactive glass and halloysite nanotubes for bone tissue regeneration. *Mater. Sci. Eng.* **2021**, *128*, 112236. [CrossRef]
71. Sandri, G.; Aguzzi, C.; Rossi, S.; Bonferoni, M.C.; Bruni, G.; Boselli, C.; Cornaglia, A.I.; Riva, F.; Viseras, C.; Caramella, C.; et al. Halloysite and chitosan oligosaccharide nanocomposite for wound healing. *Acta Biomater.* **2017**, *57*, 216–224. [CrossRef]
72. Wang, L.; You, X.; Dai, C.; Tong, T.; Wu, J. Hemostatic nanotechnologies for external and internal hemorrhage management. *Biomater. Sci.* **2020**, *8*, 4396–4412. [CrossRef]
73. Sandri, G.; Faccendini, A.; Longo, M.; Ruggeri, M.; Rossi, S.; Bonferoni, M.C.; Miele, D.; Prina-Mello, A.; Aguzzi, C.; Viseras, C.; et al. Halloysite- and Montmorillonite-Loaded Scaffolds as Enhancers of Chronic Wound Healing. *Pharmaceutics* **2020**, *12*, 179. [CrossRef]
74. Wali, A.; Gorain, M.; Inamdar, S.; Kundu, G.C.; Badiger, M.V. In vivo wound healing performance of halloysite clay and gentamicin-incorporated cellulose ether-PVA electrospun nanofiber mats. *ACS Appl. Bio Mater.* **2019**, *2*, 4324–4334. [CrossRef]
75. Cheng, C.; Gao, Y.; Song, W.; Zhao, Q.; Zhang, H.; Zhang, H. Halloysite nanotube-based H_2O_2-responsive drug delivery system with a turn on effect on fluorescence for real-time monitoring. *Chem. Eng. J.* **2020**, *380*, 122474. [CrossRef]
76. Lisuzzo, L.; Cavallaro, G.; Parisi, F.; Milioto, S.; Fakhrullin, R.; Lazzara, G. Core/shell gel beads with embedded halloysite nanotubes for controlled drug release. *Coatings* **2019**, *9*, 70. [CrossRef]
77. Akrami-Hasan-Kohal, M.; Ghorbani, M.; Mahmoodzadeh, F.; Nikzad, B. Development of reinforced aldehyde-modified kappa-carrageenan/gelatin film by incorporation of halloysite nanotubes for biomedical applications. *Int. J. Biol. Macromol.* **2020**, *160*, 669–676. [CrossRef] [PubMed]
78. Xie, M.; Huang, K.; Yang, F.; Wang, R.; Han, L.; Yu, H.; Ye, Z.; Wu, F. Chitosan nanocomposite films based on halloysite nanotubes modification for potential biomedical applications. *Int. J. Biol. Macromol.* **2019**, *151*, 1116–1125. [CrossRef]
79. Devi, N.; Dutta, J. Development and in vitro characterization of chitosan/starch/halloysite nanotubes ternary nanocomposite films. *Int. J. Biol. Macromol.* **2019**, *127*, 222–231. [CrossRef]
80. Mohebalia, A.; Abdoussa, M.; Faramarz, A.; Taromib, A. Fabrication of biocompatible antibacterial nanowafers based on HNT/PVA nanocomposites loaded with minocycline for burn wound dressing. *Mater. Sci. Eng.* **2020**, *110*, 110685. [CrossRef]
81. Weisman, J.A.; Jammalamadaka, U.; Tappa, K.; Mills, D.K. Doped Halloysite Nanotubes for Use in the 3D Printing of Medical Devices. *Bioengineering* **2017**, *4*, 96. [CrossRef] [PubMed]
82. Tappa, K.; Jammalamadaka, U.; Weisman, J.A.; Ballard, D.H.; Wolford, D.D.; Pascual-Garrido, C.; Wolford, L.M.; Woodard, P.K.; Mills, D.K. 3D Printing Custom Bioactive and Absorbable Surgical Screws, Pins, and Bone Plates for Localized Drug Delivery. *J. Funct. Biomater.* **2019**, *10*, 17. [CrossRef]
83. Wu, F.; Zheng, J.; Li, Z.; Liu, M. Halloysite nanotubes coated 3D printed PLA pattern for guiding human mesenchymal stem cells (hMSCs) orientation. *Chem. Eng. J.* **2018**, *359*, 672–683. [CrossRef]
84. Arvizo, R.R.; Bhattacharyya, S.; Kudgus, R.A.; Giri, K.; Bhattacharya, R.; Mukherjee, P. Intrinsic therapeutic applications of noble metal nanoparticles: Past, present and future. *Chem. Soc. Rev.* **2012**, *41*, 2943–2970. [CrossRef]
85. Azharuddin, M.; Zhu, G.H.; Das, D.; Ozgur, E.; Uzun, L.; Turner, A.P.F.; Patra, H.K. A repertoire of biomedical applications of noble metal nanoparticles. *Chem. Commun.* **2019**, *55*, 6964–6996. [CrossRef]
86. Massaro, M.; Cavallaro, G.; Colletti, C.G.; Lazzara, G.; Milioto, S.; Noto, R.; Riela, S. Chemical modification of halloysite nanotubes for controlled loading and release. *J. Mater. Chem. B* **2018**, *6*, 3415–3433. [CrossRef] [PubMed]
87. Boyer, C.; Mills, D.K. Method for Preparing Halloysite Supported Metal Nanoparticles through Electrolysis. U.S. Patent US9981074B1, 25 September 2015.

88. Luo, Y.; Humayun, A.; Mills, D.K. Surface Modification of 3D Printed PLA/Halloysite Composite Scaffolds with Antibacterial and Osteogenic Capabilities. *Appl. Sci.* **2020**, *10*, 3971. [CrossRef]
89. Humayun, A.; Luo, Y.; Mills, D.K. 3D printed antimicrobial PLA constructs functionalized with zinc-coated halloysite nanotubes-Ag-chitosan oligosaccharide lactate. *Mater. Technol.* **2020**, *8*, 1–8.

Review

Molecular Profile Study of Extracellular Vesicles for the Identification of Useful Small "Hit" in Cancer Diagnosis

Giusi Alberti [1,†], Christian M. Sánchez-López [2,3,†], Alexia Andres [3], Radha Santonocito [1], Claudia Campanella [1], Francesco Cappello [1,*] and Antonio Marcilla [2,3,*]

[1] Department of Biomedicine, Neurosciences and Advanced Diagnostics (BiND), University of Palermo, 90127 Palermo, Italy; giusi.alberti@unipa.it (G.A.); radha.santonocito@unipa.it (R.S.); claudia.campanella@unipa.it (C.C.)
[2] Área de Parasitología, Departamento Farmacia y Tecnología Farmacéutica y Parasitología, F. Farmacia, Universitat de València, 46100 Burjassot, Spain; christian.sanchez@uv.es
[3] Joint Unit of Endocrinology, Nutrition and Clinical Dietetics, Instituto de Investigación Sanitaria-La Fe, 46026 Valencia, Spain; alexia.andres@gmail.com
* Correspondence: francesco.cappello@unipa.it (F.C.); antonio.marcilla@uv.es (A.M.)
† Equally contributed to the manuscript.

Abstract: Tumor-secreted extracellular vesicles (EVs) are the main mediators of cell-cell communication, permitting cells to exchange proteins, lipids, and metabolites in varying physiological and pathological conditions. They contain signature tumor-derived molecules that reflect the intracellular status of their cell of origin. Recent studies have shown that tumor cell-derived EVs can aid in cancer metastasis through the modulation of the tumor microenvironment, suppression of the immune system, pre-metastatic niche formation, and subsequent metastasis. EVs can easily be isolated from a variety of biological fluids, and their content makes them useful biomarkers for the diagnosis, prognosis, monitorization of cancer progression, and response to treatment. This review aims to explore the biomarkers of cancer cell-derived EVs obtained from liquid biopsies, in order to understand cancer progression and metastatic evolution for early diagnosis and precision therapy.

Keywords: extracellular vesicles; liquid biopsy; biomarkers; tumor progression; metastasis

Citation: Alberti, G.; Sánchez-López, C.M.; Andres, A.; Santonocito, R.; Campanella, C.; Cappello, F.; Marcilla, A. Molecular Profile Study of Extracellular Vesicles for the Identification of Useful Small "Hit" in Cancer Diagnosis. *Appl. Sci.* **2021**, *11*, 10787. https://doi.org/10.3390/app112210787

Academic Editor: Neill Turner

Received: 30 September 2021
Accepted: 27 October 2021
Published: 15 November 2021

1. Introduction

Carcinogenesis is a process in which unlimited/uncontrolled cell division occurs, leading to the formation of malignant tumors. Compared to normal cells, cancer cells acquire malignant properties through genetic mutations and other aberrations that give them adaptive and proliferative advantages. This malignant transformation is a multi-stage process involving the gradual accumulation of abnormalities necessary for cell tumor progression [1,2]. Sequential selection of variant subpopulations within the neoplastic clone are responsible for the progression of the neoplasm, and the development of intra-neoplastic diversity is emerging as an important feature. In fact, heterogeneity allows cells to develop characteristics that allow them to proliferate, evade apoptosis, undergo angiogenesis, alter metabolism, and form metastases [3]. Findings suggest that cancer progression is related to altered protein expression and changes in metabolic pathways, which gives cancer cells the ability to survive and expand in a hostile microenvironment. Cancer cells use a variety of mechanisms through which they develop an extraordinary ability to adapt and expand in the microenvironment, including the release and uptake of extracellular vesicles (EVs).

EVs constitute a heterogeneous vesicle population secreted by virtually all cell types. They have been found in many body fluids, including blood [4], urine [5], saliva [6], bronchoalveolar lavage [7], and cerebrospinal fluid [8]. They are enriched with proteins, lipids, metabolites, DNA fragments, miRNA fragments, and non-coding RNAs [9]. According to their size and biogenesis, EVs are classified into the following groups: exosomes, microvesicles, and apoptotic bodies. Exosomes (30–120 nm in diameter) are initially generated inside of multivesicular bodies (MVBs), existing as intraluminal vesicles (ILVs). When these MVBs fuse with the cell membrane, vesicles are released into the extracellular space, where they become known as "exosomes" [10]. In contrast to exosomes, microvesicles are formed by a characteristic "outward blebbing" of the plasma membrane, producing a class of heterogeneous vesicles of larger size (100 nm–1 μm in diameter). The formation of microvesicles appears to occur selectively in lipid-rich membrane microdomains (e.g., in lipid rafts) [11]. These vesicles are often given other names, including ectosomes, shedding vesicles, or oncosomes (in the case of cancer cells) [12]. Apoptotic bodies (1–5 μm in diameter) are released during the last steps of apoptosis through formation of membrane protrusions, such as microtubule spikes, apoptopodia, and beaded-apoptopodia. Although these classifications have been widely used in the literature, it is advised to use this nomenclature sparingly. This is due to the fact that most techniques currently used only allow the separation of small EVs from large EVs [12]. Consequently, the International Society for Extracellular Vesicles (ISEV) recommends using terms for EV subtypes, such as size, density, instead of using terms such as "exosomes" or "microvesicles", unless there is certainty of their biogenesis [13].

Liquid biopsies are a rich source of EVs and are typically obtained using minimally invasive medical procedures that allow sampling from healthy controls, and therefore allows the detection of minimal residual disease (MRD) or recurrence and tracking of tumor evolution. It has been demonstrated that cancer cell-associated EVs carry the signatures of proteins, lipids, metabolites, and RNA of the tumor cell of origin [9]. Additionally, cancer-derived EVs are important players in cancer progression, being able to facilitate intercellular communication with cells near and far away (e.g., stromal cells), promoting the formation of pre-metastatic niches. EVs are involved in different stages of tumorigenesis and metastasis by increasing angiogenesis and remodeling the extracellular matrix. This activity allows cells to escape from immune system recognition and induces resistance to various cancer therapies [14,15]. For this reason, molecular profiling of EVs can help us understand the behavior of the cancer, in order to obtain an early diagnosis and develop targeted therapies.

In this review, we aim to cover the fields' most recent findings in proteomic, lipidomic, and metabolomic studies in EVs. We will highlight the best practices for identification of tumor biomarkers by cancer cell-derived EVs, placing an emphasis on the hallmarks of cancer, promotion of invasion, and metastasis

2. EVs Performing Multiple Functions in Cancer Formation

EVs have been found to play an important role in various physiological and pathological conditions. Factors, such as hypoxia, inflammation, and extracellular acidification in the tumor microenvironment (TME) may contribute to the increased secretion rate of EVs [16]. In fact, it has been reported that malignant cells secrete an increased number of EVs with an altered composition to those from non-malignant cells of the same type [17]. However, diagnostic and prognostic evaluations using the total abundance of EVs may be subject to confounding effects from non-malignant cells responsible for circulating EVs in some disease conditions. Even so, studies have demonstrated that EVs are enriched with a highly heterogeneous pool of biological cargo, i.e., proteins, lipids, RNA, and metabolites that modulate target cells [9]. As previously mentioned, these tumor-derived EVs are released into circulation in order to carry out functions that support tumorigenesis, such as stimulating angiogenesis and promoting metastatic diffusion [18,19].

2.1. EVs Promote Angiogenesis, Invasion, and Metastasis

EVs perform important functions throughout several stages of invasion and metastatic colonization, including angiogenesis, invasion, epithelial mesenchymal transition (EMT), migration, and pre-metastatic niche formation which are described below.

2.2. Angiogenesis

Angiogenesis is the formation of new blood vessels from an already existing vascular network. Several steps characterize angiogenic sprouting, such as enzymatic degradation of the vessel basement membrane, endothelial cell proliferation, migration, germination, branching, and formation of new vessels [19]. Angiogenesis is regulated by a precise balance between stimulatory and inhibitory signals, which can become altered in cancer pathologies [20]. The potential of EVs to induce angiogenesis is thought to be dependent on their uptake by recipient cells. For example, EVs released from glioblastomas contain a host of pro-angiogenic signals, such as proteoglycans glypican-1 and syndecan-4. These signals increase revascularization through proliferation and formation of endothelial cells and tubules [21].

EVs released by tumor cells are important mediators of the angiogenic cascade, regardless of their uptake. Tumor angiogenesis is mediated by a repertoire of membrane-bound proteins that could confer selective advantages for tumor growth and metastasis. EVs secreted by cancer cells in the brain and neck contain the ephrin type B receptor 2 (EPHB2), which is associated with tumor angiogenesis by activating the STAT3 signaling pathway via engagement of ephrin-B2 on the surface of endothelial cells [22]. Similarly, ovarian cancer cell-derived EVs contain the soluble form of E-cadherin on their surface, making them capable of stimulating tumor angiogenesis by forming a heterodimer with VE-cadherin on the surface of endothelial cells [23]. In addition, EVs secreted by tumor cells show various cytokines on their vesicular surface, such as IL-8 and VEGF, which can stimulate the growth, migration and/or formation of endothelial tubes [24]. Studies on the role of EVs in angiogenesis show that they are excellent tools for understanding the mechanisms underlying endothelial cell migration, alteration of vessel phenotype, and germination of solid tumors.

2.3. EVs in Promoting Metastasis Initiation and Progression

The metastatic process involves invasion and proliferation of tumor cells into nearby and distant tissues via the circulatory and lymphatic systems. Metastatic cells establish a microenvironment which favors angiogenic and proliferative processes, promoting cancer metastasis. EVs from initial tumor cells promote EMT by targeting EMT-related factors, such as transforming growth factor beta (TGFβ), hypoxia-inducible factor 1 alpha (HIF-1α), thus initiating metastasis. For example, EVs released by prostate cancer cells are enriched with lysosomal hyaluronidase, which promotes the mobility of stromal cells, aiding in metastasis [25]. Moreover, EVs secreted by highly metastatic cells can increase cell ability to migrate and invade low-metastatic cells by promoting an EMT process via MAPK/ERK signaling [26]. The hypoxia that characterizes the TME also influences tumor cells to release EVs enriched in matrix metalloproteinase-13 (MMP-13), with a consequent increase in vimentin and decrease in E-cadherin in normoxic cells, thus improving the metastases that occur via EMT [27].

Interestingly, Hoshino et al. demonstrated that cancer-derived EVs express a series of integrins that regulate adhesion to specific cancer cells and the extracellular matrix (ECM) molecules expressed in some organs [28]. They revealed that EVs containing ITG$\alpha6\beta4$ and ITG$\alpha6\beta1$ are associated with lung metastases through their ability to bind to lung-resident fibroblasts and epithelial cells. Similarly, integrin-expressing EVs containing $\alpha v\beta5$ are linked to liver metastases because of their ability to bind to Kupffer cells [28]. Macrophage migration inhibitory factor in EVs from pancreatic cancer can alter growth factor β production in Kupffer cells and increase fibronectin secretion of hepatic stellate cells [28]. Gastric tumor cell-derived EVs mediate the formation of a liver-like microenvironment through EGFR-mediated hepatocyte growth factor upregulation, ameliorating metastasis [29]. Many studies have also reported brain-derived secretory proteins, which can alter the brain microenvironment to promote the colonization of cancer cells, leading to brain metastasis [29]. In this respect, proteins present within EVs derived from the brain can increase the adhesive and invasive capacity of non-brain metastatic cells [30].

These results indicate that cancer-derived EVs released into the body have the potential to remodel cancerous and non-cancerous cells, leading to pre-metastatic niche formation and metastatic organotropism.

2.4. EVs in Immunomodulation

The metastatic cascade may be favored by the ability of EVs to mediate immunosuppression through the transport of inflammatory factors [31]. Importantly, tumor-derived EVs can be transmitting inflammatory factors to recipient cells to regulate their behaviors. For instance, the transfer of tumor-derived EVs into human THP-1 monocytic cells led to the production and secretion of various pro-inflammatory cytokines and tumor necrosis factor (TNF) -α, via TLR2 and TLR4 binding [32]. Moreover, EVs from breast cancer expressing Annexin A2 mediate an increase in the secretion of IL-6 and TNF-α by M1 macrophage activation [33].

Tumor-derived EVs can precondition an environment for future metastasis using other mechanisms as well. For example, EVs derived from gastric cancer express transforming growth factor (TGF)-β1, which is associated with lymph node metastasis through the induction of regulatory T cell (Treg) formation [34]. Contrarily, it has been observed that TGF-β1 in EVs derived from breast cancer cells can suppress T cell proliferation [35]. In addition, tumor-derived EVs appear to be capable of inhibiting the activation of T lymphocytes and induce apoptosis through the expression of the ligand FasL, a member of the tumor necrosis factor (TNF) family, which can promote the apoptosis of lymphocytes [36].

Cancer-derived EVs can also enhance immune evasion of cancer cells through the suppression of T cell activity and natural killer (NK) cells. For example, it has been identified that tumor-derived EVs can block IL-2-mediated activation of NK cells and their cytotoxic activity [37].

Conversely, EVs derived from immune cells can be considered attractive tools for fighting against cancer. For example, NK cell-derived EVs can induce the immune response by compromising the spread of solid tumors [38]. However, the role of EVs released by these cells is still poorly understood.

2.5. EVs in Reprogramming Energy Metabolism

The metabolic adaptations made possible by cancer cells allow them to adapt to conditions of nutritional deprivation and/or stress. TEM is heterogeneous in cancer cells, fibroblasts, endothelial cells, mesenchymal stem cells, and extracellular matrix, participating in the metabolic reprogramming crucial for cancer progression [39].

EVs released from cancer cells contribute to the transformation of normal fibroblasts into distinct functional subtypes. Cancer-associated fibroblasts (CAFs), a major component in TME, promote tumor growth and progression through various pathways, e.g., by the secretion of inflammatory factors [40]. On this note, early- and late-stage colorectal cancer (CRC) cell-derived EVs both can activate normal quiescent fibroblasts in phenotypically and functionally distinct subsets of CAFs, which could facilitate tumor progression [40]. Additionally, triple negative breast cancer (TNBC) cells overexpressing ITGB4 are capable of remodeling fibroblast metabolism and promoting glycolysis in CAFs [41].

EVs released from cancer-associated endothelial cells (CAECs) might also have an important influence on tumor progression. Analyses suggest that a decrease in levels of EC-derived EVs after chemotherapy in patients with metastatic breast cancer is associated with disease-free survival [42]. The intercellular communication between MSC and cancer cells is not well understood, and neither are the metabolic changes due to this interaction. However, TME could promote the development of cancer-associated mesenchymal stem cells (CA-MSCs) with the ability to differentiate into CAFs and influence cancer cells by secreting various metabolites by EVs. In fact, the metabolic crosstalk between tumor cells and CAFs profoundly affects TME remodeling and drives tumor growth [43]. In addition, glioma cell-derived EVs induced tumor-like phenotypes in MSCs by activating glycolysis [44] and EVs from prostate cancer promote the formation of a pre-metastatic niche through the transfer of PKM2 into MSCs [45] (Figure 1).

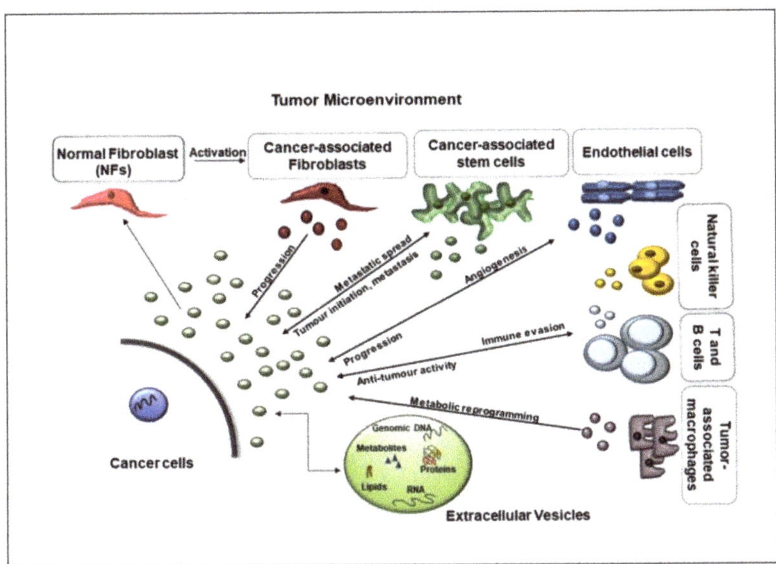

Figure 1. Schematic illustration of EV-mediated interactions between cancer cells and their TME. In cancer, EVs are involved in the process of transporting bioactive molecules between stromal and transformed cells, including endothelial, mesenchymal, immune, and CAF cells. Several subtypes of immune cells can also be found in the TME, including dendritic cells (DCs), B and T lymphocytes, and macrophages. EVs released within TME contribute to TME heterogeneity by creating a suitable environment for tumor growth, progression, and metastasis. EVs derived from immune cells which exhibit anti-cancer activity are an exception. Modified from [46].

3. EV Isolation from Different Body Fluids in Cancer

EVs can be isolated from many biofluids, including serum, plasma, urine, saliva, ascitic fluid, bile, cerebrospinal fluid (CSF), and pathological effusions from tumors [47]. The pre-analytical phase represents an important step in the analysis of EVs in terms of anatomical position, accessibility, and specific biophysical and chemical characteristics. For example, biofluids such as urine or blood are easily accessible via non-invasive/minimally invasive methods, making them ideal for longitudinal disease monitoring, and are commonly used samples for diagnostic purposes. Isolation of EVs from CSF is probably preferable over other biofluids for the study of neurological diseases. However, acquiring it is invasive and requires highly qualified professionals, creating potentially serious complications. Below is a description of some of the most commonly used biofluids.

3.1. Serum and Plasma

A crucial step in the development of biomarkers is to determine which blood sample is optimal for the study. A large limitation of serum collection is that platelets can become easily activated and release EVs, which alter the fluid composition. In addition, several variables play a significant role in the analysis of EV composition, such as patient status, syringe use, and anticoagulant use. For example, citrate and ethylenediamine tetraacetic acid (EDTA) are the most commonly used anticoagulants in preparation, in order to prevent platelet activation and release of platelet EVs. Depending on the choice of the downstream EVs analysis, all these variables must necessarily be taken into account for the standardization of the sample collection protocol [13,48]. Another important aspect is the lack of quality control (QC) for plasma samples to be used in clinical trials. Recently, researchers have agreed that human sample quality must be standardized prior to any molecular analysis [49].

3.2. Urine

Unlike serum, urine can be collected in large volumes with little or no patient discomfort. It contains a considerable number of EVs, proteins, metabolites, and cells derived from the urogenital tract and filtration of glomerular plasma. For this reason, it is ideal for the analysis of physical and pathological conditions of an individual [50]. Urine is a unique biofluid characterized by varying pH ranges, osmolarity, composition and concentration of dispersed solutes, even within the same individual [51]. Urinary EVs (uEVs) released predominantly by the kidneys, urinary tract epithelium, and male reproductive tract, constitute a heterogeneous population [52]. In addition, uEVs can be strongly influenced by factors, such as medication, exercise, food intake, and the presence of urinary pigments [53]. Consequently, careful standardization of urine collection and development of protocols compatible with downstream analyses/reproducibility of data are recommended.

3.3. Saliva

Human saliva is easy to collect, inexpensive, and can be performed in a non-invasive manner. It is also enriched in salivary extracellular vesicles (SEV) [54]. A similarity between the salivary and plasma proteome has been previously described in literature, therefore SEV should be derived partly from the salivary glands, and partly from circulation [55]. Saliva has advantages over blood as it does not coagulate, and its composition does not undergo changes brought about by the release of EVs from platelets. However, interference of high-abundance amylases and proline-rich proteins may affect the isolation and purification of SEVs, as well as the identification of low abundance proteins present inside the SEVs. Therefore, removal of amylase and other proteins from saliva prior to extraction of SEVs would be a necessary pre-analytic step for accurate downstream analysis (e.g., proteomic analysis) [56]. In addition, the composition and concentration of analytes in saliva are influenced by varying collection conditions and processing methods. Non-contaminated saliva can be collected via a buccal swab or by spitting into a test tube (which are subject to variation depending on tongue and/or cheek movements). Saliva production can be stimulated through use of chewing gum, for example, producing a volume three times greater than the volume of unstimulated saliva [57]. Whole saliva usually contains cellular debris and bacterial cells. Taking all of these variabilities into account, it is necessary to develop a standardized protocol to minimize variability when studying the composition of saliva for the identification of biomarkers.

3.4. CSF

CSF is a transparent fluid that provides a unique insight into neurological disease. CSF is the result of the ultrafiltration of blood plasma by ependymal cells in the choroid plexuses of the cerebral ventricles, containing few cells and almost devoid of proteins [58]. CSF biopsies are particularly important as they allow the collection of brain tissue from living individuals, however, sampling is more invasive than blood or urine sampling [59]. In this case, it is necessary also to pay attention to the pre-analytical procedures, since errors in the collection, storage, and exchange of biofluids are the most represented laboratory errors [59]. Cerebrospinal fluid can be collected from external ventricular drainage (i.e., during shunt or extraventricular placement) or by lumbar puncture (spinal tap). A clearly defined and consistent protocol for lumbar CSF collection ensures identical processing and the ability to compare results across institutions. Lumbar CSF typically does not contain blood, however, there are many differences between centrifuged and non-centrifuged components, and short-term temporal collections are difficult to obtain [60].

The choice of a specific isolation technique is not only dependent on the type of sample but also on the type of downstream analyses used for '-omics' characterization (e.g., proteomics).

4. Studies of the Molecular Profiling of EVs as Potential Biomarkers

EVs carry different classes of molecules, including proteins, lipids, nucleic acids, and other metabolites [9], and their content is affected by different environmental factors [61]. This, along with their ability to act as powerful cell-cell communication mediators, has contributed to their emergence as biomarkers of numerous diseases, including cancer [62]. However, mainly due to its small size, the purification of EVs is still a challenge and different isolation methods are used to obtain EVs efficiently and with high purity [63].

The techniques widely used for isolating EVs include differential ultracentrifugation (dUC), size exclusion chromatography (SEC), density gradient flotation, immunoprecipitation, and polymer-based precipitation, as well as commercially available kits [64]. dUC is the most extensively used method for isolating EVs, consisting of sequential centrifugations with increasing speed and time, to separate particles depending on their size and density [65]. This method can obtain a high yield of EVs, but it is time-expensive and requires specialized equipment [66]. Furthermore, it could damage the integrity of these vesicles, affecting their profiles of RNA, protein, and other metabolites which are a potential source of biomarkers [67]. Hence, other techniques, such as SEC and density gradient centrifugation are gaining relevance to isolate EVs [66].

On the other hand, it has been reported that the choice of a method for EVs isolation could impact down-stream analysis of their cargo, either due to the isolation of different EV populations, or the co-isolation of contaminating proteins and other possible matrix contaminants [68,69]. In addition, various challenges come into play when isolating EVs from different biological sources, such as possible co-purification of chylomicrons and lipoprotein particles with EVs in serum and plasma [70], or the presence of Tamm–Horsfall protein in urine [71], which can interfere in posterior biomarker analysis. Thus, a methodological standardization of EV isolation is still needed, in order to assure the reproducibility of the subsequent analyses, especially for clinical settings [72]. A better understanding of the protein, lipid, and metabolic composition of EVs and the extent to which EVs composition reflects the source cell composition provides a solid basis for further development of diagnostics and therapeutics [9].

In this review, we summarize the recent advances in bioactive EVs contents focusing on proteins, lipids, and metabolites that could play a significant role as diagnostic markers.

4.1. Proteome Profiling Analysis of EVs in Multiple Cancers

Proteins are an important class of molecules that are transported by EVs, and some of them are integral constituents of EV structures. Analysis of EV protein composition is crucial to understanding the mechanisms of their biogenesis and function under physiological and pathological conditions. It is well-known that EVs are highly enriched in membrane proteins, such as tetraspanins, including CD9, CD63, CD81, CD82, CD151, and Tspan8 [73,74]; proteins involved in EVs biogenesis as well as ESCRT-related proteins, such as ALIX, TSG101 (the stereotypical biomarkers for EVs characterization), and syntenin [75]. Other typical EV proteins include cell adhesion-related proteins (integrins, LFA-1, and ICAM -1) [76], and those participating in cytoskeletal construction (actin and tubulin) and vesicle trafficking (e.g., Rab family proteins). MHC class I and II complexes, involved in antigen presentation, have been also reported [77], as well as heat shock proteins, which facilitate protein folding and balance of proteostasis (i.e., Hsp60, Hsp70, and Hsp90) [78].

In addition to self-proteins, EV content reflects the physiological state of the cell from which they originated. Since EVs are normally isolated in small amounts, highly sensitive analyses are needed. In recent years, technology have been widely used to allow the massive identification and relative quantification of proteins present in EVs from different biological samples [79,80]. Moreover, many public databases have been created to share data among the scientific community, such as EVpedia, ExoCarta, and Vesiclepedia [81–83]. Current proteomic technology based on mass spectrometry (MS) is the basis for the study and discovery of specific non-invasive biomarkers in the oncology field and beyond. Its high sensitivity makes it capable of identifying low abundance proteins over wide dynamic ranges, including post-translational changes. Additionally, ELISA and Western blotting can be used to identify potential biomarkers. Although many analytical methods have been standardized, unique challenges are associated with different applications of proteomics. On this note, we have considered some studies on cancer-derived-EVs to assess the recent research progress in the field, and at the same time highlight the potential biomarkers reported.

In the literature there are many examples of potential biomarkers found in EVs derived from breast cancer (BC) from both cell lines and biological fluid. An early study on serum-derived EVs of BC patients revealed the presence of CD24, with low epithelial cell adhesion molecules (EpCAM) expression, providing evidence that EpCAM could be cleaved from EVs via serum metalloproteinases [84] (Table 1). CD24 is a glycosylphosphatidylinositol-anchored membrane protein considered to be a negative cancer stem cell marker, specifically in BC [85]. EpCAM is an epithelial cell marker which modulates biological processes, such as cell proliferation, migration, and invasion.

Human epidermal growth factor receptor-2 (HER2) positive breast cancer represents 15–20% of all breast cancers. The primary mechanism of HER2 activation in breast cancer is its gene amplification, which causes its overexpression, ultimately activating several signaling pathways [86]. It has been revealed that HER2 is overexpressed in EVs derived from plasma of BC patients [87] (Table 1).

Table 1. Summary of the potential protein biomarkers identified in EVs from different cancer types.

Cancer Type	Biomarker	Sample	Authors
Thyroid Cancer	TLN1, ITGB2, SRC and CAPNS1	serum derived-EVs	[4]
Breast Cancer	CD24 and EpCAM	serum derived-EVs	[84]
Breast Cancer	HER2	plasma derived-EVs	[87]
Breast Cancer	GPC-1, ADAM10, GLUT-1 and desintegrin	*In vitro*: MDA-MB-231 and MCF-10A cell lines	[88]
Breast Cancer	Del-1, 14-3-3 epsilon protein, β-actin, annexin A1 / 5, heat shock protein 71, and galectin-binding protein 3	*In vitro*: cell line MDA-MB-231	[89]
Pancreatic Cancer	GPC-1	serum derived-EVs	[90]
Colorectal Cancer	GPC-1	plasma derived-EVs	[91]
Colorectal Cancer	CK19, TAG72, and CA125	plasma derived-EVs, CRC cells, tumor interstitial fluid	[92]
Ovarian Cancer	CA125 and HE4	serum derived-EVs	[93,94]
Colon Cancer	CD147	*In vitro*: HCT15 and HCT116	[95]
Colon Cancer	TSPAN1	*In vitro*: HCT-116 and HT-29 CC cell lines	[96]
Colon Cancer	annexin	plasma derived-EVs	[97]
Hepatic Cancer	AMPN, VNN1, pIgR, FCN1 and NEP	serum derived-EVs	[98]
Hepatic Cancer	FIBG, A1AG1 and S100A8	serum derived-EVs	[98]
Cholangiocarcinoma	FCN2, ITIH4, FIBG; MUC1, EGFR, EpCAM, and others.	serum derived-EVs; EGI1, TFK1 cell lines and non-tumor SV40-immortalized human cholangiocytes	[98]
Cholangiocarcinoma	EpCAM, ASGPR1, annexin V and taMPs	serum derived-EVs	[99]
Cholangiocarcinoma	fetuin-A and HSP90B	*In vitro*: M213 and M213D5 cell lines	[100]
Hepatic Cancer	CAP1	*In vitro*: MHCC97-H and MHCC97-L cell lines	[101]
Gastric Cancer	PSMA3 and PSMA6	serum derived-EVs	[102]

Table 1. *Cont.*

Cancer Type	Biomarker	Sample	Authors
Prostate Cancer	FABP5, Granulin, AMBP, CHMP4A, and CHMP4C	urine derived-EVs	[103]
Bladder Cancer	MUC1, CEA, EPS8L2 and moesin	urine derived-EVs	[104]
Lung Cancer	MUC1	plasma derived-EVs	[105]
Lung Cancer	NY-ESO-1, EGFR, PLAP and EpCam	plasma derived-EVs	[106]
Lung Cancer	LBP	serum derived-EVs	[107]
Lung Cancer	BPIFA1, CRNN, MUC5B, and IQGAP1	saliva derived-Evs	[108]
Lung Cancer	LRG1	urine derived-EVs	[109]
Glioblastoma	annexin A2, vimentin, tenascin-C and others	*In vitro*: A172, Glia-Tr, Glia-L, Glia-R, and Glia-Sh cell lines	[110]
Osophageal squamous cell carcinoma	GPC1	*In vitro*: HEEpiC, Het-1A, TE-1, TE-5, TE-6, TE-8, TE-9, TE-10, TE-11, TE-14 and TE-15 and LK-2 cell lines	[111]
Nasopharyngeal carcinoma	ICAM-1, CD44v5 and TSP-1	*In vitro*: C666-1, NP69 and NP460 cell lines	[112]

Another study revealed that glypican-1 (GPC-1), disintegrin and metalloproteinase domain-containing protein 10 (ADAM10), and glucose transporter 1 (GLUT-1) were upregulated in triple negative metastatic cancer cell line MDA-MB-231 (MDA) compared to the control immortalized epithelial breast tissue cell line MCF-10A (MCF) [88] (Table 1). GPC-1 is a cell surface proteoglycan protein involved in the control of cellular growth and differentiation by activation of mitogenic signaling by heparin-binding growth factors. It has been observed to be highly expressed in tissues from BC compared to healthy patients [88]. ADAM10, a transmembrane protease protein, is very abundant in high-grade tumors, and these levels correlated with negative outcomes for the basal subtypes of BC patients [113]. Lastly, GLUT-1 is a solute carrier protein that facilitates the transport of glucose across the plasma membranes. Its expression is correlated with high-grade BC cancer and increased proliferative activity, and its absence significantly increased disease-free survival in BC patients [114].

On the other hand, proteomic analyses from cell culture EVs have also provided other potential biomarker candidates in BC, including 14-3-3 epsilon protein, β-actin, annexin A1/5, heat shock protein 71, and galectin-binding protein 3 [89] (Table 1). In addition, Del-1 has been reported as an early-stage BC EVs biomarker [115]. Del-1 was first identified as an extracellular matrix protein having 3 N-terminal epidermal growth factor-like domains and the discoidin I-like or factor V C domains, C1 and C2 [116]. The striking decrease in Del-1 concentrations in plasma after surgery suggests that this protein could be a useful surveillance biomarker to assess the response of breast cancer patients to cancer therapies [115].

Proteomics have also contributed to identifying potential biomarkers in other types of cancer, such as GPC-1, through showing increased serum EVs of pancreatic cancer (PC) patients [90] (Table 1), as well as in plasma from colorectal cancer (CRC) patients [91] (Table 1). All the studies dealing with the presence of GPC-1 in EVs confirm its potential role as a clinical cancer biomarker. In addition, CK19, TAG72, and CA125 proteins were significantly enriched in EVs derived from CRC cells, tumor interstitial fluid, and patients' plasma [92] (Table 1). CA125 is an antigenic membrane protein of unknown function, however, the release of soluble proteolytic fragments of CA125 into the extracellular space appear to be associated with the conversion from benign to cancer cells [117]. In addition, CA125 together with HE4 are candidate biomarkers for ovarian cancer. So far, the serum EV CA125 and HE4 levels can significantly identify ovarian cancer at the early stage from healthy subjects, benign ovarian disease patients, and other gastrointestinal cancer patients [93,94].

In regard to CRC, protein profiling studies of individual EVs led to the identification of CD147-positive EVs with high predictive value [95] (Table 1). CD147 is a glycoprotein released by tumor cells in a soluble form, or by EVs involved in progression, invasion, and metastasis, suggesting it as a relevant tumor biomarker for cancer diagnosis [118].

Despite the advances in colon cancer (CC) diagnosis, clinical outcomes and survival rate remain poor. Studies on EVs derived from HCT-116 and HT-29 CC cells and plasma from CC patients showed high levels of tetraspanin 1 (TSPAN1) [96] (Table 1). TSPAN 1 is a member of the tetraspanin family, which may be involved in cancer progression (e.g., proliferation, cell migration, and motility) [96].

Moreover, annexins were increased in EVs derived from plasma of CC patients compared to those from healthy controls (HCs), showing a sensitivity of 75.7% [97] (Table 1). Annexin 2A (ANXA2) is a calcium-binding cytoskeletal protein expressed on the surface of endothelial cells, macrophages, mononuclear cells, and various types of cancer cells [119]. Elevated ANXA2 expression correlates with cell migration and invasion [120].

Furthermore, high-performance analysis has identified differential proteomic profiles in EVs from the serum of intrahepatic carcinoma (iCCA), hepatocellular carcinoma (HCC), and primary sclerosing cholangitis (PSC) patients versus control donors. Reports demonstrated that aminopeptidase N (AMPN, also known as CD13), pantetheinase (VNN1), and polymeric immunoglobulin receptor (pIgR) showed good diagnostic capacity of CCA [98]. AMPN, ficolin-1 (FCN1), and neprilysin (NEP) showed good correlation with PSC compared to the healthy control group [98] (Table 1). VNN1 pantetheinase is a protein anchored to glycosylphosphatidylinositol (GPI) on the cell membrane, which participates in the synthetic pathway of pantothenic acid (vitamin B5), acting as a key regulator of tissue tolerance to stress in various diseases [121]. pIgR seems to be involved in the promotion of cell transformation and proliferation, providing new insights into the role of immunoglobulin receptors [122].

Since PSC is a major risk factor of CCA development, serum EV proteins were compared between CCA and PSC, and a selection of 10 overexpressed proteins in CCA were identified, including fibrinogen gamma chain (FIBG), alpha-1-acid glycoprotein 1 (A1AG1), and S100A8 proteins [98]. A1AG1 is a protein mainly expressed during the acute phase of the inflammatory process, that can play an important role in the tumor microenvironment, affecting immune modulation, drug resistance, and cancer progression [123], whereas S100A8 protein (calgranulin A) belongs to the S100 multigenic family of calcium-modulated proteins with roles in inflammation [124].

Discrimination between early-stage CCA (I-II) and PSC is possible through detection of ficolin-2 (FCN-2) proteins, which are involved in cancer immunity, suppression of EMT, and metastasis of HCC [125]. On this note, comparison of early-stage CCA (I-II) with PSC showed that ficolin-2, inter-alpha-trypsin inhibitor heavy chain H4 (ITIH4), and FIBG were most abundant in early-stage CCA. Further, MUC1, EGFR, EpCAM, among others, were abundant in EVs released from CCA human cell lines compared to normal human cholangiocytes (NHC) [98] (Table 1). Other enriched proteins after tumor resection found in serum EVs were EpCAM, asialoglycoprotein 1 receptor (ASGPR1), annexin V, and tumor-associated microparticles (taMPs), showing sensitivity/specificity scores and positive/negative predictive values (>78%), indicating their potential in diagnosis [99] (Table 1). Hepatic ASGPR1 is a transmembrane molecule specifically expressed on the sinusoidal and basolateral hepatocellular membranes, and not in other human tissue [126]. Furthermore, taMPs have recently emerged as novel vehicles for a horizontal crosstalk between different cells, especially in the setting of inflammatory conditions [127].

Differential exosomal phosphoproteome analysis of invasive M213 and M213D5 CCA cells has shown its potential as a biomarker in cancer. Reduced exosomal fetuin-A phosphorylation and high HSP90B phosphorylation has been detected in tissues from CCA patients with a low TNM stage, compared to those with a single high-stage TNM [100] (Table 1).

The analysis of EVs obtained from high- and low-grade metastatic HCC cell lines discriminated CCA from other tumors, highlighting adenylate cyclase associated protein 1 (CAP1) as enriched in metastatic tumor cells when compared to non-tumor/primary cell lines [101] (Table 1). CAP1 is an actin monomer binding protein involved in the reorganization of the actin filament essential for cell migration. Its overexpression in HCC is closely related to tumor metastases [128], although its role needs to be further investigated.

EVs secreted in gastric cancer (GC) are also involved in tumorigenesis. A recent study suggests that proteasome subunits PSMA3 and PSMA6 levels in patients with metastatic GC (stage III/IV) were significantly higher in serum EVs than those in healthy controls and patients with early-stage GC [102] (Table 1). PSMAs are major components of the 20S proteasome core complex with potential diagnostic utility [129]. In prostate cancer (PC), there is a need to identify new markers, due to the fact that prostate specific antigen (PSA), both in free form and in EVs derived from plasma of PC lacks specificity and sensitivity [130], and therefore could lead to overdiagnosis [131]. In the search for new biomarkers, it has been shown that levels of fatty acid binding protein 5 (FABP5) in urine-derived EVs were higher in pathological groups, as well as the levels of granulin, AMBP, CHMP4A, and others [103] (Table 1). FABP5 belongs to the family of intracellular lipid-binding proteins, and is responsible for uptake and transport of fatty acid [132]. In PC cell lines, FABP5 regulates energy metabolism via ERRα activation, suggesting that a new FABP5-ERRα signaling axis plays an important role in the regulation of AMPK activity, which is a cellular energy state sensor that directs metabolic adaptation to support cell proliferation and survival [133]. Granulins, also known as granulin-epithelial precursors, are a growth factor which regulates inflammation and tumorigenesis. They can promote migration, invasion, and proliferation in PCa [134]. CHMP4A is a subunit of the charged multivesicular body, ESCRT-III complex, which is involved in multivesicular body (MVBs) formation and sorting of endosomal cargo proteins into MVBs [135].

Although urine is an excellent source of protein biomarkers for bladder cancers, there is a high degree of variability in these samples. In this regard, the enrichment of urine EVs could reduce the variability between samples, allowing us to identify functionally relevant proteins. For instance, 1222 total proteins were detected with high confidence in EVs derived from bladder cancer, validating some of them by Western blotting, such as mucin-1 (MUC1), carcinoembryonic antigen (CEA), epidermal substrate of growth factor receptor kinase 8-protein 2 (EPS8L2), and moesin, as proteins directly associated with cancer [104] (Table 1). MUC1s are membrane glycoproteins that play important roles in cell physiology, such as mediating anti-adhesive properties between cells and the extracellular matrix (ECM). Its deregulated expression is associated with cancer progression [136]. CEA, also known as CD66, is expected to have potential value in the early diagnosis of invasive urinary bladder cancer [137]. Eps8L2 belongs to the family of epidermal growth factor receptor kinase substrate 8 (EPS8)-related proteins, which are involved in actin remodeling in response to EGF [138]. Moesin is known to be associated with an aggressive phenotype in several malignant tumors, showing predictive ability for early detection of bladder urothelial carcinoma (BUC) invasion [139]. EGFR is a transmembrane receptor whose function is to regulate both cell proliferation and apoptosis via signal transduction pathways [140], which are highly related to lung cancer [141].

A large number of biomarkers is currently being investigated in lung cancer. Consequently, it has been shown that EVs released by non-small-cell lung cancers (NSCLC) patients' plasma are also particularly enriched in MUC1 [105] (Table 1). Furthermore, EVs isolated from lung cancer cells, lung biopsies, and plasma are enriched in EGFR, making them the most powerful prognostic biomarker [106] (Table 1). Sandfeld-Paulsen and co-workers also reported an enrichment in NY-ESO-1, phospholipase A-2-activating protein (PLAP), and EpCam proteins. NY-ESO-1, also known as cancer-testis antigen (CTAs), is regularly limited in its expression to germ and placental cells, but is re-expressed in tumor cells. Similarly, NY-ESO-1 expression was found in NSCLC and has been associated with a higher risk of relapse, poorer response to treatment, and shorter survival [142]. PLAP is a member of the WD-repeat protein, G-protein-transducin superfamily, which mediates eicosanoid generation and participates in inflammatory responses [143]. On the other hand, EpCAM is a transmembrane glycoprotein that affects intercellular adhesion, and is overexpressed in various human epithelial carcinomas. It is involved in many important functions relevant to tumor progression, including cell proliferation [17]. Other works have shown that serum EVs can help distinguish patients with metastatic NSCLC from non-metastatic NSCLC by monitoring lipopolysaccharide-binding protein (LBP) levels [107] (Table 1). BPIFA1, CRNN, MUC5B, and IQGAP1 were also found highly abundant in salivary exosomes of lung cancer patients [108] (Table 1). BPIFA1 is a protein specifically expressed in the upper airways and nasopharyngeal regions. It was identified as a potential marker for the micro-metastasis of non-small cell lung cancer (NSCLC) [144]. MUC5B is a gel-forming mucin secreted from airway epithelial cells in the lung, associated with longer survival in primary EGFR mutant NSCLC [145], and IQGAP is a scaffold protein that may promote the regulation of cancer cell migration and metastasis [146]. Furthermore, proteomic analyses of urinary exosomes of NSCLC patients highlighted exosomal leucine-rich-alpha2-glycoprotein 1 (LRG1) as a candidate biomarker for non-invasive diagnosis, playing a role in epithelial-mesenchymal transition (EMT) and angiogenesis [109] (Table 1).

Analyses of the EV proteome have also provided potential new biomarkers for glioblastoma cancer (GBC). Naryzhny and colleagues (2020) provided a list of potential biomarkers by secretome profiling through LC-MS/MS, which included annexin A2, vimentin, and tenascin-C, among others [110] (Table 1). Vimentin is an intermediate filament protein that plays a central role in GBC progression [147], and tenascin-C is a non-filamentous protein that mediates cell-cell and cell-matrix interactions, which affects negatively proliferation and invasion in GBC [148].

Lastly, proteomic approaches have improved the knowledge of less known tumors thanks to the specificity of the proteins found in EVs. In this context, EVs isolated from serum of thyroid cancer lymph node metastases (LNM) patients had high amounts of talin-1 (TLN1), integrin beta-2 (ITGB2), SRC, and CAPNS1, compared with thyroid cancer without LNM [4]. TLN1 is a cytoskeletal protein involved in regulating the activity of cell adhesion proteins by coupling them to F-actin [149], it is also involved in adhesion, proliferation, survival, and tumor progression [150]. ITGB2 belongs to a family of cell surface receptors that play a key role in cell adhesion, migration, proliferation, and survival by forming physical interactions between the cell and the extracellular matrix [151]. SRC kinase activity and protein levels are elevated in several cancers, and are correlated with malignant progression [4]. Similarly, CAPNS1 belongs to a family of 15 calcium-dependent intracellular thiol proteases, whose aberrant expression or activity is involved in several diseases including cancer. Specifically, high calpain-1 levels were associated with papillary thyroid cancer, but its role in regulation of the proliferation and migration needs further investigation [4].

In esophageal squamous cell carcinoma (ESCC), GPC1 was identified as a novel biomarker [111] (Table 1). Additionally, EVs isolated either from NPC C666-1 cells or immortalized nasopharyngeal epithelial cells (NP69 and NP460) had large amounts of pro-angiogenic proteins, including intercellular adhesion molecule-1 (ICAM-1), and a variant isoform of CD44 (CD44v5), while the angio-suppressive protein thrombospondin-1 (TSP-1) was present at low levels in NPC C666-1 EVs [112] (Table 1). ICAM-1 is a member of the immunoglobulin superfamily, and CD44v5 a transmembrane glycoprotein involved in many biological activities, such as cell migration, tumor invasion, and metastasis [151].

4.2. Lipidome Profiling Analysis of EVs in Multiple Cancers

EVs represent an untapped source for the discovery of clinically relevant lipid biomarkers, which can be used in pre-clinical detection, as well as for following disease progression. Knowing the lipid profile of EVs from various tumor cell types is an important aspect, although a better understanding of how cancer cells evaluate their lipid resources to meet the metabolic demands of high proliferation rates could enhance novel anticancer therapeutic strategies. There is growing evidence about the transfer of biologically active lipids and lipid metabolites as one of the mechanisms used by cancer cells to alter the energy pathways within the tumor microenvironment [152]. It is now known that EVs transfer lipids and lipid-related proteins to influence target cell function [153,154], making the study of the EV lipidome essential for the identification of biomarkers for diagnostic purposes.

The lipidomic profiling of EVs isolated from cell cultures and biofluids includes differences in lipid composition. Based on published studies, EVs are mainly made up of membrane lipids, although small amounts of other lipids could be captured by the cytosol during ILV formation. Therefore, the lipid composition of EVs should therefore reflect the composition of a lipid bilayer. In fact, it has been shown that there is an asymmetrical distribution of lipid classes in the two leaflets of the plasma membrane, where sphingolipids and phosphatidylcholine (PC) are mainly present in the outer leaflet, while other lipid classes are mainly found in the inner leaflet [155]. Recent advances in lipid analysis by LC-MS/MS have identified lipid classes and lipid species that appear to be enriched in EVs. An important lipidomic profile study of EVs derived from single cell types was realized by Skotland and co-workers in ten exosome preparations, including cell-to-exosome enrichment factors in eight of them, showing an enrichment of 2–3 times in cholesterol (CHOL), sphingomyelin (SM), glycosphingolipids (GSLs), and phosphatidylserine (PS). Notably, the membranes of most EVs show lower phosphatidylcholine (PC) and phosphatidylinositol (PI) content than their cells of origin, but contain similar levels of phosphatidylethanolamine (PE) [156].

The synthesis of cholesterol, also known as the mevalonate pathway, is an important pathway of lipid biosynthesis, since cholesterol is a major component of membranes involved in controlling membrane fluidity and the formation of lipid rafts. Consequently, altered intracellular cholesterol levels can greatly modulate membrane architecture, promoting plasma membrane fluidity, and, therefore, contribute to cell migration and metastasis. Considering this, the active lowering of cholesterol in an advanced stage of the disease can have negative effects [157]. Conversely, the establishment of primary tumors is highly dependent on growth-stimulating signaling pathways, promoted by cholesterol concentrations on the membrane through the formation of lipid rafts. In fact, cholesterol-rich lipid rafts facilitate the accumulation of tyrosine kinase receptors, such as HER2. In this case, blocking cholesterol synthesis could inhibit the onset and proliferation of cancer in the early stages of the disease [158]. In order to use cholesterol metabolism as a therapeutic target in cancer, it is necessary first to understand why cancer cells depend on cholesterol and how this affects the progression of the disease.

Glycosphingolipids (GSLs) are a subtype of glycolipids which mediate cell-cell interactions and modulate signal transduction pathways. Aberrant expression of specific GSLs and related enzymes is strongly associated with tumor formation and malignant transformation [159]. There are several studies in literature about the expression of various glycosphingolipids in specific tumors. For example, glycosphingolipids, such as GD3 and GD2, enhance the malignant properties of cancer cells, such as cell proliferation, cell invasion, and migration [160]. In contrast, mono-sialyl gangliosides, such as GM1 and GM2 often suppress malignant properties of cancerous cells [161]. The analysis of GSLs does remain challenging due to their amphiphilic nature and inherent complexity.

Sphingomyelin (SM) is also a key component of lipid rafts involved in the regulation of several signaling pathways [162]. It has been shown that low SM levels are associated with the tumorigenic transformation [163]. However, the role of SM in cancer is yet to be understood, since other studies have reported high levels of SM and different roles in cancer [164]. Considering the glycerophospholipids (GPLs) species, i.e., phosphatidylcholine (PC), phosphatidylethanolamine (PE), and phosphatidylserine (PS) are abundant in mammalian cell membranes, there is growing evidence that these GPLs (particularly PC and PS) might be potential biomarkers of cancers. Evidence supports that the overexpression of PCs observed in cancer cells is mainly due to upregulation of choline kinase and activation of phospholipase C specific to phosphatidylcholine, the latter activated in the cycles of phosphatidylcholine induced by mitogens and oncogenes, with effects on signaling pathways, regulation of the cell cycle, and cell proliferation [165]. The phosphatidylinositol (PI) resides on the cytosolic surface of cell membranes, and differential phosphorylation generates distinct phosphoinositides, contributing to their signaling diversity, including cell growth and proliferation. The phosphoinositides have a distinct localization in the cell through which they carry out their specific function localization (e.g., PI (4,5)P2 is enriched at the plasma membrane, while PI(3)P in early endosomes). Aberrant phosphoinositide signaling has been observed in cancers, suggesting its potential role as a biomarker [166].

A pioneering work quantified 22 classes of EV lipids derived from metastatic prostate cancer cell lines, such as PC-3 cells, finding enrichment in CHOL, SM, glycosphingolipids, and phosphatidylserine, indicating a particular lipid sorting in the exosome membrane compared with the source cells by MS analysis [167] (Table 2). In contrast to healthy cells, tumor cells expose the phosphatidylserine (PS) at the cell surface. An early event in apoptosis is the appearance of PS on the cell surface, which reduces the inflammatory response by alerting phagocytic cells to engulf the cell. Macrophages recognize PS on the surface of apoptotic cells while viable cancer cells with high external PS inhibit phagocytosis by displaying CD47 [168]. In addition, EVs derived from tumor cells might expose PS, suggesting that the source of PS is mostly derived from them, thus constituting a diagnostic biomarker for cancer [169].

Research done on six different prostate cell lines observed differences in the relative abundance of the classes of glycerophospholipids between cells and their EVs [170] (Table 2). Glycerolipids comprise all glycerol-containing lipids; i.e., mono-, di- and tri-substituted glycerols (MAG, DAG, and TAG, respectively), and are important constituents of the EV membrane [171]. On the other hand, the EV lipidomes released from three prostate cell lines, i.e., RWPE1 (non-tumorigenic), NB26 (tumorigenic), and PC-3 (metastatic), have been recently published by Brzozowski et al. [74]. These authors have shown a relative enrichment of lipid species, fatty acids, glycerolipids, and prenolic lipids in EVs from RWPE1, while sterol lipids, sphingolipids, and glycerophospholipids were more abundant in EVs from NB26 and PC-3 cells. They also found that the average CHOL content of EVs derived from PC cells was three times higher than EVs derived from RWPE-1 cells [74].

Table 2. Summary of the lipids and other metabolites identified in EVs with potential as biomarkers of different cancers.

Cancer Type	Biomarker	Sample	Authors
Prostate Cancer	Glucoronate; increased creatinine, glucuronate, pantothenic acid, 4-pyridoxic acid in urina; lysine, kynurenine, threonine, tryptophan, cytidine in plasma	Urine and plasma-derived EVs	[5]
Prostate Cancer	Increased CHOL, sphingolipids and glycerophospholipids, decreased glycerolipids and prenolic lipids	*In vitro*: PC-3, RWPE1, and NB26 cell lines	[74]
Prostate Cancer	Increased glycerophospholipids and sphingolipids	*In vitro*: PC-3, DU145, VCaP, and RWPE1 cell lines	[170]
Colorectal Cancer	Increased glycerophospholipids, SM, CHOL, and PS	*In vitro*: LIM1215 cell line	[172]
Breast Cancer	Increased levels of CHOL and SM, decreased levels of PC	*In vitro*: D3H2LN and D3H1 cell lines	[173]
Glioblastoma and hepatocellular carcinoma	Increased SM and ceramides in glioblastoma than hepatocellular carcinoma	*In vitro*: Huh7 and U87 cell lines	[174]
Ovarian Cancer	Increased PS, PI, PE, and PG in HOSEPiC; Increased LPI, LPG, LPC, and LPS in SKOV-3	*In vitro*: SKOV-3 and HOSEPiC cells	[175]
Prostate Cancer	Increased PS and lactosylceramide	Urine-derived EVs	[176]
Prostate Cancer	DHEAS; acyl carnitines, citrate, and kynurenine	Urine-derived EVs: PCa and BPH patients	[177]
Endometrial adenocarcinoma	Cyclic alcohols, steroids, prenols, and amino acid conjugates	PC-1 cell line; plasma-derived EVs	[178]
Pancreatic Cancer	Alanylhistidine, 6-dimethylaminopurine, leucylproline, and methionine sulfoxide, others	Serum-derived EVs	[179]
Glioblastoma	Enrichment in glycerol, tryptophan, carnitine, and GSSG	*In vitro*: U118, LN-18, and A172 cell lines; normal human astrocytes	[180]

Additionally, differences in the relative fraction of glycerophospholipids, sphingolipids, glycerolipids, and sterol lipids were identified in LIM1215 colorectal cancer cells, and in their secreted EVs, respectively. Besides, SM, CHOL, and PS were more common in EVs derived from the colorectal cancer cell line LIM1215 than in parental cells. These authors identified a decrease in PC/PE ratios in EVs relative to LIM1215 parent cells [172] (Table 1). Moreover, a lipid composition study of EVs and cells of their origin and between

EVs derived from high and low metastasis triple negative breast cancer (TNBC) cell lines, D3H2LN and D3H1, showed an increase in the levels of CHOL, SM, and a decrease in the levels of PC in their EVs. In addition, EVs derived from D3H2LN were enriched in unsaturated diacylglycerols (DGs) compared with EVs from D3H1 [173] (Table 2).

Moreover, studies performed on glioblastoma, hepatocellular carcinoma, and human bone marrow-derived mesenchymal stem cells (MSCs) have shown high abundance of SM and ceramides in EVs released by glioblastoma cells (U87), while opposite results were reported for EVs secreted by hepatocellular carcinoma cells (Huh7). PS was only slightly enriched in EVs released by all three cell lines, while the PC and PI content were higher in cells than in EVs [174] (Table 2).

On another note, several lipid species present in EVs released from ovarian cancer cells (SKOV-3) differ when compared to those from ovarian surface epithelial cells (HOSEPiC). In particular, EVs from HOSEPiC cells were more abundant in PS, PI, PE, and phosphatidylglycerol (PG), while EVs secreted from SKOV-3 cells presented higher content in lysophosphatidylinositol (LPI), lysophosphatidylserine (LPS), lysophosphatidylinositol (LPG), lysophosphatidylcholine (LPC) [175] (Table 2). Lysophospholipids (LPLs) consist of lyso-glycerophospholipids and lysosphingolipids, which are implicated in important functional roles, e.g., through intracellular G protein-coupled receptor (GPCR)-mediated signaling. Although a comprehensive understanding of LPL levels and their distribution patterns is lacking, several studies have revealed that LPLs are associated with the development, progression, and metastasis of cancers, as in ovarian cancer [181].

On this note, analysis of the urine exosome lipid repertoire in patients with renal carcinoma suggested that lysophospholipids represent the most present lipid class than in healthy control cells [182]. Prostate cancer patients' urine exosomes (PCa) revealed up to nine differentially expressed lipid species, including lactosylceramide, with the highest patient/control ratio [176] (Table 2). To date, only a few studies have investigated the lipid composition of EVs in biofluids, due to difficulties in lipid isolation from these vesicles. There are increasing updates on the EVs' lipid composition, functionality, and potential use as biomarkers.

4.3. Metabolome Profiling Analysis of EVs in Multiple Cancers

The discovery of metabolic biomarkers in EVs is an important goal for diagnosing clinical relapses. A distinctive feature of cancer cells is their ability to perform metabolic reprogramming necessary for their high energy requirements [183]. To meet the metabolic needs associated with proliferation, a cancer cell must increase the import of nutrients from the environment. Classically, cancer metabolism has focused on carbon metabolism, including glycolysis and the tricarboxylic acid cycle (TCA cycle). Much is known about the role of glucose as a source of energy for cancer growth; however, amino acids are also important molecules in supporting cancer development. Glutamine is a non-essential amino acid abundant in circulation, which plays an important role in addition to its function as a constituent of proteins. It provides its two nitrogen atoms to synthesize hexosamines, nucleotides, and other amino acids; guides the uptake of essential amino acids; and is also a substrate for TCA, particularly under conditions of carbon diversion to the glycolytic pathways. The level of glutamine in tumor tissues *in vivo* was found to be significantly lower than in healthy surrounding tissues or plasma. Cancer cells accumulate oncogenic alterations that convey a significant degree of independence to make up for this lacking of glutamine. For instance, *c-Myc* hyperactivation results in altered levels of downstream transcriptional targets involved in glutamine uptake and metabolism. The lack of glutamine can induce apoptosis in a Myc-dependent manner [184]. On the other hand, glutamine, glycine, and aspartate serve as carbon and nitrogen donors for purine biosynthesis [185], whereas glycine, serine, and methionine provide one carbon unit through the methionine-folate cycle for nitrogenous bases [186].

Metabolites are small molecular analytes present inside the cell, which can provide real-time information on the biochemical events that occur at the time of sample collection, and, therefore, are indicative of the physiological state of the patient [187]. As a result, important clinical information on disease progression can be obtained by monitoring metabolic changes in the patient's bio-fluids, such as blood, urine, saliva, and others, as well as EVs derived from cell lines. Interestingly, during the formation of EVs, small metabolites can be packed inside the vesicle or they can be produced as a result of the activity of metabolic enzymes within the EVs [188].

It should be noted that there is a definitive crossover between EV metabolomics and lipidomics, as the size of the biologically relevant lipids make them classified as metabolites. Because their circulating levels are very low and difficult to detect, metabolites can only be distinguished through use of highly sensitive identification techniques, such as mass spectrometry (LC-MS/MS) and magnetic resonance spectroscopy.

Analyses of urine-derived EVs (uEVs) have highlighted specific metabolites, including creatinine, glucuronate, pantothenic acid, 4-pyridoxic acid, and others. Metabolites specific for plasma-derived EVs (pEVs) include lysine, kynurenine, threonine, tryptophan, cytidine. Metabolites were found to differ in abundance between uEVs and pEVs. Interestingly, the authors observed that EVs released in pre-prostatectomy present low levels of adenosine, glucuronate, isobutyryl-L-carnitine, and D-ribose 5-phosphate, compared to EVs secreted post-surgery, as well as in control and untreated samples. Specifically, they found greater differences in glucuronate between treated and untreated cancer groups when compared to a control group [5]. Another study observed a statistically significant difference in many of the key metabolites in PCa patients, including acyl carnitines, citrate, and kynurenine among benign prostate hyperplasia (BPH) samples. Importantly, they found significantly elevated levels of the steroid hormone dehydroepiandrosterone sulphate (DHEAS) in uEVs from PCa patients compared to BHP patients. There were also a few molecules differentially expressed between two subgroups of PCa patients (stages 2 and 3), as acylcarnitine [189].

A proof-of-concept study to detect EVs metabolite biomarkers from plasma of endometrial adenocarcinoma (EAC) in patients versus control subjects revealed a clear separation of metabolites. Furthermore, EVs characterized by TGF-β-treated human pancreatic cell line (PANC 1) showed marked differences compared to the control group [177] (Table 2). A metabolomic study of EVs derived from patient blood before and after chemotherapy also detected the presence of different compounds, i.e., 6-dimethylaminopurine, leucyl proline, alanyl-histidine, and methionine sulfoxide [178] (Table 2). EVs of GBM subtypes were shown to contain significantly distinguishable metabolic content from astrocytoma cells. Furthermore, a significant difference was found in the metabolic profile between GBM-derived EVs and parental cells; with enrichment in glycerol, tryptophan, carnitine, and oxidized glutathione (GSSG) [179] (Table 2).

Currently, few metabolome-oriented studies have addressed EVs under tumorigenic conditions. These studies reflect the metabolic plasticity of cancer cells and their tendency to escape dependence on canonical pathways through metabolic reprogramming. Clearly, new and innovative combinatorial analysis strategies are needed to cover the entire spectrum of the metabolome. Moreover, the method of cell culture can also impact the metabolite composition of EVs, and may need to be taken into consideration when comparing results from different studies [180] (Table 2).

4.4. miRNA Profiling Analysis of EVs in Multiple Cancers

Tumor-derived EV microRNAs (miRNAs) have received much attention as biomarker candidates for non-invasive diagnostics, given their role in tumor progression and metastasis. Analysis of serum exosomal miRNA expression profiles of CRC patients revealed that miR-19a and miR-92a were significantly upregulated compared to HCs [190] (Table 3). Moreover, let-7a, miR-1229, miR-1246, miR-150, miR-21, miR-223, and miR-23a were detectable at significantly higher levels in colon cancer cell lines and serum samples from CRC patients [191] (Table 3). A recent report demonstrated that miR-1246 and other miRNA markers were highly enriched in EVs derived from pancreatic cancer patients [192] (Table 3). In addition, miR-1246, miR-1290, miR-375 [193], miR-141 [194], and others [195] were detectable in EVs isolated from plasma and serum samples of patients with prostate cancer, respectively. Furthermore, other works highlighted miR-21-5p and miR-92a-3p as emerging diagnostic biomarkers for HCC [196] (Table 3). Analysis of plasma-derived EVs from lung cancer patients identified miR-320, miR-126 [197], as well as let-7f, miR-146, miR-203, miR-106a, and miR-20b [198].

It has been reported that EV miRNAs confer high accuracy in identifying BC; in particular, a decreased expression of miR-142-5p and miR-150-5p were significantly associated with more advanced tumor grades (grade III), while the decreased expression of miR-142-5p and miR-320a was associated with a larger tumor size (<20 mm) [199] (Table 3). Some significant miRNAs derived from EVs were also associated with the severity of BC, including miR-939 implicated in drug resistance, miR-338, and others involved in TME [200–202] (Table 3). miRNA profiling of EVs has also revealed an increased expression of miR- 200 and miR-18, while a decreased expression of miR-100 and miR-125b has been reported in different histotypes of ovarian carcinomas, as compared to normal ovarian tissue [203]. Furthermore, Let-7 miRNA family members represent a diagnostic potency marker to manage follicular nodules in the thyroid gland [204]. miR-423-5p, miR-484, miR-142-5p, and miR-17-5p were discovered to be dysregulated in GC or implicated in GC tumorigenesis/metastasis in EVs isolated from serum samples [205]. Lucero and coworkers (2020) identified eight candidate miRNAs that may mediate EV-associated angiogenesis in glioblastoma, including miR-148a and miR-9–5p [206]. Different studies involving miRNAs associated with tumor progression and diagnostics are summarized in Table 3.

Table 3. Summary of the miRNAs identified in EVs with potential as biomarkers of different cancers.

Cancer Type	Biomarker	Sample	Authors
Colorectal Cancer	Increased miR-19a and miR-92a	Serum-derived EVs	[190]
Colorectal Cancer	Increased let-7a, miR-1229, miR-1246, miR-150, miR-21, 223, and miR-23a	Serum-derived EVs	[191]
Pancreatic Cancer	Increased miR-1246, miR-4644, mir_3976, and miR-4306	Plasma-derived EVs	[192]
Prostate Cancer	Increased miR-1246, miR-1290 and miR-375	Serum-derived EVs	[193]
Prostate Cancer	Increased levels of miR-141	Serum-derived EVs	[194]
Prostate Cancer	Increased miR-21-5p and let-7a-5p	Plasma-derived EVs	[195]
Hepatocellular Carcinoma	Increased miR-21-5p, miR-92a-3p	Plasma-derived EVs	[196]
Lung cancer	Increased miR-320 and miR-126	Plasma-derived EVs	[197]
Breast Cancer	Decreased miR-142-5p and miR-150-5p	Plasma-derived EVs	[199]
Breast Cancer	miR-200a, miR-200b, miR-200c, miR-429, and miR-141	*In vitro*: 4T1, 4TO7, 67NR, and MCF10CA cell lines.	[200]
Breast Cancer	miR-338-3p, miR-340-5p, and miR124-3p	Serum-derived EVs	[201,202]
Ovarian Cancer	Decresead of miR-100 and miR-125b	*In vitro*: SKOV3, HO-8910 and U937 cell lines.	[203]
Thyroid Cancer	Let-7 miRNA family serum-derived EVs	Serum-derived EVs	[204]
Gastric Cancer	miR-423-5p, miR-484, miR-142-5p, and miR-17-5p serum-derived EVs	Serum-derived EVs	[205]
Glioblastoma	Increased miR-1246	*In vitro*: GBM8 neurospheres	[207]
Glioblastoma	Increased miR-301a	*In vitro*: U87MG and U251 cell lines	[208]
Glioblastoma	Increased miR-21	Serum-derived EVs	[209]

Several approaches are being used to create a miRNA profile to classify different cancer histotypes, nonetheless, there is a deregulated expression of specific miRNAs. Some of them have found them as regularly loaded in EVs and associated with cancer progression, which could be used as biomarkers and in therapy design. However, the development of new technology is required to advance in this aspect.

5. Future Challenges and Conclusions

Different reports have identified numerous potential EV-based biomarkers that can aid in cancer diagnosis, disease monitoring, and development of targeted therapy. Currently, studies in proteomics, lipidomics, and metabolomics have advanced our knowledge of the properties of EVs. However, limitations in sample quality, EV isolation methods, cargo analysis, and interpretation of results may contribute to the failure of them as biomarkers in achieving clinical utility.

One of the primary challenges in characterizing cancer-specific EVs in biofluids is that samples also contain large amounts of EVs secreted by healthy cells, as well as other biomolecules (e.g., albumin, lipoproteins). Methods of isolation and analysis have limited sensitivity/specificity for detecting specific tumor-secreted EVs in biofluids because they are based on a small sample size in which EVs are diluted. Therefore, it is necessary to optimize isolation methods *in vitro* in order to distinguish specific EV subgroups to address for specific clinical scenarios. In fact, the success of EV molecular profiling heavily relies on the isolation and separation process. Consequently, developing efficient isolation methods and enriching cancer-derived EVs or specific EV subpopulations from human biofluids are urgent in order to further advance in this field.

In recent years, the field of microfluidics has allowed for the development of novel exosome purification methods, starting from small sample quantities [210]. Microfluidic platforms have been shown to sort exosomes with a high level of purity and sensitivity by reducing cost, volume of reagents consumed, and time invested in the procedure. However, microfluidics platforms have the disadvantage of their manufacturing complexity [211]. The first microfluidic platform used for the isolation of EVs relied on exosome immunoaffinity and unlabeled detection. It quantified the levels of EpCAM and CD24 proteins measured in relation to CD63 (+) exosome counts, with a diagnostic accuracy of 97% in ovarian cancer ascitic fluids. Another important aspect of microfluidic platforms is their potential use in the identification of specific subpopulations of EVs. Multiplex microfluidics have enabled the selective and specific capture of IV HER2 (+) in serum from breast cancer patients [212].

EVs hold great potential for disease diagnostics, which is why it is important to further know their content and cell-targeting mechanisms. Further knowledge and classification of the proteins, lipids, and metabolites of EVs will be useful in order to gain deeper insight about their role in transmitting information between cells, as well as our understanding of disease biogenesis and progression. Online databases have been created for the purpose of cataloguing EV content. There, highly accessible websites are used to compare sequences, and upload new ones [81–83].

Despite their promising utility as cancer biomarkers, the use of EVs in the clinical setting is still far from use in everyday practice, due to difficulties in isolation through standard analysis techniques.

Author Contributions: Conceptualization, A.M. and F.C.; original draft preparation, G.A. and C.M.S.-L.; supervision, C.C. and R.S.; writing and editing, G.A., C.M.S.-L., A.A. and A.M. All authors have read and agreed to the last version of the manuscript.

Funding: G.A., R.S., C.C. and F.C. were partially supported by IMET and IEMEST. This is IMET contribution number IMET 21-001. C.M.S.-L. and A.M. are supported by grants PID2019-105713GB-I00, funded by MCIN/AEI/ 10.13039/501100011033, Spain; and grant PROMETEO 2020/071 funded by the Con-selleria de Educación, Cultura y Deporte of Generalitat Valenciana. C.M.S.L is supported by grant PRE2020-092458 funded by MCIN/AEI/ 10.13039/501100011033, Spain. The research group is part of "Red Traslacional para la Aplicación Clínica de Vesículas Extracelulares" (RED2018-102411-T), Agencia Estatal de Investigación, Spain.

Institutional Review Board Statement: Not applicable.

Informed Consent Statement: Not applicable.

Data Availability Statement: Not applicable.

Conflicts of Interest: The authors declare no conflict of interest.

References

1. Foulds, L. Tumor progression. *Cancer Res.* **1957**, *17*, 355–356.
2. Martincorena, I.; Raine, K.M.; Gerstung, M.; Dawson, K.J.; Haase, K.; Van Loo, P.; Davies, H.; Stratton, M.R.; Campbell, P.J. Universal Patterns of Selection in Cancer and Somatic Tissues. *Cell* **2017**, *171*, 1029–1041. [CrossRef]
3. Qian, M.; Wang, D.C.; Chen, H.; Cheng, Y. Detection of single cell heterogeneity in cancer. *Semin. Cell Dev. Biol.* **2017**, *64*, 143–149. [CrossRef]
4. Luo, D.; Zhan, S.; Xia, W.; Huang, L.; Ge, W.; Wang, T. Proteomics study of serum exosomes from papillary thyroid cancer patients. *Endocr. Relat. Cancer* **2018**, *25*, 879–891. [CrossRef]
5. Puhka, M.; Takatalo, M.; Nordberg, M.E.; Valkonen, S.; Nandania, J.; Aatonen, M.; Yliperttula, M.; Laitinen, S.; Velagapudi, V.; Mirtti, T.; et al. Metabolomic Profiling of Extracellular Vesicles and Alternative Normalization Methods Reveal Enriched Metabolites and Strategies to Study Prostate Cancer-Related Changes. *Theranostics* **2017**, *7*, 3824–3841. [CrossRef] [PubMed]
6. Han, Y.; Jia, L.; Zheng, Y.; Li, W. Salivary Exosomes: Emerging Roles in Systemic Disease. *Int. J. Biol. Sci.* **2018**, *14*, 633–643. [CrossRef] [PubMed]
7. Yuan, Z.; Bedi, B.; Sadikot, R.T. Bronchoalveolar Lavage Exosomes in Lipopolysaccharide-induced Septic Lung Injury. *J. Vis. Exp.* **2018**, *135*, e57737. [CrossRef]
8. Hayashi, N.; Doi, K.; Kurata, Y.; Kagawa, H.; Atobe, Y.; Funakoshi, K.; Tada, M.; Katsumoto, A.; Tanaka, K.; Kunii, M.; et al. Proteomic analysis of exosome-enriched fractions derived from cerebrospinal fluid of amyotrophic lateral sclerosis patients. *Neurosci. Res.* **2020**, *160*, 43–49. [CrossRef] [PubMed]
9. Yáñez-Mó, M.; Siljander, P.R.; Andreu, Z.; Zavec, A.B.; Borràs, F.E.; Buzas, E.I.; Buzas, K.; Casal, E.; Cappello, F.; Carvalho, J.; et al. Biological properties of extracellular vesicles and their physiological functions. *J. Extracell Vesicles* **2015**, *4*, 27066. [CrossRef] [PubMed]
10. Colombo, M.; Raposo, G.; Théry, C. Biogenesis, secretion, and intercellular interactions of exosomes and other extracellular vesicles. *Annu. Rev. Cell Dev. Biol.* **2014**, *30*, 255–289. [CrossRef]
11. Cocucci, E.; Meldolesi, J. Ectosomes and exosomes: Shedding the confusion between extracellular vesicles. *Trends Cell Biol.* **2015**, *25*, 364–372. [CrossRef] [PubMed]
12. Kowal, J.; Arras, G.; Colombo, M.; Jouve, M.; Morath, J.P.; Primdal-Bengtson, B.; Dingli, F.; Loew, D.; Tkach, M.; Théry, C. Proteomic comparison defines novel markers to characterize heterogeneous populations of extracellular vesicle subtypes. *Proc. Natl. Acad. Sci. USA* **2016**, *113*, E968–E977. [CrossRef]
13. Théry, C.; Witwer, K.W.; Aikawa, E.; Alcaraz, M.J.; Anderson, J.D.; Andriantsitohaina, R.; Antoniou, A.; Arab, T.; Archer, F.; Atkin-Smith, G.K. Minimal information for studies of extracellular vesicles 2018 (MISEV2018): A position statement of the International Society for Extracellular Vesicles and update of the MISEV2014 guidelines. *J. Extracell Vesicles* **2018**, *7*, 1535750. [CrossRef]
14. Wu, K.; Xing, F.; Wu, S.Y.; Watabe, K. Extracellular vesicles as emerging targets in cancer: Recent development from bench to bedside. *Biochim. Biophys. Acta Rev. Cancer* **2017**, *1868*, 538–563. [CrossRef] [PubMed]
15. Gopal, S.K.; Greening, D.W.; Rai, A.; Chen, M.; Xu, R.; Shafiq, A.; Mathias, R.A.; Zhu, H.J.; Simpson, R.J. Extracellular vesicles: Their role in cancer biology and epithelial-mesenchymal transition. *Biochem. J.* **2017**, *474*, 21–45. [CrossRef] [PubMed]
16. Boussadia, Z.; Lamberti, J.; Mattei, F.; Pizzi, E.; Puglisi, R.; Zanetti, C.; Pasquini, L.; Fratini, F.; Fantozzi, L.; Felicetti, F.; et al. Acidic microenvironment plays a key role in human melanoma progression through a sustained exosome mediated transfer of clinically relevant metastatic molecules. *J. Exp. Clin. Cancer Res.* **2018**, *37*, 245. [CrossRef]
17. Keller, L.; Werner, S.; Pantel, K. Biology and clinical relevance of EpCAM. *Cell Stress* **2019**, *3*, 165–180. [CrossRef] [PubMed]
18. Xavier, C.P.R.; Caires, H.R.; Barbosa, M.A.G.; Bergantim, R.; Guimarães, J.E.; Vasconcelos, M.H. The Role of Extracellular Vesicles in the Hallmarks of Cancer and Drug Resistance. *Cells* **2020**, *9*, 1141. [CrossRef] [PubMed]
19. Kuriyama, N.; Yoshioka, Y.; Kikuchi, S.; Azuma, N.; Ochiya, T. Extracellular Vesicles Are Key Regulators of Tumor Neovasculature. *Front. Cell Dev. Biol.* **2020**, *8*, 611039. [CrossRef]
20. Fallah, A.; Sadeghinia, A.; Kahroba, H.; Samadi, A.; Heidari, H.R.; Bradaran, B.; Zeinali, S.; Molavi, O. Therapeutic targeting of angiogenesis molecular pathways in angiogenesis-dependent diseases. *Biomed. Pharm.* **2019**, *110*, 775–785. [CrossRef]
21. Monteforte, A.; Lam, B.; Sherman, M.B.; Henderson, K.; Sligar, A.D.; Spencer, A.; Tang, B.; Dunn, A.K.; Baker, A.B. Glioblastoma Exosomes for Therapeutic Angiogenesis in Peripheral Ischemia. *Tissue Eng. Part. A* **2017**, *23*, 1251–1261. [CrossRef]
22. Sato, S.; Vasaikar, S.; Eskaros, A.; Kim, Y.; Lewis, J.S.; Zhang, B.; Zijlstra, A.; Weaver, A.M. EPHB2 carried on small extracellular vesicles induces tumor angiogenesis via activation of ephrin reverse signaling. *JCI Insight* **2019**, *4*, e132447. [CrossRef]
23. Tang, M.K.S.; Yue, P.Y.K.; Ip, P.P.; Huang, R.L.; Lai, H.C.; Cheung, A.N.Y.; Tse, K.Y.; Ngan, H.Y.S.; Wong, A.S.T. Soluble E-cadherin promotes tumor angiogenesis and localizes to exosome surface. *Nat. Commun.* **2018**, *9*, 2270. [CrossRef] [PubMed]
24. Feng, Q.; Zhang, C.; Lum, D.; Druso, J.E.; Blank, B.; Wilson, K.F.; Welm, A.; Antonyak, M.A.; Cerione, R.A. A class of extracellular vesicles from breast cancer cells activates VEGF receptors and tumour angiogenesis. *Nat. Commun.* **2017**, *8*, 14450. [CrossRef] [PubMed]
25. McAtee, C.O.; Booth, C.; Elowsky, C.; Zhao, L.; Payne, J.; Fangman, T.; Caplan, S.; Henry, M.D.; Simpson, M.A. Prostate tumor cell exosomes containing hyaluronidase Hyal1 stimulate prostate stromal cell motility by engagement of FAK-mediated integrin signaling. *Matrix Biol.* **2019**, *78–79*, 165–179. [CrossRef] [PubMed]

26. Chen, L.; Guo, P.; He, Y.; Chen, Z.; Chen, L.; Luo, Y.; Qi, L.; Liu, Y.; Wu, Q.; Cui, Y.; et al. HCC-derived exosomes elicit HCC progression and recurrence by epithelial-mesenchymal transition through MAPK/ERK signalling pathway. *Cell Death Dis.* **2018**, *9*, 513. [CrossRef] [PubMed]
27. Shan, Y.; You, B.; Shi, S.; Shi, W.; Zhang, Z.; Zhang, Q.; Gu, M.; Chen, J.; Bao, L.; Liu, D.; et al. Hypoxia-Induced Matrix Metalloproteinase-13 Expression in Exosomes from Nasopharyngeal Carcinoma Enhances Metastases. *Cell Death Dis.* **2018**, *9*, 382. [CrossRef] [PubMed]
28. Hoshino, A.; Costa-Silva, B.; Shen, T.L.; Rodrigues, G.; Hashimoto, A.; Tesic Mark, M.; Molina, H.; Kohsaka, S.; Di Giannatale, A.; Ceder, S.; et al. Tumour exosome integrins determine organotropic metastasis. *Nature* **2015**, *527*, 329–335. [CrossRef]
29. Zhang, H.; Deng, T.; Liu, R.; Bai, M.; Zhou, L.; Wang, X.; Li, S.; Wang, X.; Yang, H.; Li, J.; et al. Exosome-delivered EGFR regulates liver microenvironment to promote gastric cancer liver metastasis. *Nat. Commun.* **2017**, *8*, 15016. [CrossRef] [PubMed]
30. Zhang, D.X.; Vu, L.T.; Ismail, N.N.; Le, M.T.N.; Grimson, A. Landscape of extracellular vesicles in the tumour microenvironment: Interactions with stromal cells and with non-cell components, and impacts on metabolic reprogramming, horizontal transfer of neoplastic traits, and the emergence of therapeutic resistance. *Semin. Cancer Biol.* **2021**, *74*, 24–44. [CrossRef]
31. Jella, K.K.; Nasti, T.H.; Li, Z.; Malla, S.R.; Buchwald, Z.S.; Khan, M.K. Exosomes, Their Biogenesis and Role in Inter-Cellular Communication, Tumor Microenvironment and Cancer Immunotherapy. *Vaccines* **2018**, *6*, 69. [CrossRef]
32. Bretz, N.P.; Ridinger, J.; Rupp, A.K.; Rimbach, K.; Keller, S.; Rupp, C.; Marmé, F.; Umansky, L.; Umansky, V.; Eigenbrod, T.; et al. Body fluid exosomes promote secretion of inflammatory cytokines in monocytic cells via Toll-like receptor signaling. *J. Biol. Chem.* **2013**, *288*, 36691–36702. [CrossRef]
33. Maji, S.; Chaudhary, P.; Akopova, I.; Nguyen, P.M.; Hare, R.J.; Gryczynski, I.; Vishwanatha, J.K. Exosomal Annexin II Promotes Angiogenesis and Breast Cancer Metastasis. *Mol. Cancer Res.* **2017**, *15*, 93–105. [CrossRef] [PubMed]
34. Yen, E.Y.; Miaw, S.C.; Yu, J.S.; Lai, I.R. Exosomal TGF-β1 is correlated with lymphatic metastasis of gastric cancers. *Am. J. Cancer Res.* **2017**, *7*, 2199–2208.
35. Rong, L.; Li, R.; Li, S.; Luo, R. Immunosuppression of breast cancer cells mediated by transforming growth factor-β in exosomes from cancer cells. *Oncol. Lett.* **2016**, *11*, 500–504. [CrossRef]
36. Mittal, S.; Gupta, P.; Chaluvally-Raghavan, P.; Pradeep, S. Emerging Role of Extracellular Vesicles in Immune Regulation and Cancer Progression. *Cancers* **2020**, *12*, 3563. [CrossRef] [PubMed]
37. Enomoto, Y.; Li, P.; Jenkins, L.M.; Anastasakis, D.; Lyons, G.C.; Hafner, M.; Leonard, W.J. Cytokine-enhanced cytolytic activity of exosomes from NK Cells. *Cancer Gene* **2021**. [CrossRef]
38. Li, Q.; Cai, S.; Li, M.; Salma, K.I.; Zhou, X.; Han, F.; Chen, J.; Huyan, T. Tumor-Derived Extracellular Vesicles: Their Role in Immune Cells and Immunotherapy. *Int. J. Nanomed.* **2021**, *16*, 5395–5409. [CrossRef] [PubMed]
39. Giraldo, N.A.; Sanchez-Salas, R.; Peske, J.D.; Vano, Y.; Becht, E.; Petitprez, F.; Validire, P.; Ingels, A.; Cathelineau, X.; Fridman, W.H.; et al. The clinical role of the TME in solid cancer. *Br. J. Cancer* **2019**, *120*, 45–53. [CrossRef]
40. Rai, A.; Greening, D.W.; Chen, M.; Xu, R.; Ji, H.; Simpson, R.J. Exosomes Derived from Human Primary and Metastatic Colorectal Cancer Cells Contribute to Functional Heterogeneity of Activated Fibroblasts by Reprogramming Their Proteome. *Proteomics* **2019**, *19*, e1800148. [CrossRef]
41. Sung, J.S.; Kang, C.W.; Kang, S.; Jang, Y.; Chae, Y.C.; Kim, B.G.; Cho, N.H. ITGB4-mediated metabolic reprogramming of cancer-associated fibroblasts. *Oncogene* **2020**, *39*, 664–676. [CrossRef] [PubMed]
42. García Garre, E.; Luengo Gil, G.; Montoro García, S.; Gonzalez Billalabeitia, E.; Zafra Poves, M.; García Martinez, E.; Roldán Schilling, V.; Navarro Manzano, E.; Ivars Rubio, A.; Lip, G.Y.H.; et al. Circulating small-sized endothelial microparticles as predictors of clinical outcome after chemotherapy for breast cancer: An exploratory analysis. *Breast Cancer Res. Treat.* **2018**, *169*, 83–92. [CrossRef] [PubMed]
43. Atiya, H.; Frisbie, L.; Pressimone, C.; Coffman, L. Mesenchymal Stem Cells in the Tumor Microenvironment. *Adv. Exp. Med. Biol.* **2020**, *1234*, 31–42.
44. Dai, J.; Escara-Wilke, J.; Keller, J.M.; Jung, Y.; Taichman, R.S.; Pienta, K.J.; Keller, E.T. Primary prostate cancer educates bone stroma through exosomal pyruvate kinase M2 to promote bone metastasis. *J. Exp. Med.* **2019**, *216*, 2883–2899. [CrossRef]
45. Chen, F.; Chen, J.; Yang, L.; Liu, J.; Zhang, X.; Zhang, Y.; Tu, Q.; Yin, D.; Lin, D.; Wong, P.P.; et al. Extracellular vesicle-packaged HIF-1α-stabilizing lncRNA from tumour-associated macrophages regulates aerobic glycolysis of breast cancer cells. *Nat. Cell Biol.* **2019**, *21*, 498–510. [CrossRef]
46. Lucchetti, D.; Ricciardi Tenore, C.; Colella, F.; Sgambato, A. Extracellular Vesicles and Cancer: A Focus on Metabolism, Cytokines, and Immunity. *Cancers* **2020**, *12*, 171. [CrossRef]
47. Soekmadji, C.; Li, B.; Huang, Y.; Wang, H.; An, T.; Liu, C.; Pan, W.; Chen, J.; Cheung, L.; Falcon-Perez, J.M.; et al. The future of Extracellular Vesicles as Theranostics-an ISEV meeting report. *J. Extracell Vesicles* **2020**, *9*, 1809766. [CrossRef] [PubMed]
48. Geeurickx, E.; Hendrix, A. Targets, pitfalls and reference materials for liquid biopsy tests in cancer diagnostics. *Mol. Asp. Med.* **2020**, *72*, 100828. [CrossRef] [PubMed]
49. Betsou, F.; Bulla, A.; Cho, S.Y.; Clements, J.; Chuaqui, R.; Coppola, D.; De Souza, Y.; De Wilde, A.; Grizzle, W.; Guadagni, F.; et al. Assays for Qualification and Quality Stratification of Clinical Biospecimens Used in Research: A Technical Report from the ISBER Biospecimen Science Working Group. *Biopreserv. Biobank.* **2016**, *14*, 98–409. [CrossRef] [PubMed]

50. Erdbrügger, U.; Blijdorp, C.J.; Bijnsdorp, I.V.; Borràs, F.E.; Burger, D.; Bussolati, B.; Byrd, J.B.; Clayton, A.; Dear, J.W.; Falcón-Pérez, J.M.; et al. Urine Task Force of the International Society for Extracellular Vesicles. *J. Extracell Vesicles* **2021**, *10*, e12093. [CrossRef] [PubMed]
51. Palmer, B.F.; Clegg, D.J. The Use of Selected Urine Chemistries in the Diagnosis of Kidney Disorders. *Clin. J. Am. Soc. Nephrol.* **2019**, *14*, 306–316. [CrossRef]
52. Svenningsen, P.; Sabaratnam, R.; Jensen, B.L. Urinary extracellular vesicles: Origin, role as intercellular messengers and biomarkers; efficient sorting and potential treatment options. *Acta Physiol.* **2020**, *228*, e13346. [CrossRef] [PubMed]
53. Barreiro, K.; Holthofer, H. Urinary extracellular vesicles. A promising shortcut to novel biomarker discoveries. *Cell Tissue Res.* **2017**, *369*, 217–227. [CrossRef] [PubMed]
54. Sun, Y.; Xia, Z.; Shang, Z.; Sun, K.; Niu, X.; Qian, L.; Fan, L.Y.; Cao, C.X.; Xiao, H. Facile preparation of salivary extracellular vesicles for cancer proteomics. *Sci. Rep.* **2016**, *6*, 24669. [CrossRef]
55. Nonaka, T.; Wong, D.T.W. Saliva-Exosomics in Cancer: Molecular Characterization of Cancer-Derived Exosomes in Saliva. *Enzymes* **2017**, *42*, 125–151.
56. Ogawa, Y.; Miura, Y.; Harazono, A.; Kanai-Azuma, M.; Akimoto, Y.; Kawakami, H.; Yamaguchi, T.; Toda, T.; Endo, T.; Tsubuki, M.; et al. Proteomic analysis of two types of exosomes in human whole saliva. *Biol. Pharm. Bull.* **2011**, *34*, 13–23. [CrossRef] [PubMed]
57. Vijay, A.; Inui, T.; Dodds, M.; Proctor, G.; Carpenter, G. Factors That Influence the Extensional Rheological Property of Saliva. *PLoS ONE* **2015**, *10*, e0135792. [CrossRef]
58. Ghersi-Egea, J.F.; Strazielle, N.; Catala, M.; Silva-Vargas, V.; Doetsch, F.; Engelhardt, B. Molecular anatomy and functions of the choroidal blood-cerebrospinal fluid barrier in health and disease. *Acta Neuropathol.* **2018**, *135*, 337–361. [CrossRef]
59. Yan, W.; Xu, T.; Zhu, H.; Yu, J. Clinical Applications of Cerebrospinal Fluid Circulating Tumor DNA as a Liquid Biopsy for Central Nervous System Tumors. *Onco. Targets.* **2020**, *13*, 719–731. [CrossRef]
60. Witwer, K.W.; Buzás, E.I.; Bemis, L.T.; Bora, A.; Lässer, C.; Lötvall, J.; Nolte-'t Hoen, E.N.; Piper, M.G.; Sivaraman, S.; Skog, J.; et al. Standardization of sample collection, isolation and analysis methods in extracellular vesicle research. *J. Extracell Vesicles* **2013**, *2*, 20360. [CrossRef]
61. Kalluri, R. The biology and function of exosomes in cancer. *J. Clin. Invest.* **2016**, *126*, 1208–1215. [CrossRef]
62. Simeone, P.; Bologna, G.; Lanuti, P.; Pierdomenico, L.; Guagnano, M.T.; Pieragostino, D.; Del Boccio, P.; Vergara, D.; Marchisio, M.; Miscia, S.; et al. Extracellular Vesicles as Signaling Mediators and Disease Biomarkers across Biological Barriers. *Int. J. Mol. Sci.* **2020**, *21*, 2514. [CrossRef] [PubMed]
63. Konoshenko, M.Y.; Lekchnov, E.A.; Vlassov, A.V.; Laktionov, P.P. Isolation of Extracellular Vesicles: General Methodologies and Latest Trends. *Biomed. Res. Int.* **2018**, *2018*, 8545347. [CrossRef] [PubMed]
64. Veerman, R.E.; Teeuwen, L.; Czarnewski, P.; Güclüler Akpinar, G.; Sandberg, A.; Cao, X.; Pernemalm, M.; Orre, L.M.; Gabrielsson, S.; Eldh, M. Molecular evaluation of five different isolation methods for extracellular vesicles reveals different clinical applicability and subcellular origin. *J. Extracell Vesicles* **2021**, *10*, e12128. [CrossRef] [PubMed]
65. Livshits, M.A.; Khomyakova, E.; Evtushenko, E.G.; Lazarev, V.N.; Kulemin, N.A.; Semina, S.E.; Generozov, E.V.; Govorun, V.M. Isolation of exosomes by differential centrifugation: Theoretical analysis of a commonly used protocol. *Sci. Rep.* **2015**, *5*, 17319. [CrossRef]
66. Monguió-Tortajada, M.; Gálvez-Montón, C.; Bayes-Genis, A.; Roura, S.; Borràs, F.E. Extracellular vesicle isolation methods: Rising impact of size-exclusion chromatography. *Cell Mol. Life Sci.* **2019**, *76*, 2369–2382. [CrossRef]
67. Linares, R.; Tan, S.; Gounou, C.; Arraud, N.; Brisson, A.R. High-speed centrifugation induces aggregation of extracellular vesicles. *J. Extracell Vesicles* **2015**, *4*, 29509. [CrossRef] [PubMed]
68. Freitas, D.; Balmaña, M.; Poças, J.; Campos, D.; Osório, H.; Konstantinidi, A.; Vakhrushev, S.Y.; Magalhães, A.; Reis, C.A. Different isolation approaches lead to diverse glycosylated extracellular vesicle populations. *J. Extracell Vesicles* **2019**, *8*, 1621131. [CrossRef]
69. Sharma, S.; LeClaire, M.; Wohlschlegel, J.; Gimzewski, J. Impact of isolation methods on the biophysical heterogeneity of single extracellular vesicles. *Sci. Rep.* **2020**, *10*, 13327. [CrossRef]
70. Brennan, K.; Martin, K.; FitzGerald, S.P.; O'Sullivan, J.; Wu, Y.; Blanco, A.; Richardson, C.; Mc Gee, M.M. A comparison of methods for the isolation and separation of extracellular vesicles from protein and lipid particles in human serum. *Sci. Rep.* **2020**, *10*, 1039. [CrossRef]
71. Kosanović, M.; Janković, M. Isolation of urinary extracellular vesicles from Tamm- Horsfall protein-depleted urine and their application in the development of a lectin-exosome-binding assay. *Biotechniques* **2014**, *57*, 143–149. [CrossRef]
72. Gurunathan, S.; Kang, M.H.; Jeyaraj, M.; Qasim, M.; Kim, J.H. Review of the Isolation, Characterization, Biological Function, and Multifarious Therapeutic Approaches of Exosomes. *Cells* **2019**, *8*, 307. [CrossRef]
73. Xu, J.; Liao, K.; Zhou, W. Exosomes Regulate the Transformation of Cancer Cells in Cancer Stem Cell Homeostasis. *Stem Cells Int.* **2018**, *2018*, 4837370. [CrossRef] [PubMed]
74. Brzozowski, J.S.; Bond, D.R.; Jankowski, H.; Goldie, B.J.; Burchell, R.; Naudin, C.; Smith, N.D.; Scarlett, C.J.; Larsen, M.R.; Dun, M.D.; et al. Extracellular vesicles with altered tetraspanin CD9 and CD151 levels confer increased prostate cell motility and invasion. *Sci. Rep.* **2018**, *8*, 8822. [CrossRef] [PubMed]
75. Tan, C.F.; Teo, H.S.; Park, J.E.; Dutta, B.; Tse, S.W.; Leow, M.K.; Wahli, W.; Sze, S.K. Exploring Extracellular Vesicles Biogenesis in Hypothalamic Cells through a Heavy Isotope Pulse/Trace Proteomic Approach. *Cells* **2020**, *9*, 1320. [CrossRef] [PubMed]

76. Reina, M.; Espel, E. Role of LFA-1 and ICAM-1 in Cancer. *Cancers* **2017**, *9*, 153. [CrossRef]
77. Gutiérrez-Vázquez, C.; Villarroya-Beltri, C.; Mittelbrunn, M.; Sánchez-Madrid, F. Transfer of extracellular vesicles during immune cell-cell interactions. *Immunol. Rev.* **2013**, *25*, 125–142. [CrossRef]
78. Taha, E.A.; Ono, K.; Eguchi, T. Roles of Extracellular HSPs as Biomarkers in Immune Surveillance and Immune Evasion. *Int. J. Mol. Sci* **2019**, *20*, 4588. [CrossRef]
79. Kreimer, S.; Belov, A.M.; Ghiran, I.; Murthy, S.K.; Frank, D.A.; Ivanov, A.R. Mass-spectrometry-based molecular characterization of extracellular vesicles: Lipidomics and proteomics. *J. Proteome Res.* **2015**, *14*, 2367–2384. [CrossRef]
80. Shao, H.; Im, H.; Castro, C.M.; Breakefield, X.; Weissleder, R.; Lee, H. New Technologies for Analysis of Extracellular Vesicles. *Chem Rev.* **2018**, *118*, 1917–1950. [CrossRef]
81. Kim, D.K.; Kang, B.; Kim, O.Y.; Choi, D.S.; Lee, J.; Kim, S.R.; Go, G.; Yoon, Y.J.; Kim, J.H.; Jang, S.C.; et al. EVpedia: An integrated database of high-throughput data for systemic analyses of extracellular vesicles. *J. Extracell Vesicles* **2013**, *2*, 20384. [CrossRef]
82. Keerthikumar, S.; Chisanga, D.; Ariyaratne, D.; Al Saffar, H.; Anand, S.; Zhao, K.; Samuel, M.; Pathan, M.; Jois, M.; Chilamkurti, N.; et al. ExoCarta: A Web-Based Compendium of Exosomal Cargo. *J. Mol. Biol.* **2016**, *428*, 688–692. [CrossRef] [PubMed]
83. Pathan, M.; Fonseka, P.; Chitti, S.V.; Kang, T.; Sanwlani, R.; Van Deun, J.; Hendrix, A.; Mathivanan, S. Vesiclepedia 2019: A compendium of RNA, proteins, lipids and metabolites in extracellular vesicles. *Nucleic Acids Res.* **2019**, *47*, D516–D519. [CrossRef] [PubMed]
84. Rupp, A.K.; Rupp, C.; Keller, S.; Brase, J.C.; Ehehalt, R.; Fogel, M.; Moldenhauer, G.; Marmé, F.; Sültmann, H.; Altevogt, P. Loss of EpCAM expression in breast cancer derived serum exosomes: Role of proteolytic cleavage. *Gynecol. Oncol.* **2011**, *122*, 437–446. [CrossRef]
85. Chantziou, A.; Theodorakis, K.; Polioudaki, H.; de Bree, E.; Kampa, M.; Mavroudis, D.; Castanas, E.; Theodoropoulos, P.A. Glycosylation Modulates Plasma Membrane Trafficking of CD24 in Breast Cancer Cells. *Int. J. Mol. Sci* **2021**, *22*, 8165. [CrossRef] [PubMed]
86. Wuerstlein, R.; Harbeck, N. Neoadjuvant Therapy for HER2-positive Breast Cancer. *Rev. Recent Clin. Trials* **2017**, *12*, 81–92. [CrossRef]
87. Fang, S.; Tian, H.; Li, X.; Jin, D.; Li, X.; Kong, J.; Yang, C.; Yang, X.; Lu, Y.; Luo, Y.; et al. Clinical application of a microfluidic chip for immunocapture and quantification of circulating exosomes to assist breast cancer diagnosis and molecular classification. *PLoS ONE* **2017**, *12*, e0175050. [CrossRef]
88. Risha, Y.; Minic, Z.; Ghobadloo, S.M.; Berezovski, M.V. The proteomic analysis of breast cell line exosomes reveals disease patterns and potential biomarkers. *Sci. Rep.* **2020**, *10*, 13572. [CrossRef]
89. Palazzolo, G.; Albanese, N.N.; DI Cara, G.; Gygax, D.; Vittorelli, M.L.; Pucci-Minafra, I. Proteomic analysis of exosome-like vesicles derived from breast cancer cells. *Anticancer Res.* **2012**, *32*, 847–860. [PubMed]
90. Melo, S.A.; Luecke, L.B.; Kahlert, C.; Fernandez, A.F.; Gammon, S.T.; Kaye, J.; LeBleu, V.S.; Mittendorf, E.A.; Weitz, J.; Rahbari, N.; et al. Glypican-1 identifies cancer exosomes and detects early pancreatic cancer. *Nature* **2015**, *523*, 177–182. [CrossRef]
91. Li, J.; Chen, Y.; Guo, X.; Zhou, L.; Jia, Z.; Peng, Z.; Tang, Y.; Liu, W.; Zhu, B.; Wang, L.; et al. GPC1 exosome and its regulatory miRNAs are specific markers for the detection and target therapy of colorectal cancer. *J. Cell Mol. Med.* **2017**, *21*, 838–847. [CrossRef]
92. Xiao, Y.; Li, Y.; Yuan, Y.; Liu, B.; Pan, S.; Liu, Q.; Qi, X.; Zhou, H.; Dong, W.; Jia, L. The potential of exosomes derived from colorectal cancer as a biomarker. *Clin. Chim. Acta* **2019**, *490*, 186–193. [CrossRef]
93. Li, P.; Bai, Y.; Shan, B.; Zhang, W.; Liu, Z.; Zhu, Y.; Xu, X.; Chen, Q.; Sheng, X.; Deng, X.; et al. Exploration of Potential Diagnostic Value of Protein Content in Serum Small Extracellular Vesicles for Early-Stage Epithelial Ovarian Carcinoma. *Front. Oncol.* **2021**, *11*, 707658. [CrossRef] [PubMed]
94. Chen, Z.; Liang, Q.; Zeng, H.; Zhao, Q.; Guo, Z.; Zhong, R.; Xie, M.; Cai, X.; Su, J.; He, Z.; et al. Exosomal CA125 as A Promising Biomarker for Ovarian Cancer Diagnosis. *J. Cancer* **2020**, *11*, 6445–6453. [CrossRef] [PubMed]
95. Tian, Y.; Ma, L.; Gong, M.; Su, G.; Zhu, S.; Zhang, W.; Wang, S.; Li, Z.; Chen, C.; Li, L.; et al. Protein Profiling and Sizing of Extracellular Vesicles from Colorectal Cancer Patients via Flow Cytometry. *ACS Nano* **2018**, *12*, 671–680. [CrossRef] [PubMed]
96. Lee, C.H.; Im, E.J.; Moon, P.G.; Baek, M.C. Discovery of a diagnostic biomarker for colon cancer through proteomic profiling of small extracellular vesicles. *BMC Cancer* **2018**, *18*, 1058. [CrossRef]
97. Shiromizu, T.; Kume, H.; Ishida, M.; Adachi, J.; Kano, M.; Matsubara, H.; Tomonaga, T. Quantitation of putative colorectal cancer biomarker candidates in serum extracellular vesicles by targeted proteomics. *Sci. Rep.* **2017**, *7*, 12782. [CrossRef]
98. Arbelaiz, A.; Azkargorta, M.; Krawczyk, M.; Santos-Laso, A.; Lapitz, A.; Perugorria, M.J.; Erice, O.; Gonzalez, E.; Jimenez-Agüero, R.; Lacasta, A.; et al. Serum extracellular vesicles contain protein biomarkers for primary sclerosing cholangitis and cholangiocarcinoma. *Hepatology* **2017**, *66*, 1125–1143. [CrossRef]
99. Julich-Haertel, H.; Urban, S.K.; Krawczyk, M.; Willms, A.; Jankowski, K.; Patkowski, W.; Kruk, B.; Krasnodębski, M.; Ligocka, J.; Schwab, R.; et al. Cancer-associated circulating large extracellular vesicles in cholangiocarcinoma and hepatocellular carcinoma. *J. Hepatol.* **2017**, *67*, 282–292. [CrossRef]
100. Weeraphan, C.; Phongdara, A.; Chaiyawat, P.; Diskul-Na-Ayudthaya, P.; Chokchaichamnankit, D.; Verathamjamras, C.; Netsirisawan, P.; Yingchutrakul, Y.; Roytrakul, S.; Champattanachai, V.; et al. Phosphoproteome Profiling of Isogenic Cancer Cell-Derived Exosome Reveals HSP90 as a Potential Marker for Human Cholangiocarcinoma. *Proteomics* **2019**, *19*, e1800159. [CrossRef]

101. Wang, S.; Chen, G.; Lin, X.; Xing, X.; Cai, Z.; Liu, X.; Liu, J. Role of exosomes in hepatocellular carcinoma cell mobility alteration. *Oncol Lett.* **2017**, *14*, 8122–8131. [CrossRef] [PubMed]
102. Ding, X.Q.; Wang, Z.Y.; Xia, D.; Wang, R.X.; Pan, X.R.; Tong, J.H. Proteomic Profiling of Serum Exosomes from Patients With Metastatic Gastric Cancer. *Front. Oncol* **2020**, *10*, 1113. [CrossRef] [PubMed]
103. Fujita, K.; Kume, H.; Matsuzaki, K.; Kawashima, A.; Ujike, T.; Nagahara, A.; Uemura, M.; Miyagawa, Y.; Tomonaga, T.; Nonomura, N. Proteomic analysis of urinary extracellular vesicles from high Gleason score prostate cancer. *Sci. Rep.* **2017**, *7*, 42961. [CrossRef] [PubMed]
104. Lee, J.; McKinney, K.Q.; Pavlopoulos, A.J.; Niu, M.; Kang, J.W.; Oh, J.W.; Kim, K.P.; Hwang, S. Altered Proteome of Extracellular Vesicles Derived from Bladder Cancer Patients Urine. *Mol. Cells* **2018**, *41*, 179–187.
105. Jakobsen, K.R.; Paulsen, B.S.; Bæk, R.; Varming, K.; Sorensen, B.S.; Jørgensen, M.M. Exosomal proteins as potential diagnostic markers in advanced non-small cell lung carcinoma. *J. Extracell Vesicles* **2015**, *4*, 26659. [CrossRef] [PubMed]
106. Sandfeld-Paulsen, B.; Aggerholm-Pedersen, N.; Bæk, R.; Jakobsen, K.R.; Meldgaard, P.; Folkersen, B.H.; Rasmussen, T.R.; Varming, K.; Jørgensen, M.M.; Sorensen, B.S. Exosomal proteins as prognostic biomarkers in non-small cell lung cancer. *Mol. Oncol.* **2016**, *10*, 1595–1602. [CrossRef] [PubMed]
107. Wang, N.; Song, X.; Liu, L.; Niu, L.; Wang, X.; Song, X.; Xie, L. Circulating exosomes contain protein biomarkers of metastatic non-small-cell lung cancer. *Cancer Sci.* **2018**, *109*, 1701–1709. [CrossRef]
108. Sun, Y.; Huo, C.; Qiao, Z.; Shang, Z.; Uzzaman, A.; Liu, S.; Jiang, X.; Fan, L.Y.; Ji, L.; Guan, X.; et al. Comparative Proteomic Analysis of Exosomes and Microvesicles in Human Saliva for Lung Cancer. *J. Proteome Res.* **2018**, *17*, 1101–1107. [CrossRef]
109. Li, Y.; Zhang, Y.; Qiu, F.; Qiu, Z. Proteomic identification of exosomal LRG1: A potential urinary biomarker for detecting NSCLC. *Electrophoresis* **2011**, *32*, 1976–1983. [CrossRef]
110. Naryzhny, S.; Volnitskiy, A.; Kopylov, A.; Zorina, E.; Kamyshinsky, R.; Bairamukov, V.; Garaeva, L.; Shlikht, A.; Shtam, T. Proteome of Glioblastoma-Derived Exosomes as a Source of Biomarkers. *Biomedicines* **2020**, *8*, 216. [CrossRef]
111. Hara, H.; Takahashi, T.; Serada, S.; Fujimoto, M.; Ohkawara, T.; Nakatsuka, R.; Harada, E.; Nishigaki, T.; Takahashi, Y.; Nojima, S.; et al. Overexpression of glypican-1 implicates poor prognosis and their chemoresistance in oesophageal squamous cell carcinoma. *Br. J. Cancer* **2016**, *115*, 66–75. [CrossRef]
112. Chan, Y.K.; Zhang, H.; Liu, P.; Tsao, S.W.; Lung, M.L.; Mak, N.K.; Ngok-Shun Wong, R.; Ying-Kit Yue, P. Proteomic analysis of exosomes from nasopharyngeal carcinoma cell identifies intercellular transfer of angiogenic proteins. *Int. J. Cancer* **2015**, *137*, 1830–1841. [CrossRef]
113. Mullooly, M.; McGowan, P.M.; Kennedy, S.A.; Madden, S.F.; Crown, J.; O'Donovan, N.; Duffy, M.J. ADAM10: A new player in breast cancer progression? *Br. J. Cancer* **2015**, *113*, 945–951. [CrossRef]
114. Kang, S.S.; Chun, Y.K.; Hur, M.H.; Lee, H.K.; Kim, Y.J.; Hong, S.R.; Lee, J.H.; Lee, S.G.; Park, Y.K. Clinical significance of glucose transporter 1 (GLUT1) expression in human breast carcinoma. *JPN J. Cancer Res.* **2002**, *93*, 1123–1128. [CrossRef] [PubMed]
115. Moon, P.G.; Lee, J.E.; Cho, Y.E.; Lee, S.J.; Jung, J.H.; Chae, Y.S.; Bae, H.I.; Kim, Y.B.; Kim, I.S.; Park, H.Y.; et al. Identification of Developmental Endothelial Locus-1 on Circulating Extracellular Vesicles as a Novel Biomarker for Early Breast Cancer Detection. *Clin. Cancer Res.* **2016**, *22*, 1757–1766. [CrossRef]
116. Schürpf, T.; Chen, Q.; Liu, J.H.; Wang, R.; Springer, T.A.; Wang, J.H. The RGD finger of Del-1 is a unique structural feature critical for integrin binding. *FASEB J.* **2012**, *26*, 3412–3420. [CrossRef] [PubMed]
117. Felder, M.; Kapur, A.; Gonzalez-Bosquet, J.; Horibata, S.; Heintz, J.; Albrecht, R.; Fass, L.; Kaur, J.; Hu, K.; Hadi, S.; et al. MUC16 (CA125): Tumor biomarker to cancer therapy, a work in progress. *Mol. Cancer* **2014**, *13*, 29. [CrossRef]
118. Grass, G.D.; Toole, B.P. How, with whom and when: An overview of CD147-mediated regulatory networks influencing matrix metalloproteinase activity. *Biosci. Rep.* **2015**, *36*, e00283. [CrossRef]
119. Hedhli, N.; Falcone, D.J.; Huang, B.; Cesarman-Maus, G.; Kraemer, R.; Zhai, H.; Tsirka, S.E.; Santambrogio, L.; Hajjar, K.A. The annexin A2/S100A10 system in health and disease: Emerging paradigms. *J. Biomed. Biotechnol.* **2012**, *2012*, 406273. [CrossRef] [PubMed]
120. Maule, F.; Bresolin, S.; Rampazzo, E.; Boso, D.; Della Puppa, A.; Esposito, G.; Porcù, E.; Mitola, S.; Lombardi, G.; Accordi, B.; et al. Annexin 2A sustains glioblastoma cell dis Semination and proliferation. *Oncotarget* **2016**, *7*, 54632–54649. [CrossRef]
121. Naquet, P.; Pitari, G.; Duprè, S.; Galland, F. Role of the Vnn1 pantetheinase in tissue tolerance to stress. *Biochem. Soc. Trans.* **2014**, *42*, 1094–1100. [CrossRef]
122. Yue, X.; Ai, J.; Xu, Y.; Chen, Y.; Huang, M.; Yang, X.; Hu, B.; Zhang, H.; He, C.; Yang, X.; et al. Polymeric immunoglobulin receptor promotes tumor growth in hepatocellular carcinoma. *Hepatology* **2017**, *65*, 1948–1962. [CrossRef] [PubMed]
123. Bachtia, I.; Kheng, V.; Wibowo, G.A.; Gani, R.A.; Hasan, I.; Sanityoso, A.; Budhihusodo, U.; Lelosutan, S.A.; Martamala, R.; Achwan, W.A.; et al. Alpha-1-acid glycoprotein as potential biomarker for alpha-fetoprotein-low hepatocellular carcinoma. *BMC Res. Notes* **2010**, *3*, 319.
124. Xia, C.; Braunstein, Z.; Toomey, A.C.; Zhong, J.; Rao, X. S100 Proteins as an Important Regulator of Macrophage Inflammation. *Front. Immunol.* **2018**, *8*, 1908. [CrossRef] [PubMed]
125. Yang, G.; Liang, Y.; Zheng, T.; Song, R.; Wang, J.; Shi, H.; Sun, B.; Xie, C.; Li, Y.; Han, J.; et al. FCN2 inhibits epithelial-mesenchymal transition-induced metastasis of hepatocellular carcinoma via TGF-β/Smad signaling. *Cancer Lett.* **2016**, *378*, 80–86. [CrossRef] [PubMed]

126. Shi, B.; Abrams, M.; Sepp-Lorenzino, L. Expression of asialoglycoprotein receptor 1 in human hepatocellular carcinoma. *J. Histochem. Cytochem.* **2013**, *61*, 901–909. [CrossRef]
127. Kornek, M.; Schuppan, D. Microparticles: Modulators and biomarkers of liver disease. *J. Hepatol.* **2012**, *57*, 1144–1146. [CrossRef]
128. Liu, Y.; Cui, X.; Hu, B.; Lu, C.; Huang, X.; Cai, J.; He, S.; Lv, L.; Cong, X.; Liu, G.; et al. Upregulated expression of CAP1 is associated with tumor migration and metastasis in hepatocellular carcinoma. *Pathol. Res. Pract.* **2014**, *210*, 169–175. [CrossRef]
129. Li, Y.; Huang, J.; Sun, J.; Xiang, S.; Yang, D.; Ying, X.; Lu, M.; Li, H.; Ren, G. The transcription levels and prognostic values of seven proteasome alpha subunits in human cancers. *Oncotarget* **2017**, *8*, 4501–4519. [CrossRef]
130. Logozzi, M.; Angelini, D.F.; Iessi, E.; Mizzoni, D.; Di Raimo, R.; Federici, C.; Lugini, L.; Borsellino, G.; Gentilucci, A.; Pierella, F. Increased PSA expression on prostate cancer exosomes in *in vitro* condition and in cancer patients. *Cancer Lett.* **2017**, *403*, 318–329. [CrossRef]
131. Loeb, S.; Bjurlin, M.A.; Nicholson, J.; Tammela, T.L.; Penson, D.; Carter, H.B.; Carroll, P.; Etzioni, R. Overdiagnosis and overtreatment of prostate cancer. *Eur. Urol.* **2014**, *65*, 1046–1055. [CrossRef]
132. Smathers, R.L.; Petersen, D.R. The human fatty acid-binding protein family: Evolutionary divergences and functions. *Hum. Genom.* **2011**, *5*, 170–191. [CrossRef]
133. Senga, S.; Kawaguchi, K.; Kobayashi, N.; Ando, A.; Fujii, H. A novel fatty acid-binding protein 5-estrogen-related receptor α signaling pathway promotes cell growth and energy metabolism in prostate cancer cells. *Oncotarget* **2018**, *9*, 31753–31770. [CrossRef] [PubMed]
134. Monami, G.; Emiliozzi, V.; Bitto, A.; Lovat, F.; Xu, S.Q.; Goldoni, S.; Fassan, M.; Serrero, G.; Gomella, L.G.; Baffa, R.; et al. Proepithelin regulates prostate cancer cell biology by promoting cell growth, migration, and anchorage-independent growth. *Am. J. Pathol.* **2009**, *174*, 1037–1047. [CrossRef] [PubMed]
135. McCullough, J.; Frost, A.; Sundquist, W.I. Structures, Functions, and Dynamics of ESCRT-III/Vps4 Membrane Remodeling and Fission Complexes. *Annu. Rev. Cell Dev. Biol.* **2018**, *6*, 85–109. [CrossRef]
136. Kaur, S.; Momi, N.; Chakraborty, S.; Wagner, D.G.; Horn, A.J.; Lele, S.M.; Theodorescu, D.; Batra, S.K. Altered expression of transmembrane mucins, MUC1 and MUC4, in bladder cancer: Pathological implications in diagnosis. *PLoS ONE* **2014**, *9*, e92742.
137. Saied, G.M.; El-Metenawy, W.H.; Elwan, M.S.; Dessouki, N.R. Urine carcinoembryonic antigen levels are more useful than serum levels for early detection of Bilharzial and non-Bilharzial urinary bladder carcinoma: Observations of 43 Egyptian cases. *World J. Surg. Oncol.* **2007**, *5*, 4. [CrossRef] [PubMed]
138. Offenhäuser, N.; Borgonovo, A.; Disanza, A.; Romano, P.; Ponzanelli, I.; Iannolo, G.; Di Fiore, P.P.; Scita, G. The eps8 family of proteins links growth factor stimulation to actin reorganization generating functional redundancy in the Ras/Rac pathway. *Mol. Biol. Cell* **2004**, *15*, 91–98. [CrossRef]
139. Park, J.H.; Lee, C.; Han, D.; Lee, J.S.; Lee, K.M.; Song, M.J.; Kim, K.; Lee, H.; Moon, K.C.; Kim, Y.; et al. Moesin (*MSN*) as a Novel Proteome-Based Diagnostic Marker for Early Detection of Invasive Bladder Urothelial Carcinoma in Liquid-Based Cytology. *Cancers* **2020**, *12*, 1018. [CrossRef]
140. Wee, P.; Wang, Z. Epidermal Growth Factor Receptor Cell Proliferation Signaling Pathways. *Cancers* **2017**, *9*, 52. [CrossRef]
141. Gupta, R.; Dastane, A.M.; Forozan, F.; Riley-Portuguez, A.; Chung, F.; Lopategui, J.; Marchevsky, A.M. Evaluation of EGFR abnormalities in patients with pulmonary adenocarcinoma: The need to test neoplasms with more than one method. *Mod. Pathol.* **2009**, *22*, 128–133. [CrossRef]
142. Gure, A.O.; Chua, R.; Williamson, B.; Gonen, M.; Ferrera, C.A.; Gnjatic, S.; Ritter, G.; Simpson, A.J.; Chen, Y.T.; Old, L.J.; et al. Cancer-testis genes are coordinately expressed and are markers of poor outcome in non-small cell lung cancer. *Clin. Cancer Res.* **2005**, *11*, 8055–8062. [CrossRef] [PubMed]
143. Homaidan, F.R.; Chakroun, I.; Haidar, H.A.; El-Sabban, M.E. Protein regulators of eicosanoid synthesis: Role in inflammation. *Curr. Protein Pept. Sci.* **2002**, *3*, 467–484. [CrossRef]
144. Iwao, K.; Watanabe, T.; Fujiwara, Y.; Takami, K.; Kodama, K.; Higashiyama, M.; Yokouchi, H.; Ozaki, K.; Monden, M.; Tanigami, A. Isolation of a novel human lung-specific gene, LUNX, a potential molecular marker for detection of micrometastasis in non-small-cell lung cancer. *Int. J. Cancer* **2001**, *91*, 433–437. [CrossRef]
145. Wakata, K.; Tsuchiya, T.; Tomoshige, K.; Takagi, K.; Yamasaki, N.; Matsumoto, K.; Miyazaki, T.; Nanashima, A.; Whitsett, J.A.; Maeda, Y.; et al. A favourable prognostic marker for EGFR mutant non-small cell lung cancer: Immunohistochemical analysis of MUC5B. *BMJ Open* **2015**, *5*, e008366. [CrossRef] [PubMed]
146. Chuang, H.C.; Chang, C.C.; Teng, C.F.; Hsueh, C.H.; Chiu, L.L.; Hsu, P.M.; Lee, M.C.; Hsu, C.P.; Chen, Y.R.; Liu, Y.C.; et al. MAP4K3/GLK Promotes Lung Cancer Metastasis by Phosphorylating and Activating IQGAP1. *Cancer Res.* **2019**, *79*, 4978–4993. [CrossRef] [PubMed]
147. Zhao, J.; Zhang, L.; Dong, X.; Liu, L.; Huo, L.; Chen, H. High Expression of Vimentin is Associated With Progression and a Poor Outcome in Glioblastoma. *Appl. Immunohistochem. Mol. Morphol.* **2018**, *26*, 337–344. [CrossRef]
148. Xia, S.; Lal, B.; Tung, B.; Wang, S.; Goodwin, C.R.; Laterra, J. Tumor microenvironment tenascin-C promotes glioblastoma invasion and negatively regulates tumor proliferation. *Neuro-Oncol.* **2016**, *18*, 507–517. [CrossRef]
149. Desiniotis, A.; Kyprianou, N. Significance of talin in cancer progression and metastasis. *Int. Rev. Cell Mol. Biol.* **2011**, *289*, 117–147.
150. Sakamoto, S.; McCann, R.O.; Dhir, R.; Kyprianou, N. Talin1 promotes tumor invasion and metastasis via focal adhesion signaling and anoikis resistance. *Cancer Res.* **2010**, *70*, 1885–1895. [CrossRef]

151. Desgrosellier, J.S.; Cheresh, D.A. Integrins in cancer: Biological implications and therapeutic opportunities. *Nat. Rev. Cancer* **2010**, *10*, 9–22. [CrossRef]
152. Baenke, F.; Peck, B.; Miess, H.; Schulze, A. Hooked on fat: The role of lipid synthesis in cancer metabolism and tumour development. *Dis. Model. Mech.* **2013**, *6*, 1353–1363. [CrossRef] [PubMed]
153. Record, M.; Carayon, K.; Poirot, M.; Silvente-Poirot, S. Exosomes as new vesicular lipid transporters involved in cell-cell communication and various pathophysiologies. *Biochim. Biophys. Acta* **2014**, *1841*, 108–120. [CrossRef] [PubMed]
154. Beloribi-Djefaflia, S.; Siret, C.; Lombardo, D. Exosomal lipids induce human pancreatic tumoral MiaPaCa-2 cells resistance through the CXCR4-SDF-1α signaling axis. *Oncoscience* **2014**, *2*, 15–30. [CrossRef]
155. van Meer, G.; Voelker, D.R.; Feigenson, G.W. Membrane lipids: Where they are and how they behave. *Nat. Rev. Mol. Cell Biol.* **2008**, *9*, 112–124. [CrossRef] [PubMed]
156. Skotland, T.; Hessvik, N.P.; Sandvig, K.; Llorente, A. Exosomal lipid composition and the role of ether lipids and phosphoinositides in exosome biology. *J. Lipid Res.* **2019**, *60*, 9–18. [CrossRef]
157. González-Ortiz, A.; Galindo-Hernández, O.; Hernández-Acevedo, G.N.; Hurtado-Ureta, G.; García-González, V. Impact of cholesterol-pathways on breast cancer development, a metabolic landscape. *J. Cancer* **2021**, *12*, 4307–4321. [CrossRef]
158. Zhang, J.; Li, Q.; Wu, Y.; Wang, D.; Xu, L.; Zhang, Y.; Wang, S.; Wang, T.; Liu, F.; Zaky, M.Y.; et al. Cholesterol content in cell membrane maintains surface levels of ErbB2 and confers a therapeutic vulnerability in ErbB2-positive breast cancer. *Cell Commun. Signal.* **2019**, *17*, 15. [CrossRef]
159. Zhuo, D.; Li, X.; Guan, F. Biological Roles of Aberrantly Expressed Glycosphingolipids and Related Enzymes in Human Cancer Development and Progression. *Front. Physiol* **2018**, *9*, 466. [CrossRef]
160. Liang, Y.J.; Wang, C.Y.; Wang, I.A.; Chen, Y.W.; Li, L.T.; Lin, C.Y.; Ho, M.Y.; Chou, T.L.; Wang, Y.H. Interaction of glycosphingolipids GD3 and GD2 with growth factor receptors maintains breast cancer stem cell phenotype. *Oncotarget* **2017**, *8*, 47454–47473. [CrossRef]
161. Kwak, D.H.; Ryu, J.S.; Kim, C.H.; Ko, K.; Ma, J.Y.; Hwang, K.A.; Choo, Y.K. Relationship between ganglioside expression and anti-cancer effects of the monoclonal antibody against epithelial cell adhesion molecule in colon cancer. *Exp. Mol. Med.* **2011**, *43*, 693–701. [CrossRef] [PubMed]
162. Sviridov, D.; Mukhamedova, N.; Miller, Y.I. Lipid rafts as a therapeutic target. *J. Lipid Res.* **2020**, *61*, 687–695. [CrossRef] [PubMed]
163. Zheng, K.; Chen, Z.; Feng, H.; Chen, Y.; Zhang, C.; Yu, J.; Luo, Y.; Zhao, L.; Jiang, X.; Shi, F. Sphingomyelin synthase 2 promotes an aggressive breast cancer phenotype by disrupting the homoeostasis of ceramide and sphingomyelin. *Cell Death Dis.* **2019**, *10*, 157. [CrossRef] [PubMed]
164. Ogretmen, B. Sphingolipid metabolism in cancer signalling and therapy. *Nat. Rev. Cancer* **2018**, *18*, 33–50. [CrossRef] [PubMed]
165. Podo, F.; Paris, L.; Cecchetti, S.; Spadaro, F.; Abalsamo, L.; Ramoni, C.; Ricci, A.; Pisanu, M.E.; Sardanelli, F.; Canese, R.; et al. Activation of Phosphatidylcholine-Specific Phospholipase C in Breast and Ovarian Cancer: Impact on MRS-Detected Choline Metabolic Profile and Perspectives for Targeted Therapy. *Front. Oncol.* **2016**, *6*, 171. [CrossRef]
166. Raghu, P.; Joseph, A.; Krishnan, H.; Singh, P.; Saha, S. Phosphoinositides: Regulators of Nervous System Function in Health and Disease. *Front. Mol. Neurosci.* **2019**, *12*, 208. [CrossRef]
167. Llorente, A.; Skotland, T.; Sylvänne, T.; Kauhanen, D.; Róg, T.; Orłowski, A.; Vattulainen, I.; Ekroos, K.; Sandvig, K. Molecular lipidomics of exosomes released by PC-3 prostate cancer cells. *Biochim. Biophys. Acta* **2013**, *1831*, 1302–1309. [CrossRef]
168. Vallabhapurapu, S.D.; Blanco, V.M.; Sulaiman, M.K.; Vallabhapurapu, S.L.; Chu, Z.; Franco, R.S.; Qi, X. Variation in human cancer cell external phosphatidylserine is regulated by flippase activity and intracellular calcium. *Oncotarget* **2015**, *6*, 34375–34388. [CrossRef]
169. Lea, J.; Sharma, R.; Yang, F.; Zhu, H.; Ward, E.S.; Schroit, A.J. Detection of phosphatidylserine-positive exosomes as a diagnostic marker for ovarian malignancies: A proof of concept study. *Oncotarget* **2017**, *8*, 14395–14407. [CrossRef]
170. Hosseini-Beheshti, E.; Pham, S.; Adomat, H.; Li, N.; Tomlinson Guns, E.S. Exosomes as biomarker enriched microvesicles: Characterization of exosomal proteins derived from a panel of prostate cell lines with distinct AR phenotypes. *Mol. Cell Proteom.* **2012**, *11*, 863–885. [CrossRef]
171. Skotland, T.; Sagini, K.; Sandvig, K.; Llorente, A. An emerging focus on lipids in extracellular vesicles. *Adv. Drug Deliv. Rev.* **2020**, *159*, 308–321. [CrossRef] [PubMed]
172. Lydic, T.A.; Townsend, S.; Adda, C.G.; Collins, C.; Mathivanan, S.; Reid, G.E. Rapid and comprehensive 'shotgun' lipidome profiling of colorectal cancer cell derived exosomes. *Methods* **2015**, *87*, 83–95. [CrossRef]
173. Nishida-Aoki, N.; Izumi, Y.; Takeda, H.; Takahashi, M.; Ochiya, T.; Bamba, T. Lipidomic Analysis of Cells and Extracellular Vesicles from High- and Low-Metastatic Triple-Negative Breast Cancer. *Metabolites* **2020**, *10*, 67. [CrossRef]
174. Haraszti, R.A.; Didiot, M.C.; Sapp, E.; Leszyk, J.; Shaffer, S.A.; Rockwell, H.E.; Gao, F.; Narain, N.R.; DiFiglia, M.; Kiebish, M.A.; et al. High-resolution proteomic and lipidomic analysis of exosomes and microvesicles from different cell sources. *J. Extracell Vesicles* **2016**, *5*, 32570. [CrossRef]
175. Cheng, L.; Zhang, K.; Qing, Y.; Li, D.; Cui, M.; Jin, P.; Xu, T. Proteomic and lipidomic analysis of exosomes derived from ovarian cancer cells and ovarian surface epithelial cells. *J. Ovarian Res.* **2020**, *13*, 9. [CrossRef] [PubMed]
176. Skotland, T.; Ekroos, K.; Kauhanen, D.; Simolin, H.; Seierstad, T.; Berge, V.; Sandvig, K.; Llorente, A. Molecular lipid species in urinary exosomes as potential prostate cancer biomarkers. *Eur. J. Cancer* **2017**, *70*, 122–132. [CrossRef]

177. Altadill, T.; Campoy, I.; Lanau, L.; Gill, K.; Rigau, M.; Gil-Moreno, A.; Reventos, J.; Byers, S.; Colas, E.; Cheema, A.K. Enabling Metabolomics Based Biomarker Discovery Studies Using Molecular Phenotyping of Exosome-Like Vesicles. *PLoS ONE* **2016**, *11*, e0151339. [CrossRef] [PubMed]
178. Hayasaka, R.; Tabata, S.; Hasebe, M.; Ikeda, S.; Ohnuma, S.; Mori, M.; Soga, T.; Tomita, M.; Hirayama, A. Metabolomic Analysis of Small Extracellular Vesicles Derived from Pancreatic Cancer Cells Cultured under Normoxia and Hypoxia. *Metabolites* **2021**, *11*, 215. [CrossRef]
179. Čuperlović-Culf, M.; Khieu, N.H.; Surendra, A.; Hewitt, M.; Charlebois, C.; Sandhu, J.K. Analysis and Simulation of Glioblastoma Cell Lines-Derived Extracellular Vesicles Metabolome. *Metabolites* **2020**, *10*, 88. [CrossRef]
180. Williams, C.; Palviainen, M.; Reichardt, N.C.; Siljander, P.R.; Falcón-Pérez, J.M. Metabolomics Applied to the Study of Extracellular Vesicles. *Metabolites* **2019**, *9*, 276. [CrossRef]
181. Raynal, P.; Montagner, A.; Dance, M.; Yart, A. Lysophospholipides et cancer: État des lieux et perspectives [Lysophospholipids and cancer: Current status and perspectives]. *Pathol. Biol.* **2005**, *53*, 57–62. [CrossRef] [PubMed]
182. Del Boccio, P.; Raimondo, F.; Pieragostino, D.; Morosi, L.; Cozzi, G.; Sacchetta, P.; Magni, F.; Pitto, M.; Urbani, A. A hyphenated microLC-Q-TOF-MS platform for exosomal lipidomics investigations: Application to RCC urinary exosomes. *Electrophoresis* **2012**, *33*, 689–696. [CrossRef]
183. Gandhi, N.; Das, G.M. Metabolic Reprogramming in Breast Cancer and Its Therapeutic Implications. *Cells* **2019**, *8*, 89. [CrossRef] [PubMed]
184. Wise, D.R.; Thompson, C.B. Glutamine addiction: A new therapeutic target in cancer. *Trends Biochem. Sci.* **2010**, *35*, 427–433. [CrossRef] [PubMed]
185. Zhang, Y.; Morar, M.; Ealick, S.E. Structural biology of the purine biosynthetic pathway. *Cell Mol. Life Sci.* **2008**, *65*, 3699–3724. [CrossRef]
186. Locasale, J.W. Serine, glycine and one-carbon units: Cancer metabolism in full circle. *Nat. Rev. Cancer* **2013**, *13*, 572–583. [CrossRef]
187. Jacob, M.; Lopata, A.L.; Dasouki, M.; Abdel Rahman, A.M. Metabolomics toward personalized medicine. *Mass Spectrom. Rev.* **2019**, *38*, 221–238. [CrossRef]
188. Royo, F.; Moreno, L.; Mleczko, J.; Palomo, L.; Gonzalez, E.; Cabrera, D.; Cogolludo, A.; Vizcaino, F.P.; van-Liempd, S.; Falcon-Perez, J.M. Hepatocyte-secreted extracellular vesicles modify blood metabolome and endothelial function by an arginase-dependent mechanism. *Sci. Rep.* **2017**, *7*, 42798. [CrossRef]
189. Clos-Garcia, M.; Loizaga-Iriarte, A.; Zuñiga-Garcia, P.; Sánchez-Mosquera, P.; Rosa Cortazar, A.; González, E.; Torrano, V.; Alonso, C.; Pérez-Cormenzana, M.; Ugalde-Olano, A.; et al. Metabolic alterations in urine extracellular vesicles are associated to prostate cancer pathogenesis and progression. *J. Extracell Vesicles* **2018**, *7*, 1470442. [CrossRef]
190. Matsumura, T.; Sugimachi, K.; Iinuma, H.; Takahashi, Y.; Kurashige, J.; Sawada, G.; Ueda, M.; Uchi, R.; Ueo, H.; Takano, Y.; et al. Exosomal microRNA in serum is a novel biomarker of recurrence in human colorectal cancer. *Br. J. Cancer* **2015**, *113*, 275–281. [CrossRef]
191. Ogata-Kawata, H.; Izumiya, M.; Kurioka, D.; Honma, Y.; Yamada, Y.; Furuta, K.; Gunji, T.; Ohta, H.; Okamoto, H.; Sonoda, H.; et al. Circulating exosomal microRNAs as biomarkers of colon cancer. *PLoS ONE* **2014**, *9*, e92921. [CrossRef] [PubMed]
192. Madhavan, B.; Yue, S.; Galli, U.; Rana, S.; Gross, W.; Müller, M.; Giese, N.A.; Kalthoff, H.; Becker, T.; Büchler, M.W.; et al. Combined evaluation of a panel of protein and miRNA serum-exosome biomarkers for pancreatic cancer diagnosis increases sensitivity and specificity. *Int. J. Cancer* **2015**, *136*, 2616–2627. [CrossRef] [PubMed]
193. Huang, X.; Yuan, T.; Liang, M.; Du, M.; Xia, S.; Dittmar, R.; Wang, D.; See, W.; Costello, B.A.; Quevedo, F.; et al. Exosomal miR-1290 and miR-375 as prognostic markers in castration-resistant prostate cancer. *Eur. Urol.* **2015**, *67*, 33–41. [CrossRef] [PubMed]
194. Li, Z.; Ma, Y.Y.; Wang, J.; Zeng, X.F.; Li, R.; Kang, W.; Hao, X.K. Exosomal microRNA-141 is upregulated in the serum of prostate cancer patients. *Onco Targets* **2015**, *9*, 139–148.
195. Malla, B.; Aebersold, D.M.; Dal, P.A. Protocol for serum exosomal miRNAs analysis in prostate cancer patients treated with radiotherapy. *J. Transl. Med.* **2018**, *16*, 223. [CrossRef]
196. Sorop, A.; Iacob, R.; Iacob, S.; Constantinescu, D.; Chitoiu, L.; Fertig, T.E.; Dinischiotu, A.; Chivu-Economescu, M.; Bacalbasa, N.; Savu, L.; et al. Plasma Small Extracellular Vesicles Derived miR-21-5p and miR-92a-3p as Potential Biomarkers for Hepatocellular Carcinoma Screening. *Front. Genet.* **2020**, *11*, 712. [CrossRef]
197. Pontis, F.; Roz, L.; Mensah, M.; Segale, M.; Moro, M.; Bertolini, G.; Petraroia, I.; Centonze, G.; Ferretti, A.M.; Suatoni, P.; et al. Circulating extracellular vesicles from individuals at high-risk of lung cancer induce pro-tumorigenic conversion of stromal cells through transfer of miR-126 and miR-320. *J. Exp. Clin. Cancer Res.* **2021**, *40*, 237. [CrossRef]
198. Smolarz, M.; Widlak, P. Serum Exosomes and Their miRNA Load-A Potential Biomarker of Lung Cancer. *Cancers* **2021**, *13*, 1373. [CrossRef]
199. Ozawa, P.M.M.; Vieira, E.; Lemos, D.S.; Souza, I.L.M.; Zanata, S.M.; Pankievicz, V.C.; Tuleski, T.R.; Souza, E.M.; Wowk, P.F.; Urban, C.A.; et al. Identification of miRNAs Enriched in Extracellular Vesicles Derived from Serum Samples of Breast Cancer Patients. *Biomolecules* **2020**, *10*, 150. [CrossRef]
200. Le, M.T.; Hamar, P.; Guo, C.; Basar, E.; Perdigão-Henriques, R.; Balaj, L.; Lieberman, J. miR-200-containing extracellular vesicles promote breast cancer cell metastasis. *J. Clin. Invest.* **2014**, *124*, 5109–5128. [CrossRef]

201. Di Modica, M.; Regondi, V.; Sandri, M.; Iorio, M.V.; Zanetti, A.; Tagliabue, E.; Casalini, P.; Triulzi, T. Breast cancer-secreted miR-939 downregulates VE-cadherin and destroys the barrier function of endothelial monolayers. *Cancer Lett.* **2017**, *384*, 94–100. [CrossRef] [PubMed]
202. Sueta, A.; Yamamoto, Y.; Tomiguchi, M.; Takeshita, T.; Yamamoto-Ibusuki, M.; Iwase, H. Differential expression of exosomal miRNAs between breast cancer patients with and without recurrence. *Oncotarget* **2017**, *8*, 69934–69944. [CrossRef]
203. Chen, X.; Zhou, J.; Li, X.; Wang, X.; Lin, Y.; Wang, X. Exosomes derived from hypoxic epithelial ovarian cancer cells deliver microRNAs to macrophages and elicit a tumor-promoted phenotype. *Cancer Lett.* **2018**, *435*, 80–91. [CrossRef]
204. Zabegina, L.; Nazarova, I.; Knyazeva, M.; Nikiforova, N.; Slyusarenko, M.; Titov, S.; Vasilyev, D.; Sleptsov, I.; Malek, A. MiRNA let-7 from TPO(+) Extracellular Vesicles is a Potential Marker for a Differential Diagnosis of Follicular Thyroid Nodules. *Cells* **2020**, *9*, 1917. [CrossRef] [PubMed]
205. Chung, K.Y.; Quek, J.M.; Neo, S.H.; Too, H.P. Polymer-Based Precipitation of Extracellular Vesicular miRNAs from Serum Improve Gastric Cancer miRNA Biomarker Performance. *J. Mol. Diagn.* **2020**, *22*, 610–618. [CrossRef]
206. Lucero, R.; Zappulli, V.; Sammarco, A.; Murillo, O.D.; Cheah, P.S.; Srinivasan, S.; Tai, E.; Ting, D.T.; Wei, Z.; Roth, M.E.; et al. Glioma-Derived miRNA-Containing Extracellular Vesicles Induce Angiogenesis by Reprogramming Brain Endothelial Cells. *Cell Rep.* **2020**, *30*, 2065–2074.e4. [CrossRef]
207. Qian, M.; Wang, S.; Guo, X.; Wang, J.; Zhang, Z.; Qiu, W.; Gao, X.; Chen, Z.; Xu, J.; Zhao, R.; et al. Hypoxic glioma-derived exosomes deliver microRNA-1246 to induce M2 macrophage polarization by targeting TERF2IP via the STAT3 and NF-κB pathways. *Oncogene* **2020**, *39*, 428–442. [CrossRef] [PubMed]
208. Lan, F.; Qing, Q.; Pan, Q.; Hu, M.; Yu, H.; Yue, X. Serum exosomal miR-301a as a potential diagnostic and prognostic biomarker for human glioma. *Cell Oncol.* **2018**, *41*, 25–33. [CrossRef]
209. Sun, X.; Ma, X.; Wang, J.; Zhao, Y.; Wang, Y.; Bihl, J.C.; Chen, Y.; Jiang, C. Glioma stem cells-derived exosomes promote the angiogenic ability of endothelial cells through miR-21/VEGF signal. *Oncotarget* **2017**, *8*, 36137–36148. [CrossRef]
210. Gholizadeh, S.; Shehata Draz, M.; Zarghooni, M.; Sanati-Nezhad, A.; Ghavami, S.; Shafiee, H.; Akbari, M. Microfluidic approaches for isolation, detection, and characterization of extracellular vesicles: Current status and future directions. *Biosens. Bioelectron.* **2017**, *91*, 588–605. [CrossRef]
211. Contreras-Naranjo, J.C.; Wu, H.J.; Ugaz, V.M. Microfluidics for exosome isolation and analysis: Enabling liquid biopsy for personalized medicine. *Lab. Chip* **2017**, *17*, 3558–3577. [CrossRef] [PubMed]
212. Vaidyanathan, R.; Naghibosadat, M.; Rauf, S.; Korbie, D.; Carrascosa, L.G.; Shiddiky, M.J.; Trau, M. Detecting exosomes specifically: A multiplexed device based on alternating current electrohydrodynamic induced nanoshearing. *Anal. Chem.* **2014**, *86*, 11125–11132. [CrossRef] [PubMed]

Review

Extracellular Vesicles in Airway Homeostasis and Pathophysiology

Alberto Fucarino [1], Alessandro Pitruzzella [1,2,*], Stefano Burgio [1], Maria Concetta Zarcone [3], Domenico Michele Modica [4], Francesco Cappello [1,5] and Fabio Bucchieri [1,6]

1 Department of Biomedicine, Neuroscience and Advanced Diagnostics-University of Palermo, 90127 Palermo, Italy; dipartimento.bind@unipa.it (A.F.); stefano.burgio@unipa.it (S.B.); francesco.cappello@unipa.it (F.C.); fabio.bucchieri@unipa.it (F.B.)
2 Consorzio Universitario Caltanissetta, University of Palermo, 93100 Caltanissetta, Italy
3 NHLI, Imperial College London, London SW7 2AZ, UK; m.zarcone@imperial.ac.uk
4 Otorhinolaryngology Unit, Villa Sofia-Cervello Hospital, 90146 Palermo, Italy; domenicomichelemodica@gmail.com
5 Euro-Mediterranean Institute of Science and Technology (IEMEST), 90136 Palermo, Italy
6 Institute for Biomedical Research and Innovation (IRIB), National Research Council, 90146 Palermo, Italy
* Correspondence: Alessandro.pitruzzella@unipa.it

Abstract: The epithelial–mesenchymal trophic unit (EMTU) is a morphofunctional entity involved in the maintenance of the homeostasis of airways as well as in the pathogenesis of several diseases, including asthma and chronic obstructive pulmonary disease (COPD). The "muco-microbiotic layer" (MML) is the innermost layer of airways made by microbiota elements (bacteria, viruses, archaea and fungi) and the surrounding mucous matrix. The MML homeostasis is also crucial for maintaining the healthy status of organs and its alteration is at the basis of airway disorders. Nanovesicles produced by EMTU and MML elements are probably the most important tool of communication among the different cell types, including inflammatory ones. How nanovesicles produced by EMTU and MML may affect the airway integrity, leading to the onset of asthma and COPD, as well as their putative use in therapy will be discussed here.

Keywords: asthma; chronic obstructive pulmonary disease; COPD; epithelial–mesenchymal trophic unit; muco-microbiotic layer; nanovesicles; exosomes; outer membrane vesicles; microbiota

Citation: Fucarino, A.; Pitruzzella, A.; Burgio, S.; Zarcone, M.C.; Modica, D.M.; Cappello, F.; Bucchieri, F. Extracellular Vesicles in Airway Homeostasis and Pathophysiology. *Appl. Sci.* **2021**, *11*, 9933. https://doi.org/10.3390/app11219933

Academic Editor: Giuseppina Andreotti

Received: 29 September 2021
Accepted: 21 October 2021
Published: 24 October 2021

1. The Epithelial–Mesenchymal Trophic Unit and the Muco-Microbiotic Layer: Definition, Composition and Functions

Along all the lower airways, except for specific portions, the innermost and proximal layer to the lumen is composed of respiratory mucosa. From a strictly anatomical point of view, in the respiratory mucosa, the outermost layer is made up of a pseudostratified epithelium with mainly goblet and ciliated cells that lie on a basal membrane, below which there is a connective tissue layer with various cells (including fibroblasts, myofibroblasts and immune cells) of mesenchymal origin interspersed in an extracellular matrix (ECM). However, from a morphofunctional point of view, the apical epithelial tissue and the underlying connective tissue within the respiratory mucosa cannot be considered as single, separate entities of their own. Indeed, at the end of the last century, the work of Plopper and Evans focused on the close interconnection between epithelial and mesenchymal cells, providing the basis for the creation of the "epithelial–mesenchymal trophic unit" (EMTU) concept [1]. The role of the interactions between epithelial and mesenchymal elements, although known for some decades at the time of the studies of Evans and Plopper, had been exclusively relegated to airway organogenesis, erroneously assuming that this interaction was limited to intrauterine life [2]. On the contrary, the several cellular and non-cellular components of EMTU in the adult life of individuals are interconnected by a close communicative relationship that influences many other physiological and pathophysiological aspects, such as cell differentiation, tissue homeostasis, organ remodeling, reparative/regenerative processes, response to external/internal stress stimuli and participation in inflammation/autoimmunity. EMTU processes alterations even contribute to the pathogenesis of some chronic diseases of the airways, e.g., chronic obstructive pulmonary disease (COPD) and asthma [3–5].

Focusing exclusively on the structural components of EMTU, it must be emphasized that, along with its cellular components, the ECM, synthesized mainly by fibroblasts, also plays a very important role [6]. EMTU homeostasis is frequently influenced by the differences in the lymphocyte population as well as by several soluble factors and nanovesicles dispersed in the ECM that determine the outcome of reparative/regenerative processes and the establishment of pathophysiological states [7,8]. Indeed, it has already been demonstrated that variations in the composition of ECM are the basis of asthma and COPD pathogenesis [9–11].

Furthermore, many microbes reside in the mucus constantly produced by goblet cells, constituting the airways' microbiota. Nowadays, we know that this is another fundamental component for EMTU homeostasis. In the gastrointestinal tract, we already proposed the term "muco-microbiotic layer" (MML)—made by microbiota elements (bacteria, viruses, archaea and fungi) interspersed in a mucous matrix—to describe the innermost layer of the intestinal wall. The MML homeostasis is crucial for maintaining the healthy status of these organs and whose alteration is at the basis of gastrointestinal disorders [12]. An MML is also present in airways and, as in the gastrointestinal tract, it takes part in the homeostasis as well as in the pathogenesis of these organs. It is fundamental now to better characterize the constitutive elements of this MML in terms not only of microorganisms that populate it but also of nanovesicles (e.g., exosomes, microvesicles or outer membrane vesicles) that participate in the crosstalk among cells.

We want to stress here that a crucial element for the interchange of information between EMTU and MML constituents is the trafficking mediated by nanovesicles produced by both parts and called, respectively, exosomes and outer membrane vesicles (OMV). The aim of this paper is to properly highlight the roles that these nanovesicles have in airway homeostasis and disease pathophysiology, i.e., asthma and COPD.

2. Nanovesicles: Exosomes and Outer Membrane Vesicles

The paracrine communication system regulated by extracellular vesicles (EVs) and exosomes plays a key role in the communication of the EMTU, in the maintenance of tissue homeostasis of the airways as well as in many pathogenetic processes affecting the apparatus [13]. The intimate interconnection that interfaces the epithelial layer to the connective layer, synching them in their biological activities, is favored by the vesicular system that traffics nucleic acids (miRNA, siRNA) [14], growth factors, tissue-specific receptors as well as proteases [15].

As with Evans and Plopper's EMTU studies, extracellular vesicles and exosomes have long been erroneously considered as results of "garbage disposal" whose sole purpose was to eliminate waste substances from cells [16,17]. They are actively secreted by all eukaryotic and prokaryotic cells and are part of an articulated cell-to-cell communication system, both in physiological and pathological conditions [18]. In recent decades, several scientific works demonstrated the roles of EVs in a plethora of physiological processes. EVs are involved in cellular homeostasis and signaling. They can act as carriers of an enormous variety of molecules, and they express numerous signal proteins on their surface [19].

EVs are classified according to their size as (1) microvesicles (100–1000 nm in diameter); (2) apoptotic blebs (1000–5000 nm in diameter); and (3) exosomes (diameter 20–150 nm) [20].

On one side, microvesicles and apoptotic blebs originate from the outward budding of the cell membrane, unlike exosomes, which result from the invagination of endosomal membranes [21]. The understanding of the different originative pathways and the sorting mechanisms spotlighting transport, cargo packing and vesicle exocytosis find practical utility in the isolation studies of EVs for diagnostic and therapeutic purposes.

Therefore, the role of the proteins belonging to the endosomal sorting complex required for transport (ESCRT -0, -I, -II and -III), responsible for the control of the biogenetic and cargo loading processes of the EVs, is crucial [22].

As previously mentioned, the load of the EVs is tissue-specific and related to the function they can perform, i.e., EVs produced by tumor cells have a decisive impact on paracrine signaling mechanisms in support of tumor growth [23–25]; however, under physiological conditions, other EVs perform diametrically opposite roles, such as protection from traumatic tissue events or promoting the tissue healing itself [26].

A determining example might be the EVs produced in the lung microenvironment [26]. In homeostasis conditions, a broad range of cell types, such as fibroblasts, epithelial cells and endothelial cells, also actively secrete EVs: those EVs have been largely characterized, showing that epithelial cells are the main characters on the production of EVs, enriched with secretory and membrane-anchored mucins, which contribute to the mucociliary defense and boosting of innate immune defenses [27].

Not only epithelial cells but also macrophages that are present in BAL fluid play a pivotal role in the inflammatory modulation. It has been demonstrated that alveolar macrophages secrete SOCS-1 and -3 within nanoparticles, which are uptaken by lung epithelial cells. Both SOCS-1 and -3 are negative modulators of cytokine 1 and 3 biogenesis (through the STAT pathway inhibition) [28]. In normal conditions, this can modulate the inflammatory response, but at the same time, this negative modulation of IL-1 and -3 biogenesis seems to be lost in cigarette-smoking subjects, presenting a new model for the control of inflammatory response during inflammation or stress tissue [28].

Thus, with the importance of pulmonary EVs in maintaining homeostasis being confirmed, it is not difficult to think how dysregulations in this sense are closely related to the pathogenesis of various lung diseases [29]. In the following paragraphs, the relationships between microvesicles and chronic diseases of the respiratory system will be analyzed.

3. Asthma and Nanovesicles in Asthma Pathogenesis

Asthma is a chronic respiratory disease that presents several phenotypes. The global cases of asthma are estimated to be over 300 million by the World Health Organization (WHO) [30]. Asthma is commonly considered a childhood-onset disease; however, adults also can develop asthma later in life. Asthma is caused by both environmental factors such as house dust mites (HDM), particulate matter (PM), cigarette smoke (CS) [31] and genetic factors, among which allele 17q21 is the most studied [32]. Overall, the development of the disease is defined by increased mucus production, thickening of the subepithelial reticular basement membrane (RBM) of the lung mucosa, airway hyperresponsiveness and chronic inflammation [Figure 1]. All these events determine airway remodeling, which leads to the narrowing of the airways [31].

Recently, due to their high therapeutic potential, the emerging role of extracellular vesicles has been investigated in association with asthma pathogenesis [33,34]. Within the respiratory system, several cell types are involved in the release of EVs: for instance, structural cells such as epithelial cells and fibroblasts [35,36], resident immune cells such as dendritic cells (DCs) and alveolar macrophages (AMs) [28,37] as well as recruited immune cells such as eosinophils [38]. The major producer of EVs within the lung are airway epithelial cells [35]. On their membrane, epithelial cell-derived EVs expose several mucins that can neutralize virus and bacteria [27]. Furthermore, it has been shown that EVs produced by epithelial cells can trigger the proliferation of macrophages upon IL-13 release by eosinophils, thereby promoting chronic inflammation [35].

A study by Bartel and colleagues [39] revealed the role in asthma pathogenesis of miRNAs present in epithelial-derived EVs. miR-34a, miR-92b, and miR-210 in EVs can potentially lead to Th2 responses and maturation of DCs in the early development of asthma. Aberrant deposition of extracellular matrix contributes to the RMB thickening in asthma pathogenesis. On their surface, EVs secreted by fibroblasts expose fibronectin, which can trigger invasion-associated signaling pathways [40]. An in vitro study showed that exosomes produced by the fibroblasts of severe asthmatics contain a low level of TGF2, which inhibits epithelial cells proliferation, contributing to the narrowing of the airways [37]. In order to trigger an allergic response, allergens must be presented to T cells by DCs. In humans, it has been shown how DC-derived exosomes can directly present antigen to T cells [41].

In mice, exosomes, rather than microvesicles (MVs), overexpress allergens leading to the activation of allergen-specific T cells [37]. These findings highlight an important modulatory function of DC-derived exosomes in allergic responses. As epithelial cells, macrophages are a major source of EVs that display multiple functions. Macrophage-derived EVs can promote macrophage differentiation [42] via mRNA-223. During infections, both bacteria and macrophages release MVs, which have a strong proinflammatory effect. At the same time, these MVs can also induce tolerance and promote bacterial shedding [43]. EVs produced by macrophages can also activate toll-like receptors (TLRs) during infection as they contain heat-shock protein 70 (HSP70), which mediates the activation of nuclear factor kB (NFkB) [26].

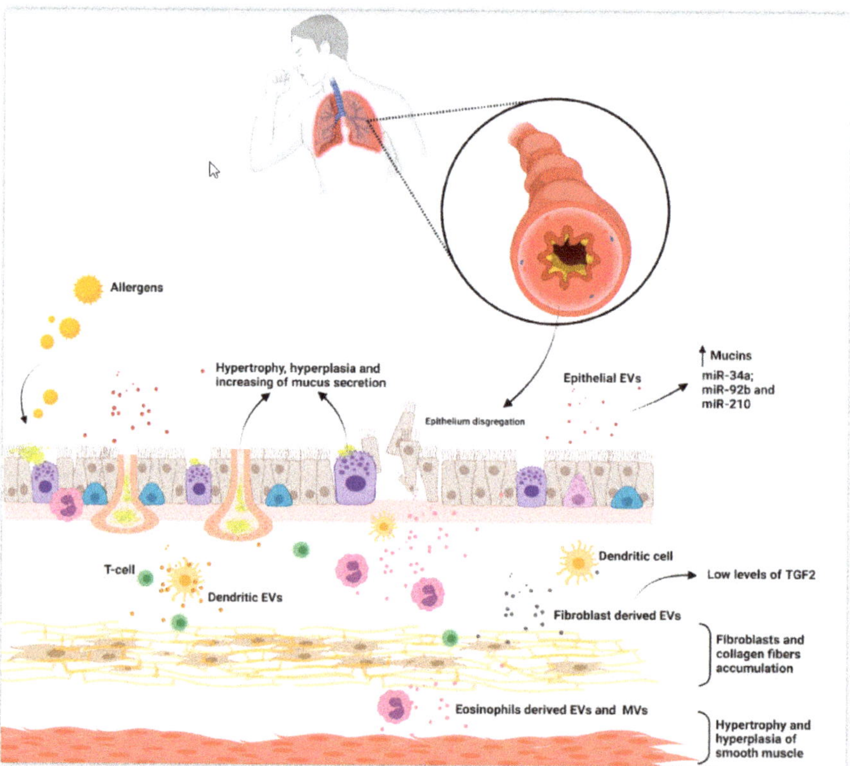

Figure 1. Scheme of allergic asthma pathogenesis. Allergic asthma is triggered by inhaling allergens. (Allergens could have different origins such as dust mites, pet dander, pollen or mold.) During the different phases and chronicization of the pathology, the tissues undergo several modifications. Among these include hyperplasia, hypertrophy of mucous cells and an increase of their secretions, continuity loss of the epithelium and variation in EV content. The extracellular vesicular release appears incremented, modified and responsible for the alterations that occurred.

Eosinophils play a major role in allergic asthma exacerbations [31]. Besides the release of potent Th2 cytokines, eosinophils release both MVs and exosomes [44]. Multiple effects of eosinophil-derived exosomes have been observed on epithelial cells and smooth muscle cells (SMCs) [45]. In asthmatic patients, these exosomes interfere with epithelial cells, wound healing and SMC proliferation. This novel research on extracellular vesicles, microvesicles and exosomes indicate that future therapies must target these components in the prevention of asthma pathogenesis.

4. Chronic Obstructive Pulmonary Disease and Nanovesicles in Its Pathogenesis

Patients with chronic obstructive pulmonary disease (COPD) face a progressive limitation in airway function. The pathology has remarkable facets and different levels of severity. Several forms of COPD have been studied and described, with a classification that reports the distinct faceting of the pathology [46,47]. The establishment of this chronic disease is due to multiple causes, although it is closely associated with the inhalation of tobacco smoke and other environmental contaminants [48,49] [Figure 2]. Several mechanisms behind COPD pathogenesis have been studied, but we are still a long way from solving the puzzle in its entirety. The number of treatments on a personal basis developed in recent years is constantly increasing, trying to counteract the effects of COPD [50–52]. Despite the progress in therapies, the settled pathology is usually connected with a condition of irreversibility. Therefore, it is even more strategic to understand any biological pathway that participates in the disease onset and its maintenance.

The onset of COPD was recently found to be closely associated with the biological airway senescence process [53]. An increase in the release of exosomes was found precisely during senescence. A variation in the molecules contained within these extracellular vesicles was observed [54]. The habits of a patient suffering from COPD can also greatly influence pathology development by directly affecting the different EVs. For example, cigarette smoke leads to a massive release of exosomes by mononucleated cells and, as a consequence, IL-8 production by the respiratory epithelium increases. The result is the construction of a microenvironment ideal for a widespread inflammatory state [55]. In particular, Fujita's group reported an increased expression of miR-210 within exosomes released by bronchial epithelial cells in smokers [56]. The direct consequence of this miR-210 overexpression is a variation in the number of myofibroblasts within the airways. This variation is due to the suppression of the ATG7 pathway biologically implicated in their autophagy phenomena. The efficiency of exosomes in transmitting long-distance messages is, in this case, a double-edged sword. CS acts on epithelial cells by upregulating CCN1 expression in exosomes. These exosomes are now able to spread inflammatory states to other distant portions of the airways as well [57]. However, CCN1, as a result of chronic exposure to cigarette smoke, is released directly into the bronchial fluids (in a truncated isoform). Extracellular matrix degradation and increased cell death are a direct consequence of this abnormal release of CCN1 outside exosomes [58]. The variation in COPD status from stable to exacerbated is also associated with a variation in exosome release. The variation in COPD status is also associated with a variation in exosome release [59]. Exosomes with CD31, CD62E and CD144 were notably reduced in stable patients than in patients with COPD exacerbation [60]. A further fascinating correlation concerns the forced expiratory volume in 1 s (FEV) and the number of exosomes present in sputum [61]. In recent years, the number of miRNAs with an assigned biomarker role for COPD is constantly increasing. miR-203, miR-4455, miR4785, miR-218-5p, miR-29c and miR-126 were analyzed by different research groups, and for each of them, a variation in patients with COPD was found. This discovery makes the miRNAs potential biomarker candidates for a more precise diagnosis and progression of COPD [62–65].

The interaction between the immune system and pathogens plays a key role in COPD as in many other complex multi-factor pathologies. Recently, the focus has also been on the EVs released by pathogenic bacteria and not only on bacteria per se. The set of extracellular vesicles released by pathogens during infection is, today, one of the mechanisms used to reveal different parameters (type of pathogen, state of infection, etc.) [66,67]. In COPD subjects, analysis of EVs showed a different lung microbiome distinct from "standard" microbiomes usually present in healthy lung tissue [67]. Outer membrane vesicles (OMVs) released by Gram-negative bacteria, for example, contain several molecules such as LPS, invasion proteins, adhesion proteins, and immunomodulatory factors [68]. Augustyniak et al. showed the potent proinflammatory effects of *Moraxella catarrhalis* OMVs in COPD. Briefly, they demonstrated how OMVs promote an inflammatory state by activating neutrophil degranulation and modifying the activation of the hBD-2 promoter in epithelial cells. An initial in vitro treatment was successfully carried out to counteract the interactions of OMVs with neutrophils and epithelial cells. This treatment is a clear example of how increasing knowledge in this specific area could, in the future, bring significant benefits to those with COPD [69]. An additional level of complexity is given by respiratory viruses that have evolved to use EVs as means of transport to spread inside the organism [70,71]. Even viruses, like bacteria, are a source of COPD exacerbation and it is crucial to monitor any viral infections to limit the course of the disease. Recent work from Roffel and her team demonstrated the role of miR-223 on the regulation of several gene expressions, providing a further example of the therapeutic potential associated with the study of different EVs [72].

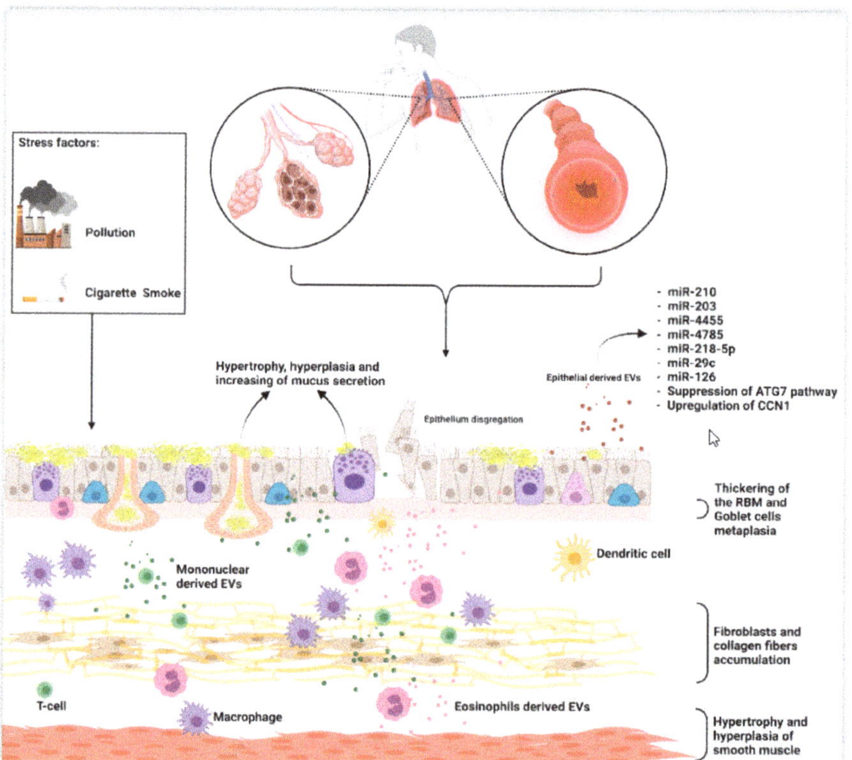

Figure 2. Scheme of COPD pathogenesis. Chronic external stimuli generate massive deregulation of lower airways. Chronic external stimuli generate massive deregulation or lower airways. EVs mediate these tissue changes directly and indirectly. Their production is also increased by the perennially present phlogistic status established with the progress of the pathology. The immune cell component contributes greatly to the release of EVs that affect all cell populations of the respiratory mucosa.

5. Microbiome Extracellular Vesicles and Chronic Respiratory Diseases

Human cells are not the only ones to produce microvesicles that affect the status of the respiratory system. Above the respiratory mucosa is a mucous layer in which all the microorganisms that make up the airway microbiota are settled [73]. The main communication pathway between the cells of our body and the microbiota is via EV [18]. Communication via EV guarantees an interaction between host and microbiome without direct contact and in a bidirectional manner. This interaction is physiologically present in a state of health, and its alteration can trigger pathogenetic processes [74]. Commensal bacterial species that constitute our microbiome are not the only EV users, but also the pathogenic infectious species could exploit this communication pathway [75]. Among the most significant aspects affected by EVs is that of immunomodulation [76,77]. One of the most studied interactions is between *Pseudomonas aeruginosa* and airway epithelial cells. *P. aeruginosa* releases outer membrane vesicles (OMVs) in the mucus layer; OMVs fuse with cellular membranes on the epithelial cells apical side and deliver a 23-nucleotide tRNA (sRNA-52320). sRNA-2320 reduces IL-8 secretion and the migration of neutrophils into the lungs and suppresses the immune response to bacterial infection by targeting several genes in the LPS-stimulated MAPK signaling pathway [76,78,79]. In recent years, specific interactions between microbiome EVs and chronic diseases have been shown. In patients with asthma, *Sphingomonas, Akkermansia, Methylophaga, Acidocella,* and *Marinobacter* were significantly more abundant. It is interesting to note how this notion was obtained indirectly by analyzing the EVs released by these bacterial species. Thus, microvesicles can be used for diagnostic purposes (in integration with other specific examinations) [80]. Urine-released microbial extracellular vesicles can be potential and novel biomarkers for chronic respiratory diseases. The analysis of EVs is, in fact, also possible through an investigation of the urine of patients [81]. By this analytic method, the different microbiome composition of asthmatic patients has been detected. The diversity also appears to correlate with IgE levels and eosinophil % [82]. These findings suggest that they may play important roles in allergic-based airway diseases. The analysis and monitoring of COPD also benefit EVs studies. Altered miRNA profiles in COPD have been discovered analyzing EVs. Sundar et al. used different EV isolation and purification methods to characterize the plasma-derived EV miRNAs from nonsmokers, smokers, and patients with COPD [83]. They analyzed plasma-derived EVs from smokers, nonsmokers and patients with COPD, discovering how EVs vary in their dimensions, distribution, concentration and phenotypic characteristics. They concluded that plasma-derived EV miRNAs are novel circulating pulmonary disease biomarkers. miR-21 to miR-181a levels have been monitored by Xie et al. In particular, heavy smokers without diagnosed COPD were taken into consideration. The levels of the two miRNAs have a dichotomous pattern: the levels of miR-21 were significantly higher in the COPD patients and asymptomatic heavy smokers than in the healthy controls (HC), while miR-18a levels were significantly lower in the COPD patients and asymptomatic heavy smokers than in the HC [84]. The ratio of these miRNA levels could be used as a potential biomarker of early COPD pathogenesis.

6. Conclusions

Asthma and COPD are, in themselves, very complex and multi-factorial diseases. These pathologies are characterized by airway inflammation, airflow reduction, and airway remodeling. For years, the scientific community has been looking for a solution to this complex puzzle without success. Probably, crucial pieces were missing before they could get the entire picture: the EVs. This carrier is used to transport different molecules and cellular material, not only by our tissues but even by any pathogens eventually present. This review tried to highlight EVs' importance as a biomarker and a potential therapeutic target in two complex chronic airways pathologies: asthma and COPD. Therefore, an accurate understanding of EVs' roles in these pathologies could lead to a more precise diagnosis and more effective treatments for patients. The contribution made by the microbiome should not be underestimated. The bacterial populations usually present, the opportunistic pathogens, and the possible infections are all able to condition the microenvironment of the airways through the EVs. The possibility of having additional biomarkers available could be essential to make an early diagnosis or analyze the state of progression of the pathologies.

Finally, in the future, the analysis of EV pathways may provide new instruments to contrast the development and progression of chronic respiratory diseases.

Author Contributions: Conceptualization, A.F. and F.C.; writing—original draft preparation, A.F., M.C.Z., F.C., S.B., A.P., D.M.M., F.B.; writing—review and editing, A.F. All authors have read and agreed to the published version of the manuscript.

Funding: This research received no external funding.

Institutional Review Board Statement: Not Applicable.

Informed Consent Statement: Not Applicable.

Conflicts of Interest: The authors declare no conflict of interest.

References

1. Evans, M.J.; Van Winkle, L.S.; Fanucchi, M.V.; Plopper, C.G. The attenuated fibroblast sheath of the respiratory tract epithelial-mesenchymal trophic unit. *Am. J. Respir. Cell Mol. Biol.* **1999**, *21*, 655–657. [CrossRef]
2. Knight, D. Does aberrant activation of the epithelial-mesenchymal trophic unit play a key role in asthma or is it an unimportant sideshow? *Curr. Opin. Pharmacol.* **2004**, *4*, 251–256. [CrossRef] [PubMed]
3. Davies, D.E. The role of the epithelium in airway remodeling in asthma. *Proc. Am. Thorac. Soc.* **2009**, *6*, 678–682. [CrossRef]
4. Bucchieri, F.; Pitruzzella, A.; Fucarino, A.; Marino Gammazza, A.; Caruso Bavisotto, C.; Marcianò, V.; Cajozzo, M.; Lo Iacono, G.; Marchese, R.; Zummo, G.; et al. Functional characterization of a novel 3D model of the epithelial-mesenchymal trophic unit. *Exp. Lung Res.* **2017**, *43*, 82–92. [CrossRef] [PubMed]
5. Pitruzzella, A.; Modica, D.M.; Burgio, S.; Gallina, S.; Manna, O.M.; Intili, G.; Bongiorno, A.; Saguto, D.; Marchese, R.; Nigro, C.L.; et al. The role of emtu in mucosae remodeling: Focus on a new model to study chronic inflammatory lung. *Dis. EuroMediterr. Biomed. J.* **2020**, *15*, 4–10. [CrossRef]
6. Hamilton, N.; Bullock, A.J.; Macneil, S.; Janes, S.M.; Birchall, M. Tissue engineering airway mucosa: A systematic review. *Laryngoscope* **2014**, *124*, 961–968. [CrossRef]
7. Reeves, S.R.; Kolstad, T.; Lien, T.Y.; Elliot, M.; Ziegler, S.F.; Wight, T.N.; Debley, J.S. Asthmatic airway epithelial cells differentially regulate fibroblast expression of extracellular matrix components. *J. Allergy Clin. Immunol.* **2014**, *134*, 663–670. [CrossRef] [PubMed]
8. Fanta, C.H. Asthma. *N. Engl. J. Med.* **2009**, *361*, 1123. [CrossRef] [PubMed]
9. Moheimani, F.; Hsu, A.C.; Reid, A.T. The genetic and epigenetic landscapes of the epithelium in asthma. *Respir. Res.* **2016**, *17*, 119. [CrossRef] [PubMed]
10. Brandsma, C.A.; Van den Berge, M.; Hackett, T.L.; Brusselle, G.; Timens, W. Recent advances in chronic obstructive pulmonary disease pathogenesis: From disease mechanisms to precision medicine. *J. Pathol.* **2020**, *250*, 624–635. [CrossRef] [PubMed]
11. Kulkarni, T.; O'Reilly, P.; Antony, V.B.; Gaggar, A.; Thannickal, V.J. Matrix Remodeling in Pulmonary Fibrosis and Emphysema. *Am. J. Respir. Cell Mol. Biol.* **2016**, *54*, 751–760. [CrossRef] [PubMed]
12. Cappello, F.; Mazzola, M.; Jurjus, A.; Zeenny, M.; Jurjus, R.; Carini, F.; Leone, A.; Bonaventura, G.; Tomasello, G.; Bucchieri, F.; et al. Hsp60 as a Novel Target in IBD Management: A Prospect. *Front. Pharmacol.* **2019**, *10*, 26. [CrossRef] [PubMed]
13. Gupta, R.; Radicioni, G.; Abdelwahab, S.; Dang, H.; Carpenter, J.; Chua, M.; Mieczkowski, P.A.; Sheridan, J.T.; Randell, S.H.; Kesimer, M. Intercellular Communication between Airway Epithelial Cells Is Mediated by Exosome-Like Vesicles. *Am. J. Respir. Cell Mol. Biol.* **2019**, *60*, 209–220. [CrossRef]
14. Alipoor, S.D.; Mortaz, E.; Garssen, J.; Movassaghi, M.; Mirsaeidi, M.; Adcock, I.M. Exosomes and Exosomal miRNA in Respiratory Diseases. *Mediat. Inflamm.* **2016**, *2016*, 5628404. [CrossRef]
15. Asef, A.; Mortaz, E.; Jamaati, H.; Velayati, A. Immunologic Role of Extracellular Vesicles and Exosomes in the Pathogenesis of Cystic Fibrosis. *Tanaffos* **2018**, *17*, 66–72. [PubMed]
16. Chargaff, E.; West, R. The biological significance of the thromboplastic protein of blood. *J. Biol. Chem.* **1946**, *166*, 189–197. [CrossRef]
17. Wolf, P. The nature and significance of platelet products in human plasma. *Br. J. Haematol.* **1967**, *13*, 269–288. [CrossRef]
18. Yanez-Mò, M.; Siljander, P.R.; Andreu, Z.; Zavec, A.B.; Borras, F.E.; Buzas, E.I.; Buzas, K.; Casal, E.; Cappello, F.; Carvalho, J.; et al. Biological Properties of extracellular vesicles and their physiology functions. *J. Extracell. Vesicles* **2015**, *4*, 27066. [CrossRef]
19. Campanella, C.; Bavisotto, C.C.; Gammazza, A.M.; Nikolic, D.; Rappa, F.; David, S.; Cappello, F.; Bucchieri, F.; Fais, S. Exosomal Heat Shock Proteins as New Players in Tumour Cell-to-Cell Communication. *JCB* **2014**, *3*, 4. [CrossRef]
20. Doyle, L.M.; Wang, M.Z. Overview of Extracellular Vesicles, Their Origin, Composition, Purpose, and Methods for Exosome Isolation and Analysis. *Cells* **2019**, *8*, 727. [CrossRef]
21. Burgio, S.; Noori, L.; Marino Gammazza, A.; Campanella, C.; Logozzi, M.; Fais, S.; Bucchieri, F.; Cappello, F.; Caruso Bavisotto, C. Extracellular Vesicles-Based Drug Delivery Systems: A New Challenge and the Exemplum of Malignant Pleural Mesothelioma. *Int. J. Mol. Sci.* **2020**, *21*, 5432. [CrossRef] [PubMed]
22. Kowal, J.; Tkach, M.; Théry, C. Biogenesis and secretion of exosomes. *Curr. Opin. Cell Biol.* **2014**, *29*, 116–125. [CrossRef] [PubMed]

23. Song, Y.H.; Warncke, C.; Choi, S.J.; Choi, S.; Chiou, A.E.; Ling, L.; Liu, H.-Y.; Daniel, S.; Antonyak, M.A.; Cerione, R.A.; et al. Breast cancer-derived extracellular vesicles stimulate myofibroblast differentiation and pro-angiogenic behavior of adipose stem cells. *Matrix Biol.* **2017**, *60–61*, 190–205. [CrossRef] [PubMed]
24. Goh, C.Y.; Wyse, C.; Ho, M.; O'Beirne, E.; Howard, J.; Lindsay, S.; Kelly, P.; Higgins, M.; McCann, A. Exosomes in triple negative breast cancer: Garbage disposals or Trojan horses? *Cancer Lett.* **2020**, *473*, 90–97. [CrossRef] [PubMed]
25. Schillaci, O.; Fontana, S.; Monteleone, F.; Taverna, S.; Di Bella, M.A.; Di Vizio, D.; Alessandro, R. Exosomes from metastatic cancer cells transfer amoeboid phenotype to non-metastatic cells and increase endothelial permeability: Their emerging role in tumor heterogeneity. *Sci. Rep.* **2017**, *7*, 4711. [CrossRef]
26. Fujita, Y.; Kosaka, N.; Araya, J.; Kuwano, K.; Ochiya, T. Extracellular vesicles in lung microenvironment and pathogenesis. *Trends Mol. Med.* **2015**, *21*, 533–542. [PubMed]
27. Kesimer, M.; Gupta, R. Physical characterization and profiling of airway epithelial derived exosomes using light scattering. *Methods* **2015**, *87*, 59–63. [CrossRef] [PubMed]
28. Bourdonnay, E.; Zasłona, Z.; Penke, L.R.K.; Speth, J.M.; Schneider, D.J.; Przybranowski, S.; Swanson, J.A.; Mancuso, P.; Freeman, C.M.; Curtis, J.L.; et al. Transcellular delivery of vesicular SOCS proteins from macrophages to epithelial cells blunts inflammatory signaling. *J. Exp. Med.* **2015**, *212*, 729–742. [CrossRef]
29. Fujita, Y.; Yoshioka, Y.; Ito, S.; Araya, J.; Kuwano, K.; Ochiya, T. Intercellular Communication by Extracellular Vesicles and Their MicroRNAs in Asthma. *Clin. Ther.* **2014**, *36*, 873–881. [CrossRef]
30. Vos, T.; Abajobir, A.A.; Abate, K.H.; Abbafati, C.; Abbas, K.M.; Abd-Allah, F.; Abdulkader, R.S.; Abdulle, A.M.; Abebo, T.A.; Abera, S.F.; et al. Global, regional, and national incidence, prevalence, and years lived with disability for 328 diseases and injuries for 195 countries, 1990–2016: A systematic analysis for the Global Burden of Disease Study 2016. *Lancet* **2017**, *390*, 1211–1259.
31. Holgate, S.; Wenzel, S.; Postma, D.; Weiss, S.T.; Renz, H.; Sly, P.D. Asthma. *Nat. Rev. Dis. Prim.* **2015**, *1*, 15025. [CrossRef]
32. Das, S.; Miller, M.; Broide, D.H. Chromosome 17q21 Genes ORMDL3 and GSDMB in Asthma and Immune Diseases. *Adv. Immunol.* **2017**, *135*, 1–52. [CrossRef] [PubMed]
33. Nagano, T.; Katsurada, M.; Dokuni, R.; Hazama, D.; Kiriu, T.; Umezawa, K.; Kobayashi, K.; Nishimura, Y. Crucial Role of Extracellular Vesicles in Bronchial Asthma. *Int. J. Mol. Sci.* **2019**, *20*, 2589.
34. Sangaphunchai, P.; Todd, I.; Fairclough, L.C. Extracellular vesicles and asthma: A review of the literature. *Clin. Exp. Allergy* **2020**, *50*, 291–307. [CrossRef] [PubMed]
35. Kulshreshtha, A.; Ahmad, T.; Agrawal, A.; Ghosh, B. Proinflammatory role of epithelial cell-derived exosomes in allergic airway inflammation. *J. Allergy Clin. Immunol.* **2013**, *131*, 1194–1203.e14. [CrossRef]
36. Haj-Salem, I.; Plante, S.; Gounni, A.S.; Rouabhia, M.; Chakir, J. Fibroblast-derived exosomes promote epithelial cell proliferation through TGF-beta2 signalling pathway in severe asthma. *Allergy* **2018**, *73*, 178–186. [CrossRef] [PubMed]
37. Wahlund, C.J.E.; Gucluler, G.; Hiltbrunner, S.; Veerman, R.E.; Naslund, T.I.; Gabrielsson, S. Exosomes from antigen-pulsed dendritic cells induce stronger antigen-specific immune responses than microvesicles in vivo. *Sci. Rep.* **2017**, *7*, 17095. [CrossRef]
38. Cañas, J.A.; Sastre, B.; Mazzeo, C.; Fernández-Nieto, M.; Rodrigo-Muñoz, J.M.; González-Guerra, A.; Izquierdo, M.; Barranco, P.; Quirce, S.; Sastre, J.; et al. Exosomes from eosinophils autoregulate and promote eosinophil functions. *J. Leukoc. Biol.* **2017**, *101*, 1191–1199. [CrossRef]
39. Bartel, S.; La Grutta, S.; Cilluffo, G.; Perconti, G.; Bongiovanni, A.; Giallongo, A.; Behrends, J.; Kruppa, J.; Hermann, S.; Chiang, D.; et al. Human airway epithelial extracellular vesicle miRNA signature is altered upon asthma development. *Allergy* **2020**, *75*, 346–356. [CrossRef]
40. Chanda, D.; Otoupalova, E.; Hough, K.P.; Locy, M.L.; Bernard, K.; Deshane, J.S.; Sanderson, R.D.; Mobley, J.A.; Thannickal, V.J. Fibronectin on the Surface of Extracellular Vesicles Mediates Fibroblast Invasion. *Am. J. Respir. Cell Mol. Biol.* **2019**, *60*, 279–288. [CrossRef]
41. Vallhov, H.; Gutzeit, C.; Hultenby, K.; Valenta, R.; Gronlund, H.; Scheynius, A. Dendritic cell-derived exosomes carry the major cat allergen Fel d 1 and induce an allergic immune response. *Allergy* **2015**, *70*, 1651–1655. [CrossRef] [PubMed]
42. Ismail, N.; Wang, Y.; Dakhlallah, D.; Moldovan, L.; Agarwal, K.; Batte, K.; Shah, P.; Wisler, J.; Eubank, T.D.; Tridandapani, S.; et al. Macrophage microvesicles induce macrophage differentiation and miR-223 transfer. *Blood* **2013**, *121*, 984–995. [CrossRef] [PubMed]
43. Volgers, C.; Benedikter, B.J.; Grauls, G.E.; Savelkoul, P.H.M.; Stassen, F.R.M. Immunomodulatory role for membrane vesicles released by THP-1 macrophages and respiratory pathogens during macrophage infection. *BMC Microbiol.* **2017**, *17*, 216. [CrossRef] [PubMed]
44. Mazzeo, C.; Canas, J.A.; Zafra, M.P.; Marco, A.R.; Fernandez-Nieto, M.; Sanz, V.; Mittelbrunn, M.; Izquierdo, M.; Baixaulli, F.; Sastre, J.; et al. Exosome secretion by eosinophils: A possible role in asthma pathogenesis. *J. Allergy Clin. Immunol.* **2015**, *135*, 1603–1613. [CrossRef] [PubMed]
45. Cañas, J.A.; Sastre, B.; Rodrigo-Muñoz, J.M.; Fernandez-Nieto, M.; Barranco, P.; Quirce, S.; Sastre, J.; Del Pozo, V. Eosinophil-derived exosomes contribute to asthma remodelling by activating structural lung cells. *Clin. Exp. Allergy* **2018**, *48*, 1173–1185. [CrossRef] [PubMed]
46. Mercado, N.; Ito, K.; Barnes, P.J. Accelerated ageing of the lung in COPD: New concepts. *Thorax* **2015**, *70*, 482–489. [CrossRef]
47. MacNee, W. Is Chronic Obstructive Pulmonary Disease an Accelerated Aging Disease? *Ann. Am. Thorac. Soc.* **2016**, *13* (Suppl. 5), S429–S437. [CrossRef] [PubMed]

48. Corlăţeanu, A.; Odajiu, I.; Botnaru, V.; Cemirtan, S. From smoking to COPD—Current approaches. *Pneumologia* **2016**, *65*, 20–23. [PubMed]
49. Mannino, D.M.; Buist, A.S. Global burden of COPD: Risk factors, prevalence, and future trends. *Lancet* **2007**, *370*, 765–773. [CrossRef]
50. Chan, S.M.H.; Selemidis, S.; Bozinovski, S.; Vlahos, R. Pathobiological mechanisms underlying metabolic syndrome (MetS) in chronic obstructive pulmonary disease (COPD): Clinical significance and therapeutic strategies. *Pharm. Ther.* **2019**, *198*, 160–188. [CrossRef]
51. Oba, Y.; Keeney, E.; Ghatehorde, N.; Dias, S. Dual combination therapy versus long-acting bronchodilators alone for chronic obstructive pulmonary disease (COPD): A systematic review and network meta-analysis. *Cochrane Database Syst. Rev.* **2018**, *12*, CD012620. [CrossRef]
52. Recio Iglesias, J.; Díez-Manglano, J.; López García, F.; Díaz Peromingo, J.A.; Almagro, P.; Varela Aguilar, J.M. Management of the COPD Patient with Comorbidities: An Experts Recommendation Document. *Int. J. Chronic Obstr. Pulm. Dis.* **2020**, *15*, 1015–1037. [CrossRef] [PubMed]
53. Barnes, P.J.; Baker, J.; Donnelly, L.E. Cellular Senescence as a Mechanism and Target in Chronic Lung Diseases. *Am. J. Respir. Crit. Care Med.* **2019**, *200*, 556–564. [CrossRef] [PubMed]
54. Kadota, T.; Fujita, Y.; Yoshioka, Y.; Araya, J.; Kuwano, K.; Ochiya, T. Emerging role of extracellular vesicles as a senescence-associated secretory phenotype: Insights into the pathophysiology of lung diseases. *Mol. Aspects Med.* **2018**, *60*, 92–103. [CrossRef]
55. Cordazzo, C.; Petrini, S.; Neri, T.; Lombardi, S.; Carmazzi, Y.; Pedrinelli, R.; Paggiaro, P.; Celi, A. Rapid shedding of proinflammatory microparticles by human mononuclear cells exposed to cigarette smoke is dependent on Ca^{2+} mobilization. *Inflamm. Res.* **2014**, *63*, 539–547. [CrossRef] [PubMed]
56. Fujita, Y.; Araya, J.; Ito, S.; Kobayashi, K.; Kosaka, N.; Yoshioka, Y.; Kadota, T.; Hara, H.; Kuwano, K.; Ochiya, T. Suppression of autophagy by extracellular vesicles promotes myofibroblast differentiation in COPD pathogenesis. *J. Extracell. Vesicles* **2015**, *4*, 28388. [CrossRef] [PubMed]
57. Moon, H.G.; Kim, S.H.; Gao, J.; Quan, T.; Qin, Z.; Osorio, J.C.; Rosas, I.O.; Wu, M.; Tesfaigzi, Y.; Jin, Y. CCN1 secretion and cleavage regulate the lung epithelial cell functions after cigarette smoke. *Am. J. Physiol. Lung Cell Mol. Physiol.* **2014**, *307*, L326–L337. [CrossRef]
58. Madhusoodanan, J. Care packages. *Nature* **2020**, *581*, S10–S11. [CrossRef]
59. Ryu, A.R.; Kim, D.H.; Kim, E.; Lee, M.Y. The Potential Roles of Extracellular Vesicles in Cigarette Smoke-Associated Diseases. *Oxid. Med. Cell. Longev.* **2018**, *2018*, 4692081. [CrossRef]
60. Takahashi, T.; Kobayashi, S.; Fujino, N.; Suzuki, T.; Ota, C.; He, M.; Yamada, M.; Suzuki, S.; Yanai, M.; Kurosawa, S.; et al. Increased circulating endothelial microparticles in COPD patients: A potential biomarker for COPD exacerbation susceptibility. *Thorax* **2012**, *67*, 1067–1074. [CrossRef]
61. Lacedonia, D.; Carpagnano, G.E.; Trotta, T.; Palladino, G.P.; Panaro, M.A.; Zoppo, L.D.; Foschino Barbaro, M.P.; Porro, C. Microparticles in sputum of COPD patients: A potential biomarker of the disease? *Int. J. Chronic Obstr. Pulm. Dis.* **2016**, *11*, 527–533. [CrossRef] [PubMed]
62. Kara, M.; Kirkil, G.; Kalemci, S. Differential expression of microRNAs in chronic obstructive pulmonary disease. *Adv. Clin. Exp. Med.* **2016**, *25*, 21–26. [CrossRef]
63. Wang, R.; Li, M.; Zhou, S.; Zeng, D.; Xu, X.; Xu, R.; Sun, G. Effect of a single nucleotide polymorphism in miR-146a on COX-2 protein expression and lung function in smokers with chronic obstructive pulmonary disease. *Int. J. Chronic Obstr. Pulm. Dis.* **2015**, *10*, 463–473. [CrossRef]
64. Shi, L.; Xin, Q.; Chai, R.; Liu, L.; Ma, Z. Ectopic expressed miR-203 contributes to chronic obstructive pulmonary disease via targeting TAK1 and PIK3CA. *Int. J. Clin. Exp. Pathol.* **2015**, *8*, 10662–10670.
65. Lewis, A.; Riddoch-Contreras, J.; Natanek, S.A.; Donaldson, A.; Man, W.D.; Moxham, J.; Hopkinson, N.S.; Polkey, M.I.; Kemp, P.R. Downregulation of the serum response factor/miR-1 axis in the quadriceps of patients with COPD. *Thorax* **2012**, *67*, 26–34. [CrossRef]
66. Rodrigues, M.; Fan, J.; Lyon, C.; Wan, M.; Hu, Y. Role of Extracellular Vesicles in Viral and Bacterial Infections: Pathogenesis, Diagnostics, and Therapeutics. *Theranostics* **2018**, *8*, 2709–2721. [CrossRef] [PubMed]
67. Kim, H.J.; Kim, Y.S.; Kim, K.H.; Choi, J.P.; Kim, Y.K.; Yun, S.; Sharma, L.; Dela Cruz, C.S.; Lee, J.S.; Oh, Y.M.; et al. The microbiome of the lung and its extracellular vesicles in nonsmokers, healthy smokers and COPD patients. *Exp. Mol. Med.* **2017**, *49*, e316. [CrossRef]
68. Kim, J.H.; Lee, J.; Park, J.; Gho, Y.S. Gram-negative and Gram-positive bacterial extracellular vesicles. *Semin. Cell Dev. Biol.* **2015**, *40*, 97–104. [CrossRef]
69. Augustyniak, D.; Roszkowiak, J.; Wiśniewska, I.; Skała, J.; Gorczyca, D.; Drulis-Kawa, Z. Neuropeptides SP and CGRP Diminish the Moraxella catarrhalis Outer Membrane Vesicle- (OMV-) Triggered Inflammatory Response of Human A549 Epithelial Cells and Neutrophils. *Mediat. Inflamm.* **2018**, *2018*, 4847205. [CrossRef] [PubMed]
70. Altan-Bonnet, N. Extracellular vesicles are the Trojan horses of viral infection. *Curr. Opin. Microbiol.* **2016**, *32*, 77–81. [CrossRef]
71. McNamara, R.P.; Dittmer, D.P. Extracellular vesicles in virus infection and pathogenesis. *Curr. Opin. Virol.* **2020**, *44*, 129–138. [CrossRef] [PubMed]

72. Roffel, M.P.; Bracke, K.R.; Heijink, I.H.; Maes, T. miR-223: A Key Regulator in the Innate Immune Response in Asthma and COPD. *Front. Med.* **2020**, *7*, 196. [CrossRef]
73. Segal, L.N.; Blaser, M.J. A brave new world: The lung microbiota in an era of change. *Ann. Am. Thorac. Soc.* **2014**, *11* (Suppl. 1), S21–S27. [CrossRef]
74. Koeppen, K.; Hampton, T.H.; Jarek, M.; Scharfe, M.; Gerber, S.A.; Mielcarz, D.W.; Demers, E.G.; Dolben, E.L.; Hammond, J.H.; Hogan, D.A.; et al. A Novel Mechanism of Host-Pathogen Interaction through sRNA in Bacterial Outer Membrane Vesicles. *PLoS Pathog.* **2016**, *12*, e1005672. [CrossRef]
75. Joshi, B.; Singh, B.; Nadeem, A.; Askarian, F.; Wai, S.N.; Johannessen, M.; Hegstad, K. Transcriptome Profiling of Staphylococcus aureus Associated Extracellular Vesicles Reveals Presence of Small RNA-Cargo. *Front. Mol. Biosci.* **2021**, *7*, 566207. [CrossRef]
76. Kim, M.R.; Hong, S.W.; Choi, E.B.; Lee, W.H.; Kim, Y.S.; Jeon, S.G.; Jang, m.H.; Gho, Y.S.; Kim, Y.K. Staphylococcus aureus-derived extracellular vesicles induce neutrophilic pulmonary inflammation via both Th1 and Th17 cell responses. *Allergy* **2012**, *67*, 1271–1281. [CrossRef]
77. Kim, Y.S.; Choi, E.J.; Lee, W.H.; Choi, S.J.; Roh, T.Y.; Park, J.; Jee, Y.K.; Zhu, Z.; Koh, Y.Y.; Gho, Y.S.; et al. Extracellular vesicles, especially derived from Gram-negative bacteria, in indoor dust induce neutrophilic pulmonary inflammation associated with both Th1 and Th17 cell responses. *Clin. Exp. Allergy* **2013**, *43*, 443–454. [CrossRef]
78. Koeppen, K.; Barnaby, R.; Jackson, A.A.; Gerber, S.A.; Hogan, D.A.; Stanton, B.A. Tobramycin reduces key virulence determinants in the proteome of Pseudomonas aeruginosa outer membrane vesicles. *PLoS ONE* **2019**, *14*, e0211290. [CrossRef]
79. Bomberger, J.M.; Ye, S.; Maceachran, D.P.; Koeppen, K.; Barnaby, R.L.; O'Toole, G.A.; Stanton, B.A. A Pseudomonas aeruginosatoxin that hijacks the host ubiquitin proteolytic system. *PLoS Pathog.* **2011**, *7*, e1001325. [CrossRef]
80. An, J.; McDowell, A.; Kim, Y.; Kim, T. Extracellular vesicle-derived microbiome obtained from exhaled breath condensate in patients with asthma. *Lett./Ann Allergy Asthma Immunol.* **2021**, *126*, 722–741.
81. Choi, Y.; Park, H.S.; Jee, Y.K. Urine Microbial Extracellular Vesicles Can Be Potential and Novel Biomarkers for Allergic Diseases. *Allergy Asthma Immunol. Res.* **2021**, *13*, 5–7. [CrossRef] [PubMed]
82. Samra, M.S.; Lim, D.H.; Han, M.Y.; Jee, H.M.; Kim, Y.K.; Kimm, J.H. Bacterial Microbiota-derived Extracellular Vesicles in Children with Allergic Airway Diseases: Compositional and Functional Features. *Allergy Asthma Immunol. Res.* **2021**, *13*, 56–74. [CrossRef] [PubMed]
83. Sundar, I.K.; Li, D.; Rahman, I. Small RNA-sequence analysis of plasma-derived extracellular vesicle miRNAs in smokers and patients with chronic obstructive pulmonary disease as circulating biomarkers. *J. Extracell. Vesicles* **2019**, *8*, 1684816. [CrossRef]
84. Xie, L.; Wu, M.; Lin, H.; Liu, C.; Yang, H.; Zhan, J.; Sun, S. An increased ratio of serum miR-21 to miR-181a levels is associated with the early pathogenic process of chronic obstructive pulmonary disease in asymptomatic heavy smokers. *Mol. Biosyst.* **2014**, *10*, 1072–1081. [CrossRef] [PubMed]

MDPI

Review

Effects of Essential Oils and Selected Compounds from *Lamiaceae* Family as Adjutants on the Treatment of Subjects with Periodontitis and Cardiovascular Risk

Giuseppa Castellino [1,2], Francisco Mesa [2], Francesco Cappello [1,3], Cristina Benavides-Reyes [4], Giuseppe Antonio Malfa [5], Inmaculada Cabello [6] and Antonio Magan-Fernandez [2,*]

1 Department of Biomedicine, Neuroscience and Advanced Diagnostics, Section of Human Anatomy, University of Palermo, 90127 Palermo, Italy; castellinogiusy@gmail.com (G.C.); francesco.cappello@unipa.it (F.C.)
2 Periodontology Department, School of Dentistry, University of Granada, 18071 Granada, Spain; fmesa@ugr.es
3 Euro-Mediterranean Institute of Science and Technology (IEMEST), 90127 Palermo, Italy
4 Department of Operative Dentistry, School of Dentistry, University of Granada, 18071 Granada, Spain; crisbr@ugr.es
5 Department of Drug and Health Sciences, Section of Biochemistry, University of Catania, 95125 Catania, Italy; g.malfa@unict.it
6 Department of Child Integrated Dentistry, Faculty of Medicine-Dentistry, University of Murcia, 30008 Murcia, Spain; icabello@um.es
* Correspondence: amaganf@ugr.es; Tel.: +34-95-8240-654

Citation: Castellino, G.; Mesa, F.; Cappello, F.; Benavides-Reyes, C.; Malfa, G.A.; Cabello, I.; Magan-Fernandez, A. Effects of Essential Oils and Selected Compounds from *Lamiaceae* Family as Adjutants on the Treatment of Subjects with Periodontitis and Cardiovascular Risk. *Appl. Sci.* **2021**, *11*, 9563. https://doi.org/10.3390/app11209563

Academic Editor: Elena G. Govorunova

Received: 23 September 2021
Accepted: 11 October 2021
Published: 14 October 2021

Abstract: Essential oils from different plant species were found to contain different compounds exhibiting anti-inflammatory effects with the potential to be a valid alternative to conventional chemotherapy that is limited in long-term use due to its serious side effects. Generally, the first mechanism by which an organism counteracts injurious stimuli is inflammation, which is considered a part of the innate immune system. Periodontitis is an infectious and inflammatory disease caused by a dysbiosis in the subgingival microbiome that triggers an exacerbated immune response of the host. The immune–inflammatory component leads to the destruction of gingival and alveolar bone tissue. The main anti-inflammation strategies negatively modulate the inflammatory pathways and the involvement of inflammatory mediators by interfering with the gene's expression or on the activity of some enzymes and so affecting the release of proinflammatory cytokines. These effects are a possible target from an effective and safe approach, suing plant-derived anti-inflammatory agents. The aim of the present review is to summarize the current evidence about the effects of essentials oils from derived from plants of the *Lamiaceae* family as complementary agents for the treatment of subjects with periodontitis and their possible effect on the cardiovascular risk of these patients.

Keywords: periodontitis; inflammation; oils; volatile; heart disease risk factors

1. Periodontitis

Periodontal disease is classically defined as a chronic inflammatory lesion, and gingivitis and periodontitis are the most common diseases derived from periodontium involvement [1]. Periodontitis is a complex chronic inflammatory disease caused by Gram-negative anaerobic bacteria located in the subgingival biofilm [2], which can induce the production of inflammatory mediators, causing the destruction and loss of dental bone support [3]. Periodontitis is also described as an infectious disease which affects the tooth-supporting tissues and leads a numerous clinical, microbiological and immunological symptoms, associated with and, probably, induced by progressive interaction among infectious agents, host immune responses, hazardous environmental exposure and genetic predisposition [4].

Anaerobic bacteria are considered as periodontal pathogens, and the following have been highlighted: *Porphyromonas gingivalis*, *Aggregatibacter actinomycetemcomitans*, *Prevotella intermedia, Tannerella forsythia*, Eikenella spp. and Capnocytophaga spp. However, it is important to highlight that those bacteria are mandatory for the disease development but are not enough and do not account for all cases of periodontitis [5]. The results of one survey in the USA indicate that chronic periodontitis affects about 46% of the adult population, with a higher prevalence among the elderly population [6]. This prevalence refers to the cohort of young adults according to the World Health Organization (WHO), aged 35 to 44 years. Interestingly, forms of periodontitis that occur at younger ages (before the age of 30 years), have other characteristics in addition to age and are known as aggressive periodontitis, with the prevalence ranging from 0.2% in Caucasians to 2.6% in Afro–Americans [7]. It is known that oral microbiome is in equilibrium between these microorganisms and the host response, which has a crucial role in both health and disease development [8]. Unfavourable modifications in the composition of the microbiota are known as dysbiosis [9], which is seen in both cases, periodontitis and CVD. In case of periodontitis, antiseptics and antibiotics such as chlorhexidine or metronidazole are delivered locally in addition to scaling and root planning procedures in order to eradicate the subgingival microbes, therefore creating a healthy subgingival environment. However, the evidence in the literature is still inconclusive [10], and future clinical trials with strict methodological criteria that will allow a more precise evaluation of the efficacy of local antimicrobials in the treatment of chronic periodontitis are required.

In the initial phase, there are no clinical signs; thus, the presence of inflammation cannot be observed. However, when the lesion progresses, vasodilation occurs locally due to the action of bacterial metabolic products, including cytokines [11]. Such initial lesion continues to progress, and a leukocyte infiltrate (mostly lymphocytes and neutrophils) is produced towards the site of inflammation. Crevicular fluid increase occurs, and clinical signs of inflammation appear [12]. In the next phase, or established injury, an inflammatory infiltrate, consisting of T and B lymphocytes, plasma cells and neutrophils, appears, followed by an increase in collagenolytic activity and more collagen-producing fibroblasts. This stage corresponds to moderate to severe gingivitis [12]. The final phase or advanced lesion is distinguished by an unresolved process, fibrosis and an irreversible loss of bone structure, characterized by clinical and histological patterns [2]. In addition, a dense inflammatory infiltrate in connective tissues and predominantly neutrophils in the epithelium are noticed, while, on the other hand, an apical migration of plasma cells to the junctional epithelium occurs to try to defend or keep the epithelial barrier intact, and, consequently, there is a continuous loss of collagen and connective tissue. Finally, if the lesion extends deep, the osteoclasts cause a resorption that affects the alveolar bone [1].

It should be highlighted that periodontitis is a multifactorial disease that requires interdisciplinary treatment concepts and the selection of a therapy that affects the microbiological nature of the disease [13]. In this regard, the recently introduced classification of periodontal diseases [14] aims to identify well-defined clinical entities using clear criteria that are able to link diagnosis with prevention and treatment, thus moving towards precision and individualized dentistry [15].

The interest in the application of natural products has been increased in the last years [16]. Several natural products and herbs have suggested that they have better properties and less side effects compared to chemical agents for irrigation. Furthermore, the use of natural extracts and essential oils (EOs) as an irrigation agent for ultrasonic instrumentation has shown to benefit slight adjunctive effect compared to chlorhexidine or water [17]. Yet, the use of natural extract in subjects with a more severe degree of periodontitis was associated with a greater improvement compared with controls [18]. Natural products in forms of oral spray have shown to be efficient against common oral pathogens, but also safe, without significant cytotoxicity in an in vitro study [19]. Thus, nutraceuticals might have the potential to prevent the infections and may be used as an adjunctive treatment to conventional therapy, as they seem to have the same or even more anti-inflammatory and antimicrobial effect without adding any chemicals. However, still there is not enough scientific evidence on this topic [20,21].

2. Cardiovascular Diseases

Cardiovascular diseases (CVDs) such as coronary heart disease, myocardial infarction, and ischemic stroke are one of the main causes of death worldwide [22]. In 2012, CVDs accounted for around 17.5 million deaths, representing almost a 31% of the worldwide mortality [23]. Atherosclerosis represents one of the main underlying processes for CVDs. It is defined as a condition characterized by formation of an atheroma plaque in the intima layer of the arterial wall; it is composed by an accumulation of lipids, cells and extracellular matrix [24]. High LDL-cholesterol (LDL-C) levels have been traditionally considered as one of the major risk factors for coronary heart disease and, together with triglycerides, are the main risk factors for atherosclerosis [25]. The current approaches for atherosclerosis management focus on the prevention of plaque growth and its destabilization through risk factor control (hypertension, lipid profile, diabetes, smoking, etc.), using lifestyle interventions (diet, physical activity, smoking cessation, etc.) and pharmacological therapies [26]. Recent advances in the understanding of the atherosclerotic process revealed that cholesterol and lipid deposition is not the only causative factor of this disease [27]. Systemic and chronic inflammation plays critical roles in the initial phases, as well as atherosclerotic plaque progression [26,28,29]. During atherosclerotic plaque formation, monocytes are recruited from the blood flow to the arterial wall and differentiate into macrophages of inflammatory phenotype and lipid-containing foam cells. These cells drive the inflammatory process and stimulate plaque maturation and thrombosis [30,31]. Interleukin 10 (IL-10) is an immunoregulatory cytokine with reported anti-inflammatory properties [32]. IL-10 has shown to play a protective role against atherogenesis by inhibiting several inflammatory mediators from activated macrophages and dendritic cells [33–35]. Intramuscular injection of IL-10-encoding plasmid DNA in IL10 knockout mice caused an increase in the cytokine level and inhibited plaque formation by 60% [33]. These findings clearly suggest that IL-10 may be a promising therapeutic target for atherosclerosis management [27]. Inflammation plays a key role in atheroma plaque formation and progression [36]. Evidence has suggested that higher levels of circulating C reactive protein (CRP) have a greater risk of suffering an acute myocardial infarction or cerebrovascular event [37]. The main factors of atheroma plaque vulnerability are the composition of the plaque core, the inflammatory process and the formation of a fibrotic layer that covers the nuclei [38]. It has been reported that inflammation of the atheroma plaque interferes with the formation of the fibrous cape, causing apoptosis and degradation of the extracellular matrix by metalloproteinase activation and increasing the risk of plaque rupture and consequent thrombotic cardiovascular events [38,39]. In those situations, alternative therapeutic approaches, such as the use of dietary supplements and nutraceuticals, may be useful [40]. It is known that the main causes of mortality of subjects with non-alcoholic fatty liver disease (NAFLD) are CVDs. Although the available data are not numerous for a final conclusion and relatively few nutraceuticals have been adequately studied for their effects on NAFLD, several nutraceuticals have been shown to contribute to the improvement of lipid infiltration of the liver and of the related anthropometric and/or biochemical parameters [41]. However, such their positive effects are associated with well-chosen dose, supplementation for a medium-long period and lifestyle changes. There are growing data in the literature demonstrating the beneficial effects of nutraceuticals in metabolic diseases and showing significant impact on different cardiometabolic risk factors (including inflammatory markers) and CVD risk [42]. However, more randomized trials as well as observational studies with specific CVD endpoints are needed.

3. Periodontitis and Cardiovascular Risk

Numerous mechanisms have been proposed as links between periodontitis and atherosclerotic CVD, but the most important include systemic inflammation, molecular mimicry and direct plaque colonization by periodontal pathogens [38]. Several systematic reviews and meta-analyses have reported an association between periodontal disease and ischaemic heart disease [43–47]. Some authors have suggested that at clinical exploration level, periodontitis and CVD have a weak association, and that, actually, systemic bacterial exposure from periodontitis could be a more plausible risk factor. In this context, Mustapha et al. [46] reported that periodontitis with increased markers of systemic bacterial exposure (periodontal bacterial burden, periodontitis-related specific serology and CRP) was associated with a greater risk of coronary heart disease compared with subjects without periodontitis [37]. It has also been shown that periodontitis patients present increased levels of inflammatory markers (tumour necrosis factor (TNF), interleukin (IL)-1, 6 and 8) [48]. Short-term adaptive response to inflammation is essential for a correct injury response and cell and tissue repair processes, while long-term consequences of a maintained inflammatory situation are often not beneficial [49]. It has also been reported that low-grade and chronic inflammation are characteristic in cardiometabolic diseases such as obesity, insulin resistance, type 2 diabetes and CVDs [50,51]. An atypical immune response has been referred in these situations, described as metabolically triggered inflammation or "metaflammation", originated by metabolic surplus, which leads to the activation of different inflammatory molecules and signalling pathways [50]. It is worth mentioning that both metabolic and immune systems are regulated by the same cellular processes, through several hormones, cytokines and bioactive lipids that have a role in the metabolic and immune response. The activation of these "metaflammatory" pathways has been related to extracellular mediators such as cytokines and lipids, especially saturated fatty acids, and also by intracellular mediators such as endoplasmic reticulum stress and elevated production of mitochondria-derived reactive oxygen species. Fatty acid-binding proteins (FABPs), a family of lipid chaperones, have shown molecular and cellular links with "metaflammation", particularly in cases of cardiometabolic diseases such as obesity, diabetes and atherosclerosis [49].

The most recent consensus document regarding association between periodontitis and CVDs was the results from the joint workshop of the European Federation of Periodontology (EFP) and the World Heart Federation (WHF) in February 2019 [52]. According to its recommendations, periodontitis was considered as an established, novel CV risk factor that influences the management of subjects suffering from CVD or at increased CVD. The management of traditional CV risk factors, such as hypertension, is also required in the presence of periodontitis, and a good periodontal health is of great relevance for achieving CV health [53]. However, it remains to be determined if periodontal treatment in subjects with hypertension would translate into a reduced CV risk. Again, inflammation still appears to be one of the main links between CVD and periodontitis [52]. It should be mentioned that the available evidence mainly comes from observational studies, assessing major CV outcomes such as myocardial infarction, stroke, heart failure or CVD death, but few studies have investigated preclinical markers of CVD in subjects with periodontitis. Very interestingly, some studies suggested that healthy subjects with periodontitis may present signs of early atherosclerosis, and thus, periodontitis may be considered as a risk factor in case of CV events that cannot be fully explained by the presence of other commonly used CV risk factors [37,52]. Future studies are needed to better understand the relationship between both diseases and to detect early stages of CVD or alterations in CV structure and function linked to periodontitis. The last published randomized controlled trials confirm a positive effect of periodontal treatment on surrogate CV measures, while its effect on the incidence of CVD events (myocardial infarction and stroke) have not been investigated in powered randomized controlled studies with adequate control of traditional CV risk factors [54,55].

Interestingly, it has been proposed that botanical products may provide a new perspective in stem cell-based periodontal regeneration thanks to their angiogenic properties that may be beneficial for bone formation and periodontal regeneration [56].

4. Plant-Derived Essential Oils

An essential oil (EO) is generally defined as a product obtained from a natural raw material of plant origin, usually by steam distillation or by mechanical processes [57].

These plant products are very complex natural mixtures of secondary volatile metabolites produced by aromatic plants, where they represent chemical defences against herbivores and pathogens such as bacteria, viruses and fungi [58]. They also exert a dual role in attracting pollinating insects and in repelling the harmful ones [59]. In plants, essential oils are synthesized by all tissues and are stored in secretory cells, epidermic cells or glandular trichomes; consequently, they can be extracted from any plant organs such as buds, stems, twigs, leaves, roots, wood, bark, flowers, fruits and seeds [59]. The particular composition of each EO depends not only on the plant species or plant tissue from which it is extracted but also on the climate, on the soil composition, on the vegetative cycle stage or age and even on the time of the harvest [60]. Moreover, the chemical profile is also affected by the extraction method carried out, thus highlighting the importance of specific extraction techniques over the steam distillation, such as solvent extraction, Soxhlet extraction, microwave-assisted hydro distillation, solvent flavour evaporation, etc., in compliance with the plant material characteristics [61]. Their peculiar chemical composition is mainly represented by various terpenoids and their oxygenated derivatives, along with aldehydes and ketones, esters and alcohols [60]. Generally, they are colourless volatile liquids soluble in organic solvents with a density lower than that of water. The most commercial EOs are extracted from various aromatic plants growing in temperate and warm regions of the Mediterranean and tropical areas, where they are historically used in traditional medicine against a wide variety of pathological conditions due to their numerous pharmacological activities including antimicrobial, antiviral, antioxidant and anti-inflammatory effects [59].

Chronic inflammation and oxidative stress are associated with most of the common chronic disorders and diseases [62,63]. It is known that the normal functions of biological molecules (such as proteins, lipids and DNA) are destabilized by oxidative stress sustained by free radicals (ROS, NRS), which also affects many inflammation-related signalling pathways, thus influencing the cellular and tissues homeostasis. On the other hand, chronic inflammation is characterized by the production of proinflammatory cytokines and chemokines, which leads to pain, redness and swelling of the involved tissue [64]. In traditional medicine, EOs have been used for the treatment of inflammatory processes [65], as they possess many beneficial properties due to the presence of several antioxidant and anti-inflammation compounds such as terpenes, the main class of compounds, and especially monoterpenes [61]. They are also present in numerous pharmaceutical products [66]. In recent studies, *Citrus bergamia* Risso and Poiteau juice (known as Bergamot) on cardiometabolic risk in dyslipidemic subjects was shown to significantly reduce plasma lipids and improve the atherogenic lipoproteins and subclinical atherosclerosis [67]. Other recent studies with chlorogenic acid and luteolin-based supplement from artichoke extract showed an improvement of two early atherosclerotic markers, carotid intima-media thickness and flow-mediated dilation, evidencing a clinical relevance, considering their beneficial nutraceutical properties, on vascular function and remodelling, including a beneficial cardiovascular and hepatoprotective effects [68].

It is widely known that EOs are recognized for their antimicrobial, antiviral and antifungal activity, but recent studies have also demonstrated potent antioxidant, anti-inflammatory and antidiabetic properties as well as cancer suppressor activity [69]. Thus, the potential of EOs as effective and safe phytotherapeutic agents should not be underestimated, although their efficacy in oral health is well documented [70]. The antibacterial activity of EOs as well as their isolated constituents and their potential applicability in novel dental formulations have been summarized in a systematic review [71], emphasizing the need for further nonclinical and clinical studies. Interestingly, in the last two decades, EOs have been extensively tested for their beneficial properties against a broad spectrum of bacterial species [72] that indicate their use in the files of dentistry and periodontal diseases. All these beneficial actions increase an interest in future investigation of these plants, including the field of periodontitis. Recent research suggests that oxygenated terpenoids found in the EOs diffuse within the bacterial cell membrane, irreversibly damaging it and causing cell death. In this regard, recently Anusha D. et al. assessed the efficacy of mouthwash containing EOs and curcumin (MEC) as an adjunct to nonsurgical periodontal therapy on the disease activity of rheumatoid arthritis (RA) among RA subjects with chronic periodontitis. They also investigated epigenetic modifications including chemical alterations of DNA and associated proteins influencing the remodelling of the chromatin and gene malfunctions, which may be related to both periodontitis and RA. The results revealed that MEC as an adjunct to SRP as an effective approach in reducing the disease activity of both RA and chronic periodontitis [73].

5. Essential Oils from *Lamiaceae* Family

Based on recent literature data, the *Lamiaceae* family seems to be one of the richest sources of a wide variety of plants with biological and medical applications [74]. It is a large group with a cosmopolitan distribution that includes 236 genera and about 7000 species that occupy different natural ecosystems [75]. The most interesting members of this family are a variety of aromatic spices traditionally harvested from spontaneous or cultivated population for their culinary use and for their characteristic and unique EOs obtained from them [76]. Numerous *Lamiaceae* species are native of the Mediterranean area and are widely used in natural medicine, pharmacology, cosmetology and aromatherapy due to the multiple therapeutic effects exerted by the components of their EOs [74]. The volatile compounds most generally found as the main ingredients in EOs among plants of this family are caryophyllene, linalool, limonene, β-pinene, 1,8-cineole, carvacrol, α-pinene, p-cymene, γ-terpinene, thymol and terpineols [77]. Several EOs from the *Lamiaceae* species are often present in toothpastes and mouthwashes as flavourings and fragrances, and recently also as antibacterial, antifungal and anti-inflammatory agents [77].

It has been proposed that high doses of EOs are needed in order to exert pharmacological effects, but recent studies indicate that this matter can be avoided using the oil formulated as nanoemulsions to improve its bioavailability [78]; thus, more investigations are needed in the future to confirm this finding. Interestingly, in addition to antioxidant, anti-inflammatory, antimicrobial, spasmolytic, antinociceptive, antitumor activity, some EOs, such as those from *Thymus vulgaris* L. may enhances cognitive function, as shown in animal models [79]. In a recent study, Carbone et al. developed nanostructured lipid carriers (NLC) using EOs from Mediterranean species *Rosmarinus officinalis* L., *Lavandula x intermedia* "Sumian", *Origanum vulgare* L. subsp. *Hirtum* (Link) Ietswaart and *Thymus capitatus* (L.) Hoffmanns. & Link (syn. *Coridothymus capitatus* (L.) Reichenb. fil., *Thymbra capitata* (L.) Cav.) with the purpose of examining the antioxidant and anti-inflammatory effects. Their results demonstrated that EOs in these NLC induced a significant and dose-dependent anti-inflammatory effect in the following order from greater to less potency: *Lavandula* L. > *Rosmarinus* L. ≥ *Origanum* L. It is worth noting that encapsulation of both *Lavandula* and *Rosmarinus* EOs in the NLC did not modify the anti-inflammatory activity of the EOs compared to the use of them as free compounds [80].

5.1. *Lavandula x intermedia (Lavender)*

The lavender EOs are generally composed mainly by camphor, terpinen-4-ol, linalool, linalyl acetate, beta-ocimene and 1,8-cineole [81]. Immunomodulatory and anti-inflammatory properties of compounds found in the lavender EOs have also been reported [82]. Many inflammatory processes are associated with leukotriene production catalysed by lipoxygenase (LOX), which can use molecular oxygen or hydrogen peroxide as oxidants. *L. x intermedia* EOs showed a moderate antioxidant activity, mainly attributed to the effect of linalool and linalyl acetate. These findings supported the use of EOs of *L. X intermedia* as natural ingredients useful for gastrointestinal disorders and for oxidative stress-related diseases [83].

5.2. *Rosmarinus officinalis* (Rosemary)

Characteristic EOs of *Rosmarinus officinalis* include 1,8-cineole, α-pinene, camphor, bornyl acetate, borneol, camphene, α-terpineol, limonene, β-pinene, β-caryophyllene and myrcene. In traditional medicine, it is used for the treatment of inflammation-related disorders [84]. Yet, this plant exerts antioxidant activity and prevents inflammatory ROS-related injury, stimulates smooth muscle relaxation and has low toxicity. Bustanji et al. performed a study to identify the effects of rosemary in blood glucose and lipid profile. They reported through an in vitro assay that rosemary extract (with the component gallic acid: IC50 14.5 μg/mL) inhibited in a dose-dependent manner the activity of cyclic adenosine monophosphate of gluconeogenic genes, cytosolic phosphoenol-pyruvate carboxykinase and glucose-6-phosphatase. Rosemary extract showed hypoglycaemic and hypolipidemic effects by activation of signalling pathways including AMP-activated protein kinase (which induces glycolysis) and proliferator-activated receptor gamma (PPAR-γ). It upregulated the expression of LDL-C receptor (responsible of the endocytosis of LDL-C from blood plasma to liver hepatocytes), sirtuin-1 (increasing the oxidation of fatty acids) and PGC1α (activates PPAR-γ) [85]. Neutrophils are rapidly mobilized, and they are one of the first and main cells to arrive at the inflammation site. Oral treatment with *R. officinalis* aqueous extract reduced the neutrophil influx, the release of cytokines and the oxidative stress on inflamed exudates [86]. The study of Borges et al. [87] showed that all nanoemulsions showed no toxicity and also showed the ability to potentiate the anti-inflammatory action of essential oils by exerting immunomodulatory activity by inhibiting the production of the proinflammatory mediator nitric oxide.

Plant extracts and their compounds have proven to be an alternative for treating periodontal diseases, since rosemary has shown antibacterial and anti-inflammatory activities in toothpaste presentation, which reduced biofilm formation and improved gingival bleeding [88]. Rasooli et al. also reported an in vivo reduction of biofilm by rosemary EO and suggested its potential use as an anticaries agent [89]. Bernardes et al. also confirmed this antimicrobial activity against oral bacteria in their study. These authors used common bacterial species from the oral cavity (*Streptococcus mutans*, *Streptococcus mitis*, *Streptococcus sanguinis*, *Streptococcus salivarius*, *Streptococcus sobrinus* and *Enterococcus faecalis*) in planktonic form, and the greatest antimicrobial activity of rosemary EO was shown against *Streptococcus mitis* [90]. The results of Smullen et al. have shown that rosemary and other plant-derived extracts inhibited growth and adhesion of oral bacteria to glass, inhibited both glucosyltransferase activity and glucan production by *S. mutans* and prevented plaque formation in vitro and on bovine teeth [91].

5.3. *Thymus capitatus* (Thyme)

The main single constituents of EOs from *Thymus* genus are thymol, carvacrol, linalool, a-terpineol, 1,8-cineole and borneol [92]. Iauk et al. [93] have investigated the hypoglycaemic activity of *T. capitatus* (L.) Hoffsgg. & Link via the inhibition of α-amylase and α-glucosidase, inhibitors that offer an attractive strategy to control postprandial hyperglycaemia for type 2 diabetes management. Manconi et al. [94] had suggested that formulations on *Thymus capitatus* EO in phospholipid vesicles might be used as an antibacterial–antioxidant mouthwash for the treatment of oral cavity diseases. Finally, Valerio et al. [95] indicated the potential use of thyme as biopreservative for bakery products due to its antimicrobial properties. The antioxidant activity of the formulations was evaluated as a protector of keratinocytes against the damage induced by hydrogen peroxide. They were capable of favouring wound repair in keratinocytes. The antibacterial activity of the EO was demonstrated against cariogenic *Streptococcus mutans*, *Lactobacillus acidophilus* and commensal *Streptococcus sanguinis* [94]. Alvarez-Echazú et al. used thymol–chitosan hydrogels to protect the dental biofilm from breakdown and treat inflammation [96]. *Thymus zygis* has been studied more extensively from an immunological point of view. In a cellular model with human macrophages the gene expression for IL-1β, TNFα and Il-6 were significantly reduced, and anti-inflammatory cytokines such as IL-10 dose-dependently highly increased [97]. *Thymus zygis* has also been tested in vitro as an antibacterial agent against *E. coli*, *S. enteritidis*, *S. essen* and other bacterial species [98]. Interestingly, in addition to antioxidant, anti-inflammatory, antimicrobial, spasmolytic, antinociceptive and antitumor activity, some EOs, such as those from *Thymus vulgaris* L. may enhances cognitive function, as shown in animal models [79].

5.4. Selected Compounds as Essential Oils from Lamiaceae Family

The following table presents a summary of the main compounds of EOs from the *Lamiaceae* family and their effects on inflammatory processes (Table 1).

Table 1. Anti-inflammatory effects of selected compounds of Essential Oils from *Lamiaceae* family.

Compound	Experimental Model	Effects	References
Terpineols	in vitro	↓ inflammatory mediators	Hart et al. 2000 [99]
	in vitro	↓inflammatory cytokines	Nogueira et al. 2014 [100]
	in vivo	↓ inflammation	Zhang et al. 2018 [101]
Linalool	in vitro	↓ inflammatory mediators	Huo et al. 2013 [102]
	in vivo	↓ acute lung inflammation	Ma et al. 2015 [103]
Limonene	in vitro	↓ inflammatory mediators	Kummer et al. 2013 [104]
	in vitro	↓ inflammatory mediators	Yoon et al. 2010 [105]
	in vitro	↓ inflammatory mediators	Rufino et al. 2015 [106]
Carvacrol	in vivo	↓ inflammation, antibacterial	Botelho et al. 2008 [107]
	in vitro/in vivo	↓ inflammatory mediators	Guimaraes et al. 2012 [108]
	in vitro	↓ inflammatory mediators	Hotta et al. 2010 [109]
	in vitro	↓ inflammatory mediators	Landa et al. 2009 [110]
Eucalyptol	in vivo	↓ inflammation	Yalçin et al. 2007 [111]
	in vitro	↓ inflammatory mediators	Bastos et al. 2011 [112]
	in vitro	↓ inflammatory mediators	Kennedy-Feitosa et al. 2016 [113]
	in vitro	↓ inflammatory mediators	Juergens et al. 2004 [114]
	in vitro	↓ inflammatory mediators	Kim et al. 2015 [115]
Tymol	in vitro	↓ inflammatory mediators	Chauhan et al. 2014 [116]
	in vitro	↓ inflammatory mediators	Vigo et al. 2004 [117]
	in vitro	↓ inflammatory mediators	Marsik et al. 2005 [118]
	in vitro	↓ inflammatory mediators	Liang et al. 2014 [119]

↓: Decrease of levels.

Terpineols are isomers of monocyclic monoterpene alcohol naturally present in different plants, among which the most common are α-terpineol and terpinen-4-ol [120] (Figure 1). In particular, the latter has been shown in vitro to suppress inflammatory mediator production by activation of monocytes [99] and to inhibit inflammatory cytokine generation in LPS-stimulated human macrophages [100] but also in animal models to attenuate inflammation in dextran sulphate sodium-induced colitis [121], prevent LPS-induced acute lung injury by decreasing LPS-induced NF-κB activation and trigger peroxisome PPAR-γ [122] (Figure 2).

Linalool (3,7-dimethyl-1,6-octadien-3-ol) is an acyclic monoterpene found in EOs of hundreds of plants widely spread worldwide and principally in *Lamiaceae* family [74] (Figure 3A). Several in vitro and in vivo studies demonstrated different anti-inflammatory effects of this monoterpene also interfering with the mediators of the inflammation pathways. In detail, in RAW 264.6 monocyte/macrophage-like cells linalool decreased the generation of lipopolysaccharide (LPS)-induced TNF-α and IL-6 and inhibited the activation of the nuclear factor-κB (NF-κB) and mitogen-activated protein kinase (MAPK) pathways [102]. In addition, in animal models, it has been shown that linalool attenuated acute lung inflammation by reducing TNF-α, IL-6, IL-8, IL-1β and monocyte chemoattractant protein-1 (MCP-1) production [103], further supporting linalool as a promising tool to treat inflammatory related diseases.

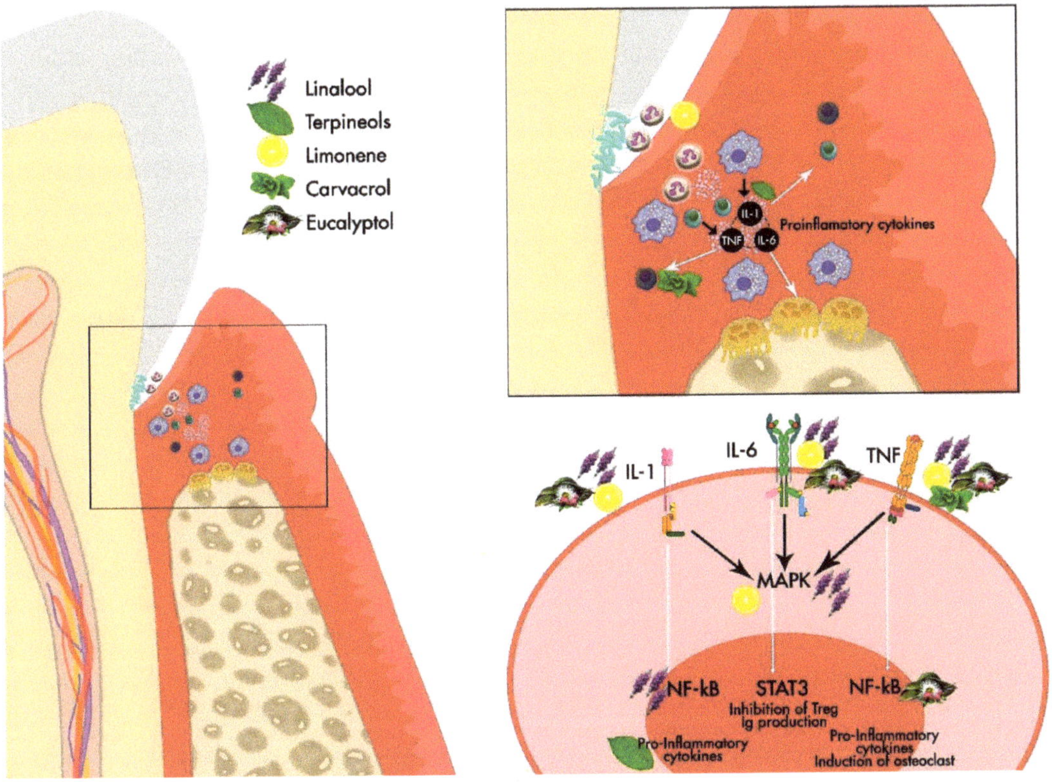

Figure 1. General Scheme of the effect of the main EOs derived from *Lamiaceae* family on the periodontium.

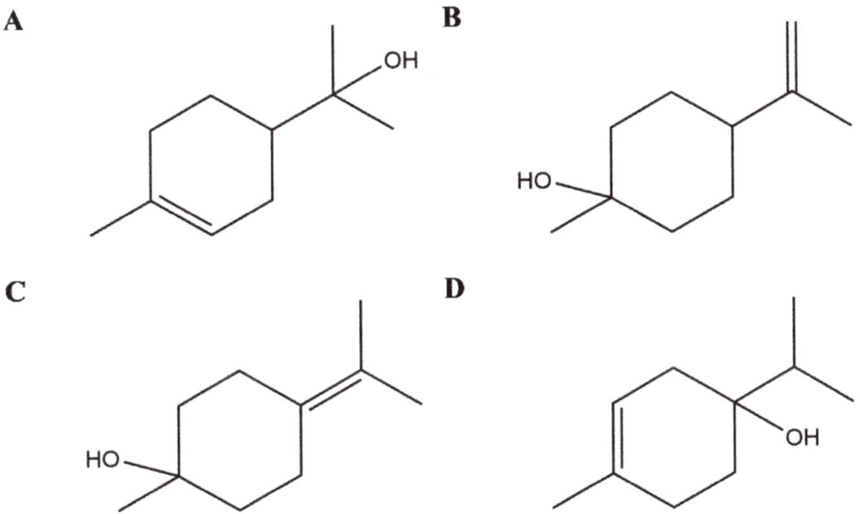

Figure 2. Selected compounds from Essential Oils from Lamiaceae family. Terpineols: α- terpineol (**A**), β- terpineol (**B**), γ-terpineol (**C**), and terpinen-4-ol (**D**).

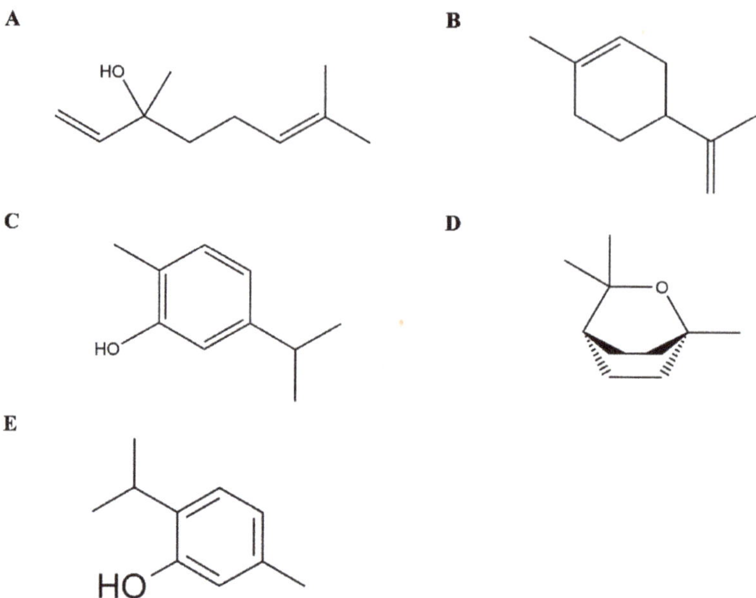

Figure 3. Selected compounds from Essential Oils from *Lamiaceae* family. Linalool (**A**), Limonene (**B**), Carvacrol (**C**), Eucalyptol (**D**), Thymol (**E**).

Limonene (1-Methyl-4-(prop-1-en-2-yl)cyclohex-1-ene) is a cyclic monoterpene and one of the most common terpenes in nature as well as the main constituent of citrus Eos (Figure 3B). Its anti-inflammatory effects are principally linked to the modulation of cytokines and the interference with the inflammatory-related pathways, as demonstrated by in vitro and in vivo assays. Limonene decreases leukocytes infiltration and neutrophils migration, as well as the levels of TNF-α in cell derived from the peritoneal cavity and in the peritoneal exudate of zymosan-induced peritonitis BALB/C mice [104]. In LPS inflammation-induced RAW 264.7 macrophages, limonene reduced in a dose-dependent manner the levels of proinflammatory cytokines TNF-α, IL-6 and IL-1β, together with the expression of inducible nitric oxide synthase (iNOS), cyclooxygenase (COX) and prostaglandin E2 (PGE2) [105]. Similarly, in vitro model of osteoarthritis with IL-1β-stimulated human chondrocytes, limonene negatively modulated nitric oxide (NO) production by decreasing iNOS, matrix metalloproteinase (MMP)-1 and MMP-13expression, besides NF-κB and p38 activation [106].

Carvacrol (5-isopropyl-2-methylphenol) is a cyclic monoterpene mainly present in the EO of plants from *Lamiaceae* family (Figure 3C). In an experimental rat model of periodontal disease, carvacrol maintained alveolar bone resorption and decreased tissue lesion at histopathology, with preservation of the gingival tissue also demonstrating anti-inflammatory and antibacterial activities [107]. In addition, in vitro in murine macrophages carvacrol (1, 10, and 100 μg/mL) reduced the LPS-induced nitrite production as well as in vivo in a model of carrageenan-induced pleurisy with a pretreatment 50 or 100 mg/kg, i.e., carvacrol reduced the levels of TNF-α and suppressed leukocytes recruitment in pleural lavage [108]. In human macrophage-like U937 cells, carvacrol suppressed LPS-induced COX-2 expression, activating PPARγ, indicating its anti-inflammatory properties [109] and having no selectivity for both COX-1 and COX-2 enzyme isoforms [110]. Furthermore, it has been shown that carvacrol induces Nav blockade in DRG neurons [123,124], and Gonçalves et al. also reported significant analgesic activity and dose-dependency of *T. capitatus* EO. It is suggest that it is the main active molecule behind the antinociceptive effects of *T. capitatus* through peripheral nervous excitability blockade [125]. Other study showed that carvacrol and thymol have the most potent antimicrobial activity against *Escherichia coli, Sta. aureus, Str. epidermidis, Enterococcus faecalis, Yersinia enterocolitica, Candida albicans, Bacillus cereus, Listeria monocytogenes, Salmonella typhimurium* and *Saccharomyces cerevisiae*, with the exception of *Pseudomonas aeruginosa* [126]. Taking into account that *T. capitatus* fractions are characterized by the presence of carvacrol as dominant constituent, the antibacterial properties may be attributed to this oxygenated monoterpene.

Thymol (2-isopropyl-5-methylphenol), a monoterpene phenol, is a typical compound in EOs from thyme species (Figure 3E). Thymol also ameliorates inflammation in vitro in LPS-induced inflammation in murine macrophage cells [116] as well as in LPS- and interferon (IFN)-γ-induced macrophage inflammation, besides inhibition of the iNO RNA expression in J774A.1 cells [117]. Thymol may also modify prostaglandin catalysed biosynthesis by the inhibition of COX-1 and COX-2 isoforms [118]. Moreover, in mouse mammary epithelial cells, LPS-induced inflammatory response was decreased after thymol treatment (40 μg/mL) by the downregulation of MAPK and NF-κB signalling pathways [119]. Recently, Perrino et al. [127] have reported a high bioactivity also found in endemic wild species as *spinulosus* Ten., indicating its potential use in organic agriculture, since thymol may also serve as natural agent against phytopathogenic microorganisms.

Eucalyptol or 1,8-cineole (1,3,3-trimethyl-2-oxabicyclo[2.2.2]octane) is a bicyclic monoterpene isolated from EOs from numerous plants [111] (Figure 3D). Its anti-inflammatory properties have also been investigated in human and animal models of respiratory diseases such as asthma, Chronic Obstructive Pulmonary Disease and bronchitis [112,113,128]. In vitro studies on LPS-induced human lymphocytes and monocytes showed a reduced expression of cytokines including TNFα, IL-6 and IL-1β accompanied with a decrease in NFκB activated form [114,115].

6. Conclusions

The available published evidence provides substantial data to consider several species of the *Lamiaceae* family (*Rosmarinus officinalis, Lavandula x intermedia, Thymus capitatus*) as potential agents for the treatment of inflammatory diseases. The studies reviewed support the use of EOs of these plants against inflammation-related diseases, and the mechanisms described here provided pathways that explain these anti-inflammatory effects. Furthermore, the wide variety of species of the *Lamiaceae* family should be considered. Different chemotypes can be found, even within the same species, depending on the region and the environment in which the species have been grown. These factors could affect the percentage of EOs. The phenological phase of the same individual can also change the chemical composition and therefore the EOs obtained from it. Several nutraceutical products based on EOs as an adjuvant therapeutic agent for periodontal treatment might have some additional beneficial effect on periodontitis variables, preventing progression of disease when used as an irrigation solution and/or mouthwash. The utilization of such agents may reduce the intake of the drugs and consequently minimize the risk of a possible appearance of drug resistance as well as other risk factors. The plant extracts may act beneficially on periodontitis possibly through the known anti-inflammatory effects of each compound that remain to be clarified by future studies. Therefore, EOs from the *Lamiaceae* family can be considered as potential therapeutic agents for the complementary treatment periodontitis, as well as for other immune-inflammatory diseases. The potential applications role of each specific EO needs to be further studied to allow a full understanding of the spectrum of potential applications.

Author Contributions: Conceptualization, G.C., F.M. and A.M.-F.; writing—original draft preparation, G.C., F.C., C.B.-R., G.A.M., I.C. and A.M.-F.; writing—review and editing, A.M.-F. and F.M.; supervision, F.M., G.C. and G.A.M. All authors have read and agreed to the published version of the manuscript.

Funding: This research received no external funding.

Institutional Review Board Statement: Not applicable.

Informed Consent Statement: Not applicable.

Data Availability Statement: Not applicable.

Conflicts of Interest: The authors declare no conflict of interest.

References

1. Cekici, A.; Kantarci, A.; Hasturk, H.; Van Dyke, T.E. Inflammatory and immune pathways in the pathogenesis of periodontal disease. *Periodontol. 2000* **2014**, *64*, 57–80. [CrossRef] [PubMed]
2. Hasturk, H.; Kantarci, A. Activation and resolution of periodontal inflammation and its systemic impact. *Periodontol. 2000* **2015**, *69*, 255–273. [CrossRef] [PubMed]
3. Teng, Y.T. The role of acquired immunity and periodontal disease progression. *Crit. Rev. Oral Biol. Med.* **2003**, *14*, 237–252. [CrossRef] [PubMed]
4. Kononen, E.; Gursoy, M.; Gursoy, U.K. Periodontitis: A Multifaceted Disease of Tooth-Supporting Tissues. *J. Clin. Med.* **2019**, *8*, 1135. [CrossRef] [PubMed]
5. Tatakis, D.N.; Trombelli, L. Modulation of clinical expression of plaque-induced gingivitis. I. Background review and rationale. *J. Clin. Periodontol.* **2004**, *31*, 229–238. [CrossRef]
6. Eke, P.I.; Wei, L.; Borgnakke, W.S.; Thornton-Evans, G.; Zhang, X.; Lu, H.; McGuire, L.C.; Genco, R.J. Periodontitis prevalence in adults $\geq$ 65 years of age, in the USA. *Periodontol. 2000* **2016**, *72*, 76–95. [CrossRef]
7. Susin, C.; Haas, A.N.; Albandar, J.M. Epidemiology and demographics of aggressive periodontitis. *Periodontol. 2000* **2014**, *65*, 27–45. [CrossRef]
8. Kilian, M.; Chapple, I.L.; Hannig, M.; Marsh, P.D.; Meuric, V.; Pedersen, A.M.; Tonetti, M.S.; Wade, W.G.; Zaura, E. The oral microbiome—An update for oral healthcare professionals. *Br. Dent. J.* **2016**, *221*, 657–666. [CrossRef]
9. Hajishengallis, G.; Lamont, R.J. Beyond the red complex and into more complexity: The polymicrobial synergy and dysbiosis (PSD) model of periodontal disease etiology. *Mol. Oral Microbiol.* **2012**, *27*, 409–419. [CrossRef]
10. Ramanauskaite, E.; Machiulskiene, V. Antiseptics as adjuncts to scaling and root planing in the treatment of periodontitis: A systematic literature review. *BMC Oral Health* **2020**, *20*, 143. [CrossRef]
11. Kinane, D.F.; Lappin, D.F. Clinical, pathological and immunological aspects of periodontal disease. *Acta Odontol. Scand.* **2001**, *59*, 154–160. [CrossRef]
12. Barros, S.P.; Williams, R.; Offenbacher, S.; Morelli, T. Gingival crevicular fluid as a source of biomarkers for periodontitis. *Periodontol. 2000* **2016**, *70*, 53–64. [CrossRef]
13. Haffajee, A.D.; Socransky, S.S. Microbial etiological agents of destructive periodontal diseases. *Periodontol. 2000* **1994**, *5*, 78–111. [CrossRef] [PubMed]
14. Caton, J.G.; Armitage, G.; Berglundh, T.; Chapple, I.L.C.; Jepsen, S.; Kornman, K.S.; Mealey, B.L.; Papapanou, P.N.; Sanz, M.; Tonetti, M.S. A new classification scheme for periodontal and peri-implant diseases and conditions—Introduction and key changes from the 1999 classification. *J. Periodontol.* **2018**, *89* (Suppl. 1), S1–S8. [CrossRef] [PubMed]
15. Tonetti, M.S.; Sanz, M. Implementation of the new classification of periodontal diseases: Decision-making algorithms for clinical practice and education. *J. Clin. Periodontol.* **2019**, *46*, 398–405. [CrossRef]
16. Harvey, A.L. Natural products in drug discovery. *Drug Discov. Today* **2008**, *13*, 894–901. [CrossRef]
17. Yilmaz, H.G.; Bayindir, H. Clinical evaluation of chlorhexidine and essential oils for adjunctive effects in ultrasonic instrumentation of furcation involvements: A randomized controlled clinical trial. *Int. J. Dent. Hyg.* **2012**, *10*, 113–117. [CrossRef] [PubMed]
18. Varela-Lopez, A.; Navarro-Hortal, M.D.; Giampieri, F.; Bullon, P.; Battino, M.; Quiles, J.L. Nutraceuticals in Periodontal Health: A Systematic Review on the Role of Vitamins in Periodontal Health Maintenance. *Molecules* **2018**, *23*, 1226. [CrossRef]
19. Nittayananta, W.; Limsuwan, S.; Srichana, T.; Sae-Wong, C.; Amnuaikit, T. Oral spray containing plant-derived compounds is effective against common oral pathogens. *Arch. Oral Biol.* **2018**, *90*, 80–85. [CrossRef]
20. Basu, A.; Masek, E.; Ebersole, J.L. Dietary Polyphenols and Periodontitis—A Mini-Review of Literature. *Molecules* **2018**, *23*, 1786. [CrossRef]
21. Zhu, F.; Du, B.; Xu, B. Anti-inflammatory effects of phytochemicals from fruits, vegetables, and food legumes: A review. *Crit. Rev. Food Sci. Nutr.* **2018**, *58*, 1260–1270. [CrossRef]
22. Weber, C.; Noels, H. Atherosclerosis: Current pathogenesis and therapeutic options. *Nat. Med.* **2011**, *17*, 1410–1422. [CrossRef]
23. World Health Organization. Cardiovascular Diseases (CVDs). Available online: https://www.who.int/en/news-room/fact-sheets/detail/cardiovascular-diseases-(cvds) (accessed on 25 January 2021).
24. Wong, B.W.; Meredith, A.; Lin, D.; McManus, B.M. The biological role of inflammation in atherosclerosis. *Can. J. Cardiol.* **2012**, *28*, 631–641. [CrossRef] [PubMed]
25. Rizzo, M.; Berneis, K.; Koulouris, S.; Pastromas, S.; Rini, G.B.; Sakellariou, D.; Manolis, A.S. Should we measure routinely oxidised and atherogenic dense low-density lipoproteins in subjects with type 2 diabetes? *Int. J. Clin. Pract.* **2010**, *64*, 1632–1642. [CrossRef]
26. Raggi, P.; Genest, J.; Giles, J.T.; Rayner, K.J.; Dwivedi, G.; Beanlands, R.S.; Gupta, M. Role of inflammation in the pathogenesis of atherosclerosis and therapeutic interventions. *Atherosclerosis* **2018**, *276*, 98–108. [CrossRef] [PubMed]
27. Kim, M.; Sahu, A.; Hwang, Y.; Kim, G.B.; Nam, G.H.; Kim, I.S.; Chan Kwon, I.; Tae, G. Targeted delivery of anti-inflammatory cytokine by nanocarrier reduces atherosclerosis in Apo E($-/-$) mice. *Biomaterials* **2020**, *226*, 119550. [CrossRef]

28. Viola, J.; Soehnlein, O. Atherosclerosis—A matter of unresolved inflammation. *Semin. Immunol.* **2015**, *27*, 184–193. [CrossRef] [PubMed]
29. Paoletti, R.; Gotto, A.M., Jr.; Hajjar, D.P. Inflammation in atherosclerosis and implications for therapy. *Circulation* **2004**, *109*, III20–III26. [CrossRef] [PubMed]
30. Tabas, I.; Bornfeldt, K.E. Macrophage Phenotype and Function in Different Stages of Atherosclerosis. *Circ. Res.* **2016**, *118*, 653–667. [CrossRef]
31. Moore, K.J.; Sheedy, F.J.; Fisher, E.A. Macrophages in atherosclerosis: A dynamic balance. *Nat. Rev. Immunol.* **2013**, *13*, 709–721. [CrossRef]
32. De Vries, J.E. Immunosuppressive and anti-inflammatory properties of interleukin 10. *Ann. Med.* **1995**, *27*, 537–541. [CrossRef]
33. Mallat, Z.; Besnard, S.; Duriez, M.; Deleuze, V.; Emmanuel, F.; Bureau, M.F.; Soubrier, F.; Esposito, B.; Duez, H.; Fievet, C.; et al. Protective role of interleukin-10 in atherosclerosis. *Circ. Res.* **1999**, *85*, e17–e24. [CrossRef]
34. Pinderski Oslund, L.J.; Hedrick, C.C.; Olvera, T.; Hagenbaugh, A.; Territo, M.; Berliner, J.A.; Fyfe, A.I. Interleukin-10 blocks atherosclerotic events in vitro and in vivo. *Arterioscler. Thromb. Vasc. Biol.* **1999**, *19*, 2847–2853. [CrossRef]
35. Murray, P.J. The primary mechanism of the IL-10-regulated antiinflammatory response is to selectively inhibit transcription. *Proc. Natl. Acad. Sci. USA* **2005**, *102*, 8686–8691. [CrossRef]
36. Libby, P.; Ridker, P.M.; Hansson, G.K. Inflammation in atherosclerosis: From pathophysiology to practice. *J. Am. Coll. Cardiol.* **2009**, *54*, 2129–2138. [CrossRef] [PubMed]
37. Carrizales-Sepulveda, E.F.; Ordaz-Farias, A.; Vera-Pineda, R.; Flores-Ramirez, R. Periodontal Disease, Systemic Inflammation and the Risk of Cardiovascular Disease. *Heart Lung Circ.* **2018**, *27*, 1327–1334. [CrossRef] [PubMed]
38. Lockhart, P.B.; Bolger, A.F.; Papapanou, P.N.; Osinbowale, O.; Trevisan, M.; Levison, M.E.; Taubert, K.A.; Newburger, J.W.; Gornik, H.L.; Gewitz, M.H.; et al. Periodontal disease and atherosclerotic vascular disease: Does the evidence support an independent association? A scientific statement from the American Heart Association. *Circulation* **2012**, *125*, 2520–2544. [CrossRef] [PubMed]
39. Golia, E.; Limongelli, G.; Natale, F.; Fimiani, F.; Maddaloni, V.; Pariggiano, I.; Bianchi, R.; Crisci, M.; D'Acierno, L.; Giordano, R.; et al. Inflammation and cardiovascular disease: From pathogenesis to therapeutic target. *Curr. Atheroscler. Rep.* **2014**, *16*, 435. [CrossRef] [PubMed]
40. Patti, M.G.; Allaix, M.E.; Fisichella, P.M. Analysis of the Causes of Failed Antireflux Surgery and the Principles of Treatment: A Review. *JAMA Surg.* **2015**, *150*, 585–590. [CrossRef] [PubMed]
41. Cicero, A.F.G.; Colletti, A.; Bellentani, S. Nutraceutical Approach to Non-Alcoholic Fatty Liver Disease (NAFLD): The Available Clinical Evidence. *Nutrients* **2018**, *10*, 1153. [CrossRef]
42. Patti, A.M.; Al-Rasadi, K.; Giglio, R.V.; Nikolic, D.; Mannina, C.; Castellino, G.; Chianetta, R.; Banach, M.; Cicero, A.F.G.; Lippi, G.; et al. Natural approaches in metabolic syndrome management. *Arch. Med. Sci. AMS* **2018**, *14*, 422–441. [CrossRef]
43. Blaizot, A.; Vergnes, J.N.; Nuwwareh, S.; Amar, J.; Sixou, M. Periodontal diseases and cardiovascular events: Meta-analysis of observational studies. *Int. Dent. J.* **2009**, *59*, 197–209.
44. Zeng, X.T.; Leng, W.D.; Lam, Y.Y.; Yan, B.P.; Wei, X.M.; Weng, H.; Kwong, J.S. Periodontal disease and carotid atherosclerosis: A meta-analysis of 17,330 participants. *Int. J. Cardiol.* **2016**, *203*, 1044–1051. [CrossRef] [PubMed]
45. Humphrey, L.L.; Fu, R.; Buckley, D.I.; Freeman, M.; Helfand, M. Periodontal disease and coronary heart disease incidence: A systematic review and meta-analysis. *J. Gen. Intern. Med.* **2008**, *23*, 2079–2086. [CrossRef] [PubMed]
46. Mustapha, I.Z.; Debrey, S.; Oladubu, M.; Ugarte, R. Markers of systemic bacterial exposure in periodontal disease and cardiovascular disease risk: A systematic review and meta-analysis. *J. Periodontol.* **2007**, *78*, 2289–2302. [CrossRef] [PubMed]
47. Bahekar, A.A.; Singh, S.; Saha, S.; Molnar, J.; Arora, R. The prevalence and incidence of coronary heart disease is significantly increased in periodontitis: A meta-analysis. *Am. Heart J.* **2007**, *154*, 830–837. [CrossRef]
48. Loos, B.G. Systemic markers of inflammation in periodontitis. *J. Periodontol.* **2005**, *76*, 2106–2115. [CrossRef] [PubMed]
49. Furuhashi, M.; Ishimura, S.; Ota, H.; Miura, T. Lipid chaperones and metabolic inflammation. *Int. J. Inflam.* **2011**, *2011*, 642612. [CrossRef] [PubMed]
50. Hotamisligil, G.S. Inflammation and metabolic disorders. *Nature* **2006**, *444*, 860–867. [CrossRef]
51. Gregor, M.F.; Hotamisligil, G.S. Inflammatory mechanisms in obesity. *Annu. Rev. Immunol.* **2011**, *29*, 415–445. [CrossRef]
52. Sanz, M.; Marco Del Castillo, A.; Jepsen, S.; Gonzalez-Juanatey, J.R.; D'Aiuto, F.; Bouchard, P.; Chapple, I.; Dietrich, T.; Gotsman, I.; Graziani, F.; et al. Periodontitis and cardiovascular diseases: Consensus report. *J. Clin. Periodontol.* **2020**, *47*, 268–288. [CrossRef] [PubMed]
53. Del Pinto, R.; Pietropaoli, D.; Munoz-Aguilera, E.; D'Aiuto, F.; Czesnikiewicz-Guzik, M.; Monaco, A.; Guzik, T.J.; Ferri, C. Periodontitis and Hypertension: Is the Association Causal? *High Blood Press. Cardiovasc. Prev.* **2020**, *27*, 281–289. [CrossRef] [PubMed]
54. Orlandi, M.; Graziani, F.; D'Aiuto, F. Periodontal therapy and cardiovascular risk. *Periodontol. 2000* **2020**, *83*, 107–124. [CrossRef]
55. Priyamvara, A.; Dey, A.K.; Bandyopadhyay, D.; Katikineni, V.; Zaghlol, R.; Basyal, B.; Barssoum, K.; Amarin, R.; Bhatt, D.L.; Lavie, C.J. Periodontal Inflammation and the Risk of Cardiovascular Disease. *Curr. Atheroscler. Rep.* **2020**, *22*, 28. [CrossRef] [PubMed]
56. Xue, W.; Yu, J.; Chen, W. Plants and Their Bioactive Constituents in Mesenchymal Stem Cell-Based Periodontal Regeneration: A Novel Prospective. *Biomed. Res. Int.* **2018**, *2018*, 7571363. [CrossRef]
57. Ríos, J.-L. Essential Oils. In *Essential Oils in Food Preservation, Flavor and Safety*; Preedy, V.R., Ed.; Academic Press: San Diego, CA, USA, 2016; pp. 3–10.

58. Ul Hassan, M.N.; Zainal, Z.; Ismail, I. Green leaf volatiles: Biosynthesis, biological functions and their applications in biotechnology. *Plant Biotechnol. J.* **2015**, *13*, 727–739. [CrossRef]
59. Sharifi-Rad, J.; Sureda, A.; Tenore, G.C.; Daglia, M.; Sharifi-Rad, M.; Valussi, M.; Tundis, R.; Sharifi-Rad, M.; Loizzo, M.R.; Ademiluyi, A.O.; et al. Biological Activities of Essential Oils: From Plant Chemoecology to Traditional Healing Systems. *Molecules* **2017**, *22*, 70. [CrossRef]
60. Sangwan, N.S.; Farooqi, A.H.A.; Shabih, F.; Sangwan, R.S. Regulation of essential oil production in plants. *Plant Growth Regul.* **2001**, *34*, 3–21. [CrossRef]
61. Aziz, Z.A.A.; Ahmad, A.; Setapar, S.H.M.; Karakucuk, A.; Azim, M.M.; Lokhat, D.; Rafatullah, M.; Ganash, M.; Kamal, M.A.; Ashraf, G.M. Essential Oils: Extraction Techniques, Pharmaceutical And Therapeutic Potential—A Review. *Curr. Drug Metab.* **2018**, *19*, 1100–1110. [CrossRef]
62. Khansari, N.; Shakiba, Y.; Mahmoudi, M. Chronic inflammation and oxidative stress as a major cause of age-related diseases and cancer. *Recent Pat. Inflamm. Allergy Drug Discov.* **2009**, *3*, 73–80. [CrossRef]
63. Dandekar, A.; Mendez, R.; Zhang, K. Cross talk between ER stress, oxidative stress, and inflammation in health and disease. *Methods Mol. Biol.* **2015**, *1292*, 205–214. [CrossRef]
64. Arulselvan, P.; Fard, M.T.; Tan, W.S.; Gothai, S.; Fakurazi, S.; Norhaizan, M.E.; Kumar, S.S. Role of Antioxidants and Natural Products in Inflammation. *Oxid. Med. Cell Longev.* **2016**, *2016*, 5276130. [CrossRef]
65. Bonesi, M.; Loizzo, M.R.; Acquaviva, R.; Malfa, G.A.; Aiello, F.; Tundis, R. Anti-inflammatory and Antioxidant Agents from Salvia Genus (Lamiaceae): An Assessment of the Current State of Knowledge. *Antiinflamm. Antiallergy Agents Med. Chem.* **2017**, *16*, 70–86. [CrossRef] [PubMed]
66. Dajic Stevanovic, Z.; Sieniawska, E.; Glowniak, K.; Obradovic, N.; Pajic-Lijakovic, I. Natural Macromolecules as Carriers for Essential Oils: From Extraction to Biomedical Application. *Front. Bioeng. Biotechnol.* **2020**, *8*, 563. [CrossRef] [PubMed]
67. Toth, P.P.; Patti, A.M.; Nikolic, D.; Giglio, R.V.; Castellino, G.; Biancucci, T.; Geraci, F.; David, S.; Montalto, G.; Rizvi, A.; et al. Bergamot Reduces Plasma Lipids, Atherogenic Small Dense LDL, and Subclinical Atherosclerosis in Subjects with Moderate Hypercholesterolemia: A 6 Months Prospective Study. *Front. Pharmacol.* **2015**, *6*, 299. [CrossRef]
68. Castellino, G.; Nikolic, D.; Magan-Fernandez, A.; Malfa, G.A.; Chianetta, R.; Patti, A.M.; Amato, A.; Montalto, G.; Toth, P.P.; Banach, M.; et al. Altilix((R)) Supplement Containing Chlorogenic Acid and Luteolin Improved Hepatic and Cardiometabolic Parameters in Subjects with Metabolic Syndrome: A 6 Month Randomized, Double-Blind, Placebo-Controlled Study. *Nutrients* **2019**, *11*, 2580. [CrossRef] [PubMed]
69. Leyva-Lopez, N.; Gutierrez-Grijalva, E.P.; Vazquez-Olivo, G.; Heredia, J.B. Essential Oils of Oregano: Biological Activity beyond Their Antimicrobial Properties. *Molecules* **2017**, *22*, 989. [CrossRef] [PubMed]
70. Dagli, N.; Dagli, R.; Mahmoud, R.S.; Baroudi, K. Essential oils, their therapeutic properties, and implication in dentistry: A review. *J. Int. Soc. Prev. Community Dent.* **2015**, *5*, 335–340. [CrossRef] [PubMed]
71. Freires, I.A.; Denny, C.; Benso, B.; de Alencar, S.M.; Rosalen, P.L. Antibacterial Activity of Essential Oils and Their Isolated Constituents against Cariogenic Bacteria: A Systematic Review. *Molecules* **2015**, *20*, 7329–7358. [CrossRef]
72. Hammer, K.A.; Carson, C.F.; Riley, T.V. Antifungal effects of *Melaleuca alternifolia* (tea tree) oil and its components on *Candida albicans*, *Candida glabrata* and *Saccharomyces cerevisiae*. *J. Antimicrob. Chemother.* **2004**, *53*, 1081–1085. [CrossRef]
73. Anusha, D.; Chaly, P.E.; Junaid, M.; Nijesh, J.E.; Shivashankar, K.; Sivasamy, S. Efficacy of a mouthwash containing essential oils and curcumin as an adjunct to nonsurgical periodontal therapy among rheumatoid arthritis patients with chronic periodontitis: A randomized controlled trial. *Indian J. Dent. Res.* **2019**, *30*, 506–511. [CrossRef]
74. Uritu, C.M.; Mihai, C.T.; Stanciu, G.D.; Dodi, G.; Alexa-Stratulat, T.; Luca, A.; Leon-Constantin, M.M.; Stefanescu, R.; Bild, V.; Melnic, S.; et al. Medicinal Plants of the Family Lamiaceae in Pain Therapy: A Review. *Pain Res. Manag.* **2018**, *2018*, 7801543. [CrossRef]
75. Li, B.; Cantino, P.D.; Olmstead, R.G.; Bramley, G.L.; Xiang, C.L.; Ma, Z.H.; Tan, Y.H.; Zhang, D.X. A large-scale chloroplast phylogeny of the Lamiaceae sheds new light on its subfamilial classification. *Sci. Rep.* **2016**, *6*, 34343. [CrossRef]
76. Mamadalieva, N.Z.; Akramov, D.K.; Ovidi, E.; Tiezzi, A.; Nahar, L.; Azimova, S.S.; Sarker, S.D. Aromatic Medicinal Plants of the Lamiaceae Family from Uzbekistan: Ethnopharmacology, Essential Oils Composition, and Biological Activities. *Medicines* **2017**, *4*, 8. [CrossRef]
77. Karpinski, T.M. Essential Oils of Lamiaceae Family Plants as Antifungals. *Biomolecules* **2020**, *10*, 103. [CrossRef] [PubMed]
78. Borges, R.S.; Ortiz, B.L.S.; Pereira, A.C.M.; Keita, H.; Carvalho, J.C.T. Rosmarinus officinalis essential oil: A review of its phytochemistry, anti-inflammatory activity, and mechanisms of action involved. *J. Ethnopharmacol.* **2019**, *229*, 29–45. [CrossRef]
79. Capatina, L.; Todirascu-Ciornea, E.; Napoli, E.M.; Ruberto, G.; Hritcu, L.; Dumitru, G. Thymus vulgaris Essential Oil Protects Zebrafish against Cognitive Dysfunction by Regulating Cholinergic and Antioxidants Systems. *Antioxidants* **2020**, *9*, 1083. [CrossRef] [PubMed]
80. Carbone, C.; Martins-Gomes, C.; Caddeo, C.; Silva, A.M.; Musumeci, T.; Pignatello, R.; Puglisi, G.; Souto, E.B. Mediterranean essential oils as precious matrix components and active ingredients of lipid nanoparticles. *Int. J. Pharm.* **2018**, *548*, 217–226. [CrossRef] [PubMed]
81. Cavanagh, H.M.; Wilkinson, J.M. Biological activities of lavender essential oil. *Phytother. Res.* **2002**, *16*, 301–308. [CrossRef] [PubMed]

82. Silva, G.L.; Luft, C.; Lunardelli, A.; Amaral, R.H.; Melo, D.A.; Donadio, M.V.; Nunes, F.B.; de Azambuja, M.S.; Santana, J.C.; Moraes, C.M.; et al. Antioxidant, analgesic and anti-inflammatory effects of lavender essential oil. *An. Acad. Bras. Cienc.* **2015**, *87*, 1397–1408. [CrossRef]
83. Carrasco, A.; Martinez-Gutierrez, R.; Tomas, V.; Tudela, J. Lavandin (*Lavandula x intermedia* Emeric ex Loiseleur) essential oil from Spain: Determination of aromatic profile by gas chromatography-mass spectrometry, antioxidant and lipoxygenase inhibitory bioactivities. *Nat. Prod. Res.* **2016**, *30*, 1123–1130. [CrossRef]
84. Andrade, J.M.; Faustino, C.; Garcia, C.; Ladeiras, D.; Reis, C.P.; Rijo, P. *Rosmarinus officinalis* L.: An update review of its phytochemistry and biological activity. *Future Sci. OA* **2018**, *4*, FSO283. [CrossRef] [PubMed]
85. Bustanji, Y.; Issa, A.; Mohammad, M.; Hudaib, M.; Tawah, K.; Alkhatib, H.; Almasri, I.; Al-Khalidi, B. Inhibition of hormone sensitive lipase and pancreatic lipase by *Rosmarinus officinalis* extract and selected phenolic constituents. *J. Med. Plants Res.* **2010**, *4*, 2235–2242.
86. Silva, A.M.; Machado, I.D.; Santin, J.R.; de Melo, I.L.; Pedrosa, G.V.; Genovese, M.I.; Farsky, S.H.; Mancini-Filho, J. Aqueous extract of *Rosmarinus officinalis* L. inhibits neutrophil influx and cytokine secretion. *Phytother. Res.* **2015**, *29*, 125–133. [CrossRef] [PubMed]
87. Borges, R.S.; Keita, H.; Ortiz, B.L.S.; Dos Santos Sampaio, T.I.; Ferreira, I.M.; Lima, E.S.; de Jesus Amazonas da Silva, M.; Fernandes, C.P.; de Faria Mota Oliveira, A.E.M.; da Conceicao, E.C.; et al. Anti-inflammatory activity of nanoemulsions of essential oil from *Rosmarinus officinalis* L.: In vitro and in zebrafish studies. *Inflammopharmacology* **2018**, *26*, 1057–1080. [CrossRef] [PubMed]
88. Valones, M.A.A.; Silva, I.C.G.; Gueiros, L.A.M.; Leao, J.C.; Caldas, A.F., Jr.; Carvalho, A.A.T. Clinical Assessment of Rosemary-based Toothpaste (*Rosmarinus officinalis* Linn.): A Randomized Controlled Double-blind Study. *Braz. Dent. J.* **2019**, *30*, 146–151. [CrossRef] [PubMed]
89. Rasooli, I.; Shayegh, S.; Taghizadeh, M.; Astaneh, S.D. Phytotherapeutic prevention of dental biofilm formation. *Phytother. Res.* **2008**, *22*, 1162–1167. [CrossRef] [PubMed]
90. Bernardes, W.A.; Lucarini, R.; Tozatti, M.G.; Flauzino, L.G.; Souza, M.G.; Turatti, I.C.; Andrade e Silva, M.L.; Martins, C.H.; da Silva Filho, A.A.; Cunha, W.R. Antibacterial activity of the essential oil from *Rosmarinus officinalis* and its major components against oral pathogens. *Z. Nat. C* **2010**, *65*, 588–593. [CrossRef]
91. Smullen, J.; Finney, M.; Storey, D.M.; Foster, H.A. Prevention of artificial dental plaque formation in vitro by plant extracts. *J. Appl. Microbiol.* **2012**, *113*, 964–973. [CrossRef]
92. Salehi, B.; Mishra, A.P.; Shukla, I.; Sharifi-Rad, M.; Contreras, M.D.M.; Segura-Carretero, A.; Fathi, H.; Nasrabadi, N.N.; Kobarfard, F.; Sharifi-Rad, J. Thymol, thyme, and other plant sources: Health and potential uses. *Phytother. Res.* **2018**, *32*, 1688–1706. [CrossRef]
93. Iauk, L.; Acquaviva, R.; Mastrojeni, S.; Amodeo, A.; Pugliese, M.; Ragusa, M.; Loizzo, M.R.; Menichini, F.; Tundis, R. Antibacterial, antioxidant and hypoglycaemic effects of *Thymus capitatus* (L.) Hoffmanns. et Link leaves' fractions. *J. Enzyme Inhib. Med. Chem.* **2015**, *30*, 360–365. [CrossRef] [PubMed]
94. Manconi, M.; Petretto, G.; D'Hallewin, G.; Escribano, E.; Milia, E.; Pinna, R.; Palmieri, A.; Firoznezhad, M.; Peris, J.E.; Usach, I.; et al. Thymus essential oil extraction, characterization and incorporation in phospholipid vesicles for the antioxidant/antibacterial treatment of oral cavity diseases. *Colloids Surf. B Biointerfaces* **2018**, *171*, 115–122. [CrossRef] [PubMed]
95. Valerio, F.; Mezzapesa, G.N.; Ghannouchi, A.; Mondelli, D.; Logrieco, A.F.; Perrino, E.V. Characterization and Antimicrobial Properties of Essential Oils from Four Wild Taxa of Lamiaceae Family Growing in Apulia. *Agronomy* **2021**, *11*, 1431. [CrossRef]
96. Alvarez Echazu, M.I.; Olivetti, C.E.; Anesini, C.; Perez, C.J.; Alvarez, G.S.; Desimone, M.F. Development and evaluation of thymol-chitosan hydrogels with antimicrobial-antioxidant activity for oral local delivery. *Mater. Sci. Eng. C Mater. Biol. Appl.* **2017**, *81*, 588–596. [CrossRef]
97. Ocana, A.; Reglero, G. Effects of Thyme Extract Oils (from *Thymus vulgaris*, *Thymus zygis*, and *Thymus hyemalis*) on Cytokine Production and Gene Expression of oxLDL-Stimulated THP-1-Macrophages. *J. Obes.* **2012**, *2012*, 104706. [CrossRef]
98. Penalver, P.; Huerta, B.; Borge, C.; Astorga, R.; Romero, R.; Perea, A. Antimicrobial activity of five essential oils against origin strains of the Enterobacteriaceae family. *APMIS* **2005**, *113*, 1–6. [CrossRef]
99. Hart, P.H.; Brand, C.; Carson, C.F.; Riley, T.V.; Prager, R.H.; Finlay-Jones, J.J. Terpinen-4-ol, the main component of the essential oil of *Melaleuca alternifolia* (tea tree oil), suppresses inflammatory mediator production by activated human monocytes. *Inflamm. Res.* **2000**, *49*, 619–626. [CrossRef]
100. Nogueira, M.N.; Aquino, S.G.; Rossa Junior, C.; Spolidorio, D.M. Terpinen-4-ol and alpha-terpineol (tea tree oil components) inhibit the production of IL-1beta, IL-6 and IL-10 on human macrophages. *Inflamm. Res.* **2014**, *63*, 769–778. [CrossRef]
101. Zhang, Y.; Li, D.; Wang, Z.; Zang, W.; Rao, P.; Liang, Y.; Mei, Y. Alpha-terpineol affects synthesis and antitumor activity of triterpenoids from *Antrodia cinnamomea* mycelia in solid-state culture. *Food Funct.* **2018**, *9*, 6517–6525. [CrossRef] [PubMed]
102. Huo, M.; Gao, R.; Jiang, L.; Cui, X.; Duan, L.; Deng, X.; Guan, S.; Wei, J.; Soromou, L.W.; Feng, H.; et al. Suppression of LPS-induced inflammatory responses by gossypol in RAW 264.7 cells and mouse models. *Int. Immunopharmacol.* **2013**, *15*, 442–449. [CrossRef] [PubMed]
103. Ma, J.; Xu, H.; Wu, J.; Qu, C.; Sun, F.; Xu, S. Linalool inhibits cigarette smoke-induced lung inflammation by inhibiting NF-kappaB activation. *Int. Immunopharmacol.* **2015**, *29*, 708–713. [CrossRef]

104. Kummer, R.; Fachini-Queiroz, F.C.; Estevao-Silva, C.F.; Grespan, R.; Silva, E.L.; Bersani-Amado, C.A.; Cuman, R.K. Evaluation of Anti-Inflammatory Activity of Citrus latifolia Tanaka Essential Oil and Limonene in Experimental Mouse Models. *Evid. Based Complement. Alternat. Med.* **2013**, *2013*, 859083. [CrossRef]
105. Yoon, W.J.; Lee, N.H.; Hyun, C.G. Limonene suppresses lipopolysaccharide-induced production of nitric oxide, prostaglandin E2, and pro-inflammatory cytokines in RAW 264.7 macrophages. *J. Oleo Sci.* **2010**, *59*, 415–421. [CrossRef]
106. Rufino, A.T.; Ribeiro, M.; Sousa, C.; Judas, F.; Salgueiro, L.; Cavaleiro, C.; Mendes, A.F. Evaluation of the anti-inflammatory, anti-catabolic and pro-anabolic effects of E-caryophyllene, myrcene and limonene in a cell model of osteoarthritis. *Eur. J. Pharmacol.* **2015**, *750*, 141–150. [CrossRef] [PubMed]
107. Botelho, M.A.; Rao, V.S.; Montenegro, D.; Bandeira, M.A.; Fonseca, S.G.; Nogueira, N.A.; Ribeiro, R.A.; Brito, G.A. Effects of a herbal gel containing carvacrol and chalcones on alveolar bone resorption in rats on experimental periodontitis. *Phytother. Res.* **2008**, *22*, 442–449. [CrossRef] [PubMed]
108. Guimaraes, A.G.; Xavier, M.A.; de Santana, M.T.; Camargo, E.A.; Santos, C.A.; Brito, F.A.; Barreto, E.O.; Cavalcanti, S.C.; Antoniolli, A.R.; Oliveira, R.C.; et al. Carvacrol attenuates mechanical hypernociception and inflammatory response. *Naunyn Schmiedebergs Arch. Pharmacol.* **2012**, *385*, 253–263. [CrossRef] [PubMed]
109. Hotta, M.; Nakata, R.; Katsukawa, M.; Hori, K.; Takahashi, S.; Inoue, H. Carvacrol, a component of thyme oil, activates PPARalpha and gamma and suppresses COX-2 expression. *J. Lipid Res.* **2010**, *51*, 132–139. [CrossRef] [PubMed]
110. Landa, P.; Kokoska, L.; Pribylova, M.; Vanek, T.; Marsik, P. In vitro anti-inflammatory activity of carvacrol: Inhibitory effect on COX-2 catalyzed prostaglandin E(2) biosynthesis. *Arch. Pharm. Res.* **2009**, *32*, 75–78. [CrossRef]
111. Yalcin, H.; Anik, M.; Sanda, M.A.; Cakir, A. Gas chromatography/mass spectrometry analysis of Laurus nobilis essential oil composition of northern Cyprus. *J. Med. Food* **2007**, *10*, 715–719. [CrossRef]
112. Bastos, V.P.; Gomes, A.S.; Lima, F.J.; Brito, T.S.; Soares, P.M.; Pinho, J.P.; Silva, C.S.; Santos, A.A.; Souza, M.H.; Magalhaes, P.J. Inhaled 1,8-cineole reduces inflammatory parameters in airways of ovalbumin-challenged Guinea pigs. *Basic Clin. Pharmacol. Toxicol.* **2011**, *108*, 34–39. [CrossRef]
113. Kennedy-Feitosa, E.; Okuro, R.T.; Pinho Ribeiro, V.; Lanzetti, M.; Barroso, M.V.; Zin, W.A.; Porto, L.C.; Brito-Gitirana, L.; Valenca, S.S. Eucalyptol attenuates cigarette smoke-induced acute lung inflammation and oxidative stress in the mouse. *Pulm. Pharmacol. Ther.* **2016**, *41*, 11–18. [CrossRef] [PubMed]
114. Juergens, U.R.; Engelen, T.; Racke, K.; Stober, M.; Gillissen, A.; Vetter, H. Inhibitory activity of 1,8-cineol (eucalyptol) on cytokine production in cultured human lymphocytes and monocytes. *Pulm. Pharmacol. Ther.* **2004**, *17*, 281–287. [CrossRef] [PubMed]
115. Kim, K.Y.; Lee, H.S.; Seol, G.H. Eucalyptol suppresses matrix metalloproteinase-9 expression through an extracellular signal-regulated kinase-dependent nuclear factor-kappa B pathway to exert anti-inflammatory effects in an acute lung inflammation model. *J. Pharm. Pharmacol.* **2015**, *67*, 1066–1074. [CrossRef] [PubMed]
116. Chauhan, A.K.; Jakhar, R.; Paul, S.; Kang, S.C. Potentiation of macrophage activity by thymol through augmenting phagocytosis. *Int. Immunopharmacol.* **2014**, *18*, 340–346. [CrossRef]
117. Vigo, E.; Cepeda, A.; Gualillo, O.; Perez-Fernandez, R. In-vitro anti-inflammatory effect of Eucalyptus globulus and Thymus vulgaris: Nitric oxide inhibition in J774A.1 murine macrophages. *J. Pharm. Pharmacol.* **2004**, *56*, 257–263. [CrossRef]
118. Marsik, P.; Kokoska, L.; Landa, P.; Nepovim, A.; Soudek, P.; Vanek, T. In vitro inhibitory effects of thymol and quinones of Nigella sativa seeds on cyclooxygenase-1- and -2-catalyzed prostaglandin E2 biosyntheses. *Planta Med.* **2005**, *71*, 739–742. [CrossRef] [PubMed]
119. Liang, D.; Li, F.; Fu, Y.; Cao, Y.; Song, X.; Wang, T.; Wang, W.; Guo, M.; Zhou, E.; Li, D.; et al. Thymol inhibits LPS-stimulated inflammatory response via down-regulation of NF-kappaB and MAPK signaling pathways in mouse mammary epithelial cells. *Inflammation* **2014**, *37*, 214–222. [CrossRef]
120. Khaleel, C.; Tabanca, N.; Buchbauer, G.J.O.C. α-Terpineol, a natural monoterpene: A review of its biological properties. *Open Chem.* **2018**, *16*, 349–361. [CrossRef]
121. Zhang, H.; Hua, R.; Zhang, B.; Zhang, X.; Yang, H.; Zhou, X. Serine Alleviates Dextran Sulfate Sodium-Induced Colitis and Regulates the Gut Microbiota in Mice. *Front. Microbiol.* **2018**, *9*, 3062. [CrossRef]
122. Peng, L.Y.; Shi, H.T.; Yuan, M.; Li, J.H.; Song, K.; Huang, J.N.; Yi, P.F.; Shen, H.Q.; Fu, B.D. Madecassoside Protects Against LPS-Induced Acute Lung Injury via Inhibiting TLR4/NF-kappaB Activation and Blood-Air Barrier Permeability. *Front. Pharmacol.* **2020**, *11*, 807. [CrossRef]
123. Joca, H.C.; Vieira, D.C.; Vasconcelos, A.P.; Araujo, D.A.; Cruz, J.S. Carvacrol modulates voltage-gated sodium channels kinetics in dorsal root ganglia. *Eur. J. Pharmacol.* **2015**, *756*, 22–29. [CrossRef] [PubMed]
124. Joca, H.C.; Cruz-Mendes, Y.; Oliveira-Abreu, K.; Maia-Joca, R.P.; Barbosa, R.; Lemos, T.L.; Lacerda Beirao, P.S.; Leal-Cardoso, J.H. Carvacrol decreases neuronal excitability by inhibition of voltage-gated sodium channels. *J. Nat. Prod.* **2012**, *75*, 1511–1517. [CrossRef] [PubMed]
125. Goncalves, J.C.; de Meneses, D.A.; de Vasconcelos, A.P.; Piauilino, C.A.; Almeida, F.R.; Napoli, E.M.; Ruberto, G.; de Araujo, D.A. Essential oil composition and antinociceptive activity of Thymus capitatus. *Pharm. Biol.* **2017**, *55*, 782–786. [CrossRef] [PubMed]
126. Cosentino, S.; Tuberoso, C.I.; Pisano, B.; Satta, M.; Mascia, V.; Arzedi, E.; Palmas, F. In-vitro antimicrobial activity and chemical composition of Sardinian Thymus essential oils. *Lett. Appl. Microbiol.* **1999**, *29*, 130–135. [CrossRef]

127. Perrino, E.V.; Valerio, F.; Jallali, S.; Trani, A.; Mezzapesa, G.N. Ecological and Biological Properties of Satureja cuneifolia Ten. and *Thymus spinulosus* Ten.: Two Wild Officinal Species of Conservation Concern in Apulia (Italy). A Preliminary Survey. *Plants* **2021**, *10*, 1952. [CrossRef] [PubMed]
128. Juergens, U.R.; Dethlefsen, U.; Steinkamp, G.; Gillissen, A.; Repges, R.; Vetter, H. Anti-inflammatory activity of 1.8-cineol (eucalyptol) in bronchial asthma: A double-blind placebo-controlled trial. *Respir. Med.* **2003**, *97*, 250–256. [CrossRef]

MDPI

Systematic Review

Protective Treatments against Endothelial Glycocalyx Degradation in Surgery: A Systematic Review and Meta-Analysis

Hasnain Q. R. B. Khan [1,*] and Gwendolen C. Reilly [2]

[1] Academic Unit of Medical Education, Faculty of Medicine, Dentistry and Health, Sheffield Medical School, University of Sheffield, Beech Hill Road, Broomhall, Sheffield S10 2RX, UK
[2] Department of Material Sciences and Engineering, INSIGNEO Institute for In Silico Medicine University of Sheffield, Mappin Street, Sheffield S1 3JD, UK; g.reilly@sheffield.ac.uk
* Correspondence: hqrkhan1@sheffield.ac.uk; Tel.: +44-7888657373

Abstract: The aim was to explore the body of literature focusing on protective treatments against endothelial glycocalyx degradation in surgery. A comprehensive systematic review of relevant articles was conducted across databases. Inclusion criteria: (1) treatments for the protection of the endothelial glycocalyx in surgery; (2) syndecan-1 used as a biomarker for endothelial glycocalyx degradation. Outcomes analysed: (1) mean difference of syndecan-1 (2) correlation between glycocalyx degradation and inflammation; (3) correlation between glycocalyx degradation and extravasation. A meta-analysis was used to present mean differences and 95% confidence intervals. Seven articles with eight randomised controlled trials were included. The greatest change from baseline values in syndecan-1 concentrations was generally from the first timepoint measured post-operatively. Interventions looked to either dampen the inflammatory response or fluid therapy. Methylprednisolone had the highest mean difference in plasma syndecan-1 concentrations. Ulinastatin showed correlations between alleviation of degradation and preserving vascular permeability. In this systematic review of 385 patients, those treated were more likely than those treated with placebo to exhibit less shedding of the endothelial glycocalyx. Methylprednisolone has been shown to specifically target the transient increase of glycocalyx degradation immediately post-operation and has displayed anti-inflammatory effects. We have proposed suggestions for improved uniformity and enhanced confidence for future randomised controlled trials.

Keywords: endothelial glycocalyx; inflammatory response; fluid loading; surgery; post-operative; albumin extravasation

Citation: Khan, H.Q.R.B.; Reilly, G.C. Protective Treatments against Endothelial Glycocalyx Degradation in Surgery: A Systematic Review and Meta-Analysis. *Appl. Sci.* **2021**, *11*, 6994. https://doi.org/10.3390/app11156994

Academic Editor: Francesco Cappello

Received: 9 July 2021
Accepted: 26 July 2021
Published: 29 July 2021

1. Introduction

The vascular endothelium plays several important roles including haemostatic balance, endothelial integrity, and blood flow regulation [1]. It comprises a single layer of endothelial cells, lining every blood vessel in the body and is understood to be 4000–7000 m^2 [2]. These cells play a vital role as a semipermeable membrane to allow the exchange of nutrients and the removal of waste to and from the blood. A key structure involved in these actions is the endothelial glycocalyx (EG), which coats cells' extracellular matrix. EG thickness ranges from 0.2 μm in capillaries to 4.5 μm in the carotid artery [3]. The glycocalyx enables changes by a process known as mechanotransduction.

Mechanotransduction is the mechanism by which external mechanical stimuli are converted into cellular responses through signalling pathways [4]. The apical surface of the glycocalyx consists of glycosaminoglycans (GAGs); those commonly associated with the vasculature are heparan sulphate (HS), chondroitin sulphate (CS) and hyaluronic acid (HA) [5]. Additionally present are syndecans, which provide sites on the apical surface to highly regulate proteolytic cleavage. The best conceptual theory for glycocalyx to detect changes in blood flow was described by Squire [6] as the "wind in the trees". The wind (fluid flow) is sensed by the branches (GAGs) of which the drag force is transmitted through the trees/forest (glycocalyx) and stimulates a response. Twenty-four-hour exposure of fluid shear-stress (FSS) in an in vitro model of EG has been shown to enhance synthesis and distribution of the prevalent components (HS, CS, and HA) of the glycocalyx with nearly normal uniform spatial distribution, similar to baseline levels compared to only 30 min of exposure [7].

Surgery is associated with EG degradation [8]. A suggested mechanism for this is that the thickness of the glycocalyx, and thus the surface layer, is reduced by ischaemia/reperfusion [9]. This increases capillary permeability and, thereby, contributes to tissue oedema (Figure 1) [10]. EG degradation is known to disrupt the equilibrium between pro-inflammatory cytokines and adhesion molecules, and vasodilators and vasoconstrictors, leading to endothelial dysfunction [11]. Glycocalyx components are released, such as syndecan-1 and heparan sulphate, during surgery. However, the steps in which diseases, trauma, and surgery thin the EG are not well understood. Johansson et al. [12] undertook a prospective 75-patient double-blind cohort study and found trauma patients to have raised plasma syndecan-1 levels associated with 16 other markers for inflammation, and tissue and endothelial cell damage. Patients with significant EG degradation were shown to have a three-fold increase in mortality compared to those with lower syndecan-1 levels. A mechanism to explain the strong association between EG degradation and patients with trauma is through the extensive activation of the inflammatory and coagulation pathways. Protection against EG degradation could help treat against known diseases to shed major EG constituents, such as diabetes and hyperglycaemia, atherosclerosis and chronic kidney disease [13].

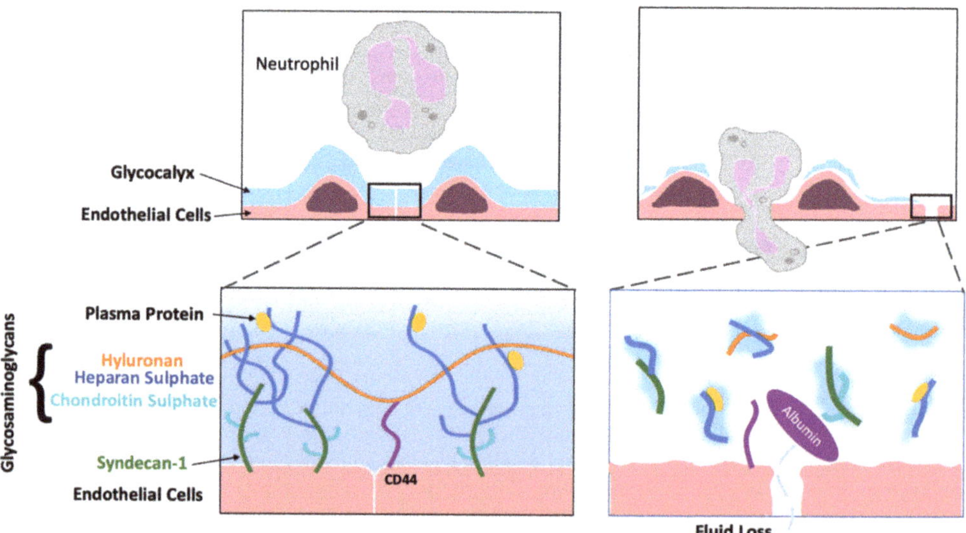

Figure 1. EG structure during health (**left**) and degradation (**right**). Disease causes thinning of the glycocalyx. Glycocalyx constituents are released into the plasma. An inflammatory response is produced with the recruitment of leucocytes and the loss of fluid.

Recent studies have looked into pharmacological interventions to prevent EG degradation during surgery. With the limited studies, interventions have looked into two therapeutic routes: (1) dampening the inflammatory response and (2) fluid-therapy. (1) The EG plays an important role in the post-injury inflammatory response [14]. An increase in C-reactive protein (CRP) has been associated with a decreased thickness of the EG and impaired vasoreactivity [15]. (2) Infused fluid therapy has been associated with impaired microcirculation, resulting in tissue oedema. Moreover, fluid infusion could cause EG degradation and further fluid loss into the lymphatic system through albumin and other plasma proteins moving across the vascular wall [16].

This systematic review and meta-analysis looks to examine the body of literature into protective treatments for EG degradation in surgery from most of the published clinical trials. The aim is to identify trends in treatments which may address the mechanism for which surgery causes EG degradation.

2. Materials and Methods

2.1. Search Strategy

A comprehensive review of the literature was conducted in May 2020, using databases, such as Cochrane Library of Systematic Reviews, Cochrane Central Register of Clinical Trials (CENTRAL), MEDLINE, PubMed, and Clinical Trials.org. Article selection was limited to publications in the English Language between 1 January 1950 and 20 May 2020. Search terms used were a combination of the following: "endothelial", "glycocalyx", and "degradation". Reference lists of the selected articles and other related studies were assessed for eligibility. Cross-referencing from identified articles and conference abstracts were also performed. Clinical trials which are currently in progress with no data available were not included.

2.2. Inclusion Criteria

One researcher (H.K.) performed the review process for inclusion in the systematic review. Specific inclusion criteria mandated systematic reviews and clinical trials that were from retrospective or prospective investigations which met the following criteria; (1) papers published in the English Language; (2) clinical investigations into protective treatments for EG; (3) syndecan-1 used as a biomarker for EG degradation (specific marker for significant degradation as opposed to HS for minor disturbances); (4) clinical investigations in which patients were followed from pre-operative/induction of anaesthesia to the end of surgical treatment [17].

2.3. Exclusion Criteria

From the search strategy, 987 studies were selected. The first screen excluded articles if they had not conducted investigations on surgical patients. A further screen excluded articles that investigated combined treatments and those in which more than one substance was administered during surgery.

After exclusion, 112 articles formed the inclusion of the initial review. Following this, an independent abstract review was conducted by the same author for the title review. Three were removed as they investigated the role of the EG within surgery without the use of an intervention. One assessed the microperfusion abilities of endothelial cells following surgery. Twenty-nine articles investigated the effects of surgery (cardiothoracic, abdominal, hysterectomy, etc.) on the EG and whether surgery is a stimulatory factor for EG degradation. Twelve articles looked into protective biological factors and signalling pathways of the body that are used to protect against EG degradation in surgery. Twenty-one articles were trials conducted on animals/in-vitro models. Twenty-nine articles investigated the effects of different surgical procedures had on the structural integrity of the EG. Six articles were related to the effects that degradation of the EG had on other systems.

The final screen of 11 articles in their entirety was completed by the same researcher (Khan, H) to ensure adequate data content for inclusion in the systematic review. One had been centred on post-operative interventions which focuses more into regeneration/recovery of the EG [18]. Three articles were excluded as they were clinical trials still undergoing with no preliminary results [19–21].

After the full text review, the final search included 7 articles that met the inclusion criteria and thus were used for analysis and data were extracted (Brettner et al. [22];

Kim et al. [23]; Lindberg-Larsen et al. [24]; Mennander et al. [25]; Nemme et al. [26]; Pesonen et al. [27]; Wang et al. [28]). The search strategy has been depicted in Figure 2.

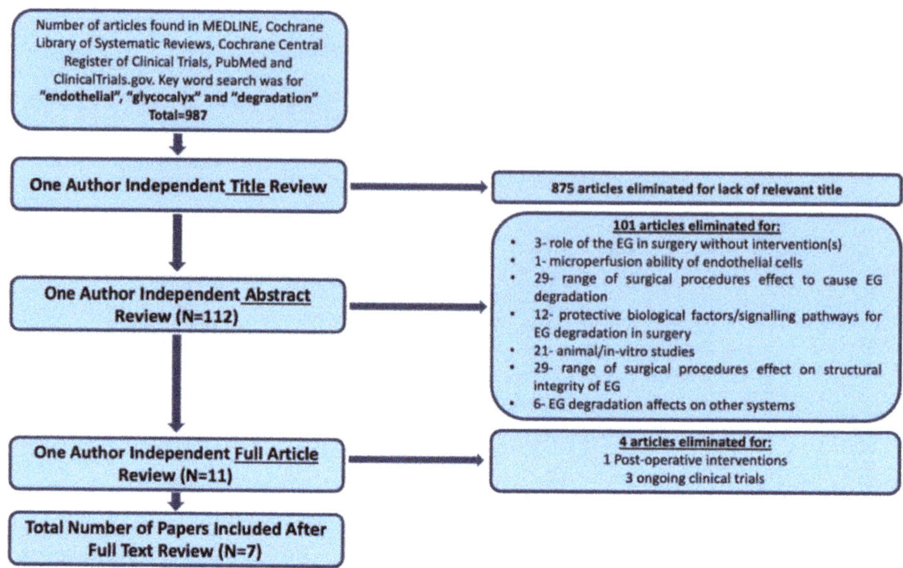

Figure 2. PRISMA flow chart of search strategy to identify pivotal publications of protective treatments against EG degradation in surgery for meta-analysis.

2.4. Data Extraction

Data were extracted and then reviewed, and all reported results were summarised to include comparable and clinically relevant outcomes. The following data were extracted from each clinical trial and used for descriptive comparison; author, year, study design, sample size, surgery, type of treatment, time points for administration of treatment (and/or placebo), biomarkers used to measure endothelial cell glycocalyx degradation, time points used for biomarker measurements (Table 1), and details regarding degree of endothelial cell glycocalyx degradation and post-surgical complications (Table 2).

2.5. Statistical Analyses

Using GraphPad Prism version 8, a line graph was produced to assess the change in plasma syndecan-1 concentrations in the control groups of the included articles. This enabled identification of the timepoints at which to expect interventions to target. This analysis formed the basis of a forest plot. Some articles provided checkpoints in surgery for when blood samples were taken rather than averaged time-points of each sample. The provided averaged times for each stage in surgery were converted to averaged timepoints when blood samples were taken. Meta-analysis was conducted using RevMan 5.4 from Cochrane Review to derive pooled effect estimates as a mean difference associated with 95% confidence intervals. The effect on the analysis of within and between study heterogeneity was quantified by calculating I^2. Overall, 4 studies with 5 randomised controlled trials were included for the forest plot.

Table 1. Publications included for review of protective treatments for EG degradation in surgery.

Reference	Sample Size	Study	Controlled	Comparative	Patients Age	Syndecan-1 Marker	Surgery	Anaesthesia	Treatment (Drug Class)
Pesonen et al. (2016) [Neonates] Pesonen et al. (2016) [VSD]	40 45	Prospective	No	Yes	7 (1–27) 0.37 (0.15–1.36)	Yes Yes	CPB	Sufentanil, Pancuronium, S-ketamine Maintained-> Sevoflurane	Methylprednisolone (Corticosteroid)
Brettner et al. (2019)	30	Prospective	No	No	65 (57.3–74)	Yes	CPB	Midazolam, Sufentanil, Pancuronium Maintained-> Sufentanil	Hydrocortisone (Corticosteroid)
Wang et al. (2017)	50	Prospective	No	Yes	58.56	Yes	VATS Lobectomy	Midazolam, Propofol, Sufentanil-> Propofol, Remifentanil, Rocuronium	Ulinastatin (UTI, anti-inflammatory agent)
Lindberg-Larsen et al. (2017)	63	Prospective	No	Yes	63.27	Yes	Unilateral Total Knee Arthroplasty	Standard procedure	Methylprednisolone (Corticosteroid)
Nemme et al. (2019)	24	Prospective	No	No	47(5) 46(4)	Yes	Hysterectomy	Midazolam, Fentanyl, Propofol	Ringers Lactate (Fluid Therapy)
Mennander et al. (2012)	13	Prospective	No	Yes	Not stated	Yes	CPB	Propofol, Sufentanil, cis-atracurium Maintained-> Sevoflurane	Diazoxide (Thiazide)
Kim et al. (2017)	120	Prospective	No	No	67.8 (9.9) 65.3 (10.5)	Yes	CPB	Midazolam, Sufentanil, Vecuronium Miantained0> Remifentanil, Propofol	Hydroxyl Starch (Crystalloid Starch Fluid Therapy)

Table 2. Outcome on EG degradation.

Reference	Timepoints	Syndecan-1 Levels	Timepoints of Statistical Significance	Heparan Sulphate	Inflammatory Markers	HGB and Albumin	Hyluronan
Pesonen et al. (2016) [Neonates] Pesonen et al. (2016) [VSD]	T1: induction of anaesthesia T2: 30-min on CPB T3: weaning of CPB T4: 6-h post-operative	Significant lowering in intervention group to none in control group, when comparing to baseline values None	T1->T2 N/A	N/A N/A	N/A N/A	N/A N/A	N/A N/A
Brettner et al. (2019)	T0: preoperative T1: induction of anaesthesia T2: 30 min after onset of CPB T3: weaning of CPB T4: 1-h post-operative T5: 4-h post-operative	None	N/A	T2–3	Higher CRP post-operatively on Days 1,2,3 in control to intervention group, when comparing to baseline values Higher IL-6 on Day 2 post-op in control to intervention group, when comparing to baseline values	N/A	N/A
Wang et al. (2017)	T0: preoperative T1: end of surgery	Control group showed a significant increase with none in intervention group, when comparing to baseline values	T0->T1	None	N/A	Significant lowering of albumin in control to intervention group, when comparing to baseline values	N/A
Lindberg-Larsen et al. (2017)	T0: pre-operative T1: 2-h post-operative T2: 6-h post-operative T3: 24-h post-operative	Control group showed a significant increase with none in intervention group	T0->T3	N/A	No effect on sE, prevent significant drop in thrombomodulin, Reduced increase in VEGF in intervention to control, CRP increased less so in intervention group, when comparing to baseline values	N/A	N/A

Table 2. *Cont.*

Reference	Timepoints	Syndecan-1 Levels	Timepoints of Statistical Significance	Heparan Sulphate	Inflammatory Markers	HGB and Albumin	Hyluronan
Nemme et al. (2019)	T0: pre-operative T1: 30-min intra-operative T2: 60-min intra-operative T3: 90-min intra-operative T4: 2-h post-operative	Significant increase from baseline values in both groups	T3->T4	T3->T4	Some patients showed raised CRP post-operatively However, only results for some were significant	Albumin showed similar trends to HB but lower No significant differences at time points The greatest difference of values was at 20 min but lowered significantly at 90 min	N/A
Mennander et al. (2012)	T1: induction of anaesthesia T2: after aortic clamp removal T3: 60-min intra-operative T4: closure of skin wound	Significant drop found in intervention group of which control group did not show, when comparing to baseline values	T2->T3 T3->T4	N/A	N/A	N/A	Similar changes to syndecan-1 levels
Kim et al. (2017)	T1: induction of anaesthesia T2: 60-min after coronary artery anastomosis T3: upon infusion of HES/crystalloid T4: skin closure T5: 12-h after ICU admission	None	N/A	N/A	N/A	N/A	N/A

3. Results

3.1. Description of Included Studies

The articles included were published in 2012–2019. Sample sizes varied from 16 patients [25] to 120 patients [23]. From the remaining 6 articles, patients' median age ranged from 7 months [27] to 67.8 years-old [21]. One article did not provide any information for patient average age [25]. Four articles did not include a range for the patients' ages [23,24,27,28] but a standard deviation. From the remaining 2 articles that included patient's age range, the range was from 1 month to 72 years-old.

Two articles described two treatments using different sample groups with no control group [23,25]. The remaining 5 articles investigated their treatment using two patient groups (control vs. intervention). One article conducted two separate randomised, prospective clinical trials by testing the intervention on two different patient cohorts based on age and surgical procedure [27]. All articles had delivered their drug following the administration of anaesthesia. Of the articles to investigate protective interventions for EG degradation, 4 looked into CPB [22–24,26], 1 into abdominal hysterectomy [26], 1 in knee replacement surgery [24] and 1 in pulmonary lobe resection [28].

3.2. Outcomes and Results

When analysing the included articles of our systematic review, the focus was on (1) the comparison between control and intervention groups to reduce levels of EG degradation in surgery; (2) correlations between glycocalyx degradation and an inflammatory response; and (3) correlation between glycocalyx degradation and extravasation. For each study, we assessed biomarkers for glycocalyx degradation (syndecan-1, and those that measured HS), as well as those that measured inflammatory markers and capillary leakage.

Four studies investigated the protective treatments against degradation of the EG in CPB [22,23,25,27]. Only 2 studies [25,27] showed a significant lowering of plasma syndecan-1, which in both cases occurred during surgery. The Mennander [25] study, which investigated the effects of diazoxide, found similar changes in hyaluronan ($p < 0.04$ and $p < 0.04$ respectively). The Brettner study [22] investigated the effects of hydrocortisone. There were no significant differences in plasma syndecan-1 levels between intervention and control groups. However, they did establish significant lowering of HS levels intra-operatively. The Kim study [23] differed from the other clinical trials by comparing two interventions; hydroxyl starch (HES) vs. crystalloid. It has also differed from the other studies as patients had undergone off-pump CPB, which is of clinical significance as studies have shown differences in plasma syndecan-1 and plasma heparan sulphate concentrations between surgeries [29]. Following infusion of 20 mL/kg of the study fluids, median syndecan-1 levels were higher in the HES group than the crystalloid group [(79.7 (46.6–176.6) vs. 62.7 (30.1–103)]. However, overall peri-operative changes in syndecan-1 were not significantly different between the groups.

Of the 4 selected articles, the Brettner study [22] was the only to investigate the inflammatory response in addition to glycocalyx degradation. There was significant lowering of CRP levels on days 1, 2 and 3 post-operatively in the hydrocortisone group (p is 0.014, 0.012 and 0.022 respectively). Interleukin-6 (IL-6) was significantly lower in the intervention group to the control group ($p < 0.05$).

One study [27] used urinary trypsin inhibitor (UTI) to treat EG degradation in video-assisted thoracoscopy (VATS) lobectomy. There were no significant differences in baseline values of syndecan-1, HS, HGB and serum albumin levels in the control and UTI group. However, syndecan-1 levels were elevated at T1 in the control group (3.77 ± 3.15 versus 4.28 ± 3.30, $p = 0.022$), whereas the UTI group showed no significant increase at T1 (3.98 ± 3.04 versus 4.24 ± 3.12, $p = 0.160$). There were no obvious changes in HS levels between groups.

One study looked into the effects of pre-operative methylprednisolone treatment in total knee arthroplasty [24]. Syndecan-1 concentrations remained stable in the control group with a statistically significant drop from baseline to 24-h post-operative (14.1 ± 1.4

versus 12.4 ± 12.4, $p = 0.001$). Vascular endothelial growth factor (VEGF) increased in both groups with only a transient increase in the methylprednisolone group (T0->T6: 42.4 (2.5) to 54.8 (4.1), $p = 0.008$ versus 37.7 (2.5) to 47.7 (4.9), $p = 0.019$). Soluble thrombomodulin (sTE) increased in both groups but only transiently in the methylprednisolone group (T0->T6: 5.0 (0.2) to 5.4 (0.3), $p = 0.008$ versus 5.0 (0.2) to 5.2 (0.4), $p = 0.022$). CRP increased in both groups but less so in the methylprednisolone group (T0->T24: 4.5 (1.1) to 36.6 (3.6) versus 6.9 (2.2) to 74.4 (5.0)). Overall, effects of methylprednisolone were more pronounced at higher base values in sTM, sE-Selectin and VEGF ($p = 0.012$, $p = 0.009$ and $p < 0.001$, respectively) but this was not the case for syndecan-1.

One study looked into the effects of ringer's lactate (a form of fluid therapy) to prevent EG degradation in abdominal hysterectomy [26]. This article differs from other included studies, as the two groups were based on different anaesthetic medications (propofol or sevoflurane), with no control group. Plasma syndecan-1 and HS levels showed minimal variations during surgery but significantly increased 2-h post-operatively ($p < 0.05$ and $p < 0.001$, respectively, between 90 min and 2 h, compared using Wilcoxon's matched-pair test). Plasma concentrations for brain natriuretic peptide (BNP) showed small changes but no significant differences between groups.

There was a common significant increase in plasma syndecan-1 concentrations between baseline values and the first time-point post-operation. Data from the articles plotted (Figure 3) showed similar baseline values compared with their corresponding intervention group. Hence, we used the first timepoint post-operation from each article as the basis for the forest plot (Figure 4). Patients receiving diazoxide in CPB, methylprednisolone in CPB for neonates, methylprednisolone in VSD trial, methylprednisolone in total-knee arthroplasty, UTI in VATS lobectomy, had pooled mean differences of −1.4 (95% CI: −3.67, 0.87), −83.1 (95% CI: −150.4, −15.8), −59.0 (95% CI: −96.97, −21.03), −10 (95% CI: −20.79, 0.79) and −0.1 (95% CI: −9.0, −9.80), respectively.

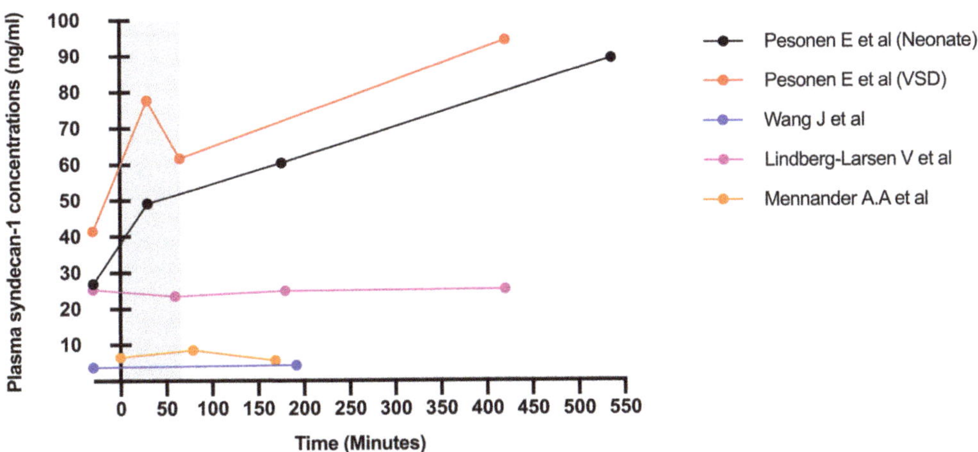

Figure 3. Changes in plasma syndecan-1 concentration in the control groups of the included articles. Values lower than x = 0 represent blood samples taken pre-operative. Values within the grey region are samples recorded intra-operatively.

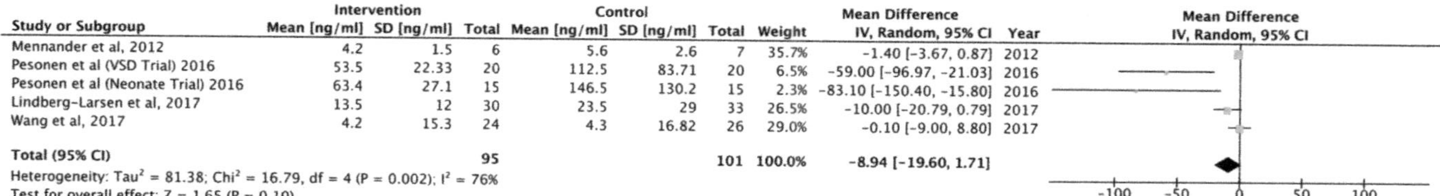

Study or Subgroup	Intervention Mean [ng/ml]	Intervention SD [ng/ml]	Intervention Total	Control Mean [ng/ml]	Control SD [ng/ml]	Control Total	Weight	Mean Difference IV, Random, 95% CI	Year
Mennander et al, 2012	4.2	1.5	6	5.6	2.6	7	35.7%	-1.40 [-3.67, 0.87]	2012
Pesonen et al (VSD Trial) 2016	53.5	22.33	20	112.5	83.71	20	6.5%	-59.00 [-96.97, -21.03]	2016
Pesonen et al (Neonate Trial) 2016	63.4	27.1	15	146.5	130.2	15	2.3%	-83.10 [-150.40, -15.80]	2016
Lindberg-Larsen et al, 2017	13.5	12	30	23.5	29	33	26.5%	-10.00 [-20.79, 0.79]	2017
Wang et al, 2017	4.2	15.3	24	4.3	16.82	26	29.0%	-0.10 [-9.00, 8.80]	2017
Total (95% CI)			**95**			**101**	**100.0%**	**-8.94 [-19.60, 1.71]**	

Heterogeneity: Tau² = 81.38; Chi² = 16.79, df = 4 (P = 0.002); I² = 76%
Test for overall effect: Z = 1.65 (P = 0.10)

Figure 4. Forest plot of the mean difference of plasma syndecan-1 concentration changes for treatments to protect against EG degradation in surgery.

4. Discussion

To the authors' knowledge, there has been no systematic review to investigate treatments to protect against EG degradation in surgery. In our study, the greatest change in plasma syndecan-1 concentrations from baseline values, was generally found at the first timepoint measured post-operatively. Syndecan-1 is a marker for greater trauma of EG shedding, so more time may be required for this extent of damage to occur and to be detected. The transient increase can be explained by the proteolytic degradation of the glycocalyx with subsequent rapid clearance, especially via the kidneys [30]. A slow systemic degradation could be explained by a general activation of leucocytes and platelets with an associated release of enzymes that shed the glycocalyx [31]

Pooled data from 196 patients in 5 randomised controlled trials for protective treatments against EG degradation in surgery indicated that patients treated with experimental drugs were more likely to respond than those treated with placebo, with a pooled mean difference of −8.94 (95% CI: −19.60, 1.61). Randomised controlled trials using methylprednisolone were shown to have the greatest effect to inhibit elevation of syndecan-1 levels. This may be important as it provides evidence for the development of personalised interventions, targeting patients in selected groups with altered risk profiles, for example, using bolus I.V. methylprednisolone in patients with high baseline values of endothelial activation and damage. Lindberg-Larsen study [24] found that the effect of methylprednisolone on syndecan-1, sTM and VEGF concentrations was dependant on the time of sampling, with these outcomes increasing with time. Pesonen study [27] showed similar changes in which methylprednisolone failed to inhibit the increase of syndecan-1 shedding at the early time-points, yet showed significant lowering of syndecan-1 levels to its corresponding control group for later time-points. A combination of the previous work on interleukins by Keski-Nisula [32] and these present syndecan-1 results suggest that methylprednisolone probably mediates the conservation of glycocalyx by an anti-inflammatory action. This is operative at the time-points when interleukin production is regulated via de novo protein synthesis due to glucocorticoid receptor activation. Of the three randomised clinical trials, with only one of them to have their upper-confidence interval to go beyond the line of no effect by 0.79 ng/mL, future work on methylprednisolone could be promising to protect the EG in surgery.

Brettner [22] showed hydrocortisone to reduce minor disturbances and not significant degradation of the EG. This shows that mechanistic pathways involved with shedding of heparan sulphate side-chains evoked by the combined stimuli of surgery were more susceptible to inhibition by hydrocortisone than the one leading to cleavage of the transmembrane core protein, syndecan-1. Whilst inflammatory markers (IL-6 and CRP) had shown significantly higher levels in the control group to the hydrocortisone group post-operatively, it had no relevant influence on inflammatory clinical parameters. As highlighted previously, the Lindberg-Larsen study [24] showed the effect of methylprednisolone on syndecan-1, sTM and VEGF concentrations was dependant on the time of sampling, with these outcomes increasing with time. However, this study showed no correlation with changes in CRP and any of the EG degradation markers. Analysis of the Nemme study [26] was challenging with no control group, however, there were no marked changes in BNP, syndecan-1, heparan sulphate and CRP between the two groups. There was no correlation between inflammatory markers (BNP and CRP) and glycocalyx shedding products.

The following results from the three articles differed from their previous research. Experimental studies had shown hydrocortisone to provide EG protective properties, most probably due to the stabilisation of mast cells, and, therefore, the amelioration of histamine, cytokines, lysases and protease production [9,33,34]. Stress doses of hydrocortisone administered before cardiac surgery in high risk patients attenuated systemic inflammation and improved early outcomes [35]. Glucocorticoids, such as methylprednisolone, have been shown in animal studies to reduce oedema formation and shedding of the glycocalyx by reducing the systemic inflammatory response in surgery [36]. Nemme [26] was unable to replicate the typical acute increase in BNPs seen previously experimentally [37,38] and post-operatively [39]. The included articles did not investigate the association between the degradation of the EG and cytokine production. Further surgical studies should look to understand the association and mechanism between EG shedding and cytokines' influence on endothelial permeability.

A common problem found in each randomised controlled trial was minimal shedding. The two reasons in which minimal shedding could have occurred are (1) the small patient sample and (2) the low operative stress in each clinical trial. Johansson study [12] showed that patients undergoing surgery from significant trauma had elevated syndecan-1 concentrations associated with raised inflammatory markers. This could highlight that significant EG degradation is needed to stimulate an inflammatory response, and hence, allow the action of anti-inflammatory medication to protect and treat against the symptoms.

Wang et al. [28] found a significantly greater decrease in plasma albumin compared with HGB at POD1, suggesting that serum albumin was not only lost as a result of blood loss but also by extravasation and attributed this to increased vascular permeability. Furthermore, serum albumin levels were significantly lower at POD1 in the control group compared to the UTI group, yet HGB levels were similar in both groups. This suggests that serum albumin leakage was reduced as a result of a decrease in microvascular permeability in the UTI group. With the UTI group showing no significant increase in syndecan-1, not only does this provide evidence of association between alleviation of degradation and preserving vascular permeability, it also suggests the development of personalised interventions against oedema formation as a result of surgery. With these findings, it may suggest a mechanism whereby UTI acts to protects the vascular endothelial barrier function during surgery, and therefore reduce tissue oedema. The Nemme study [26], also used the latter approach to investigate fluid maintenance in surgery to protect against EG degradation. A comparison between variables could only be made from the first hour post-infusion, which suggested intravascular albumin leakage out of the bloodstream at a normal rate. However, the data did not support previous work that suggested significant capillary leakage of albumin in response to hypervolemia [40].

Whilst this analysis provides useful information for clinicians for future work, the following limitations should be noted. As with all meta-analyses, the precision of pooled effect estimates is dependent on the sample size. Therefore, of all the included articles which had small sample sizes, greater patient cohorts are needed to be truly representative of the clinical outcomes experienced in the population. We have seen the contradictory evidence of EG degradation in surgery, yet it is well-understood that greater operative-stress does have an impact on endothelial integrity. Where there was a lack of ability to disrupt EG structure, this produced varying degrees of inflammatory responses and albumin leakage. Correlation between EG degradation and inflammatory markers as well as pronounced extravasation were difficult to analyse. Included articles did not investigate the source of EG degradation products produced during surgical operations. Overall, the included trials showed no significant difference in clinical signs and symptoms between intervention and control groups, differing to the work of Johansson et al. [12].

Despite the relative similarity of the study design of the randomised controlled trials (RCTs) included in the systematic review, there are always some levels of inconsistency due to variability in factors, such as treatment duration, outcomes assessed, patient age, severity of illness for need of surgery and co-morbid conditions. This is of clinical significance as plasma concentration of syndecan-1 and hyaluronic acid can vary dependant on the kidney's ability to excrete them [41]. These facts have not been reported consistently and cannot be assessed in this study. Finally, the authors were unable to find universal dosing strategies identified from official bodies for guidelines to treat EG degradation in surgery. Therefore, the dosing strategies in the RCTs were variable.

With the relatively-high level of heterogeneity indicated by I^2, these limitations highlight a real need for a set design for clinical trials to cohesively investigate protective treatments against EG degradation in surgery. Guidelines and reporting standards should be used to increase uniformity and reduced heterogeneity between RCTs, therefore maximising confidence, validity and comparative analysis for subsequent systematic reviews. Following is a list of suggestions for future RCTs on this topic;

- Study design: trials should be a placebo-controlled, double-blind, and a parallel design, with treatment administered pre-operatively or following the induction of anaesthesia but before the start of the surgical procedure. Time-points should be assessed at all stages of the surgical operation (pre-operative, intra-operative and post-operative) to assess trends in which the treatments affect the production of glycocalyx products.
- Patient population: RCTs should evaluate the work of Johansson study [12] to assess EG degradation in patients with greater operative stress. Mean age, concomitant medication, incidence and common co-morbidities should be collected and reported for each treatment group.
- Clinical outcomes: syndecan-1 and heparan sulphate should be used as primary markers to measure the degree of EG degradation. In addition, inflammatory markers (CRP, white blood cells) as well as capillary leakage of albumin (HGB and albumin) should be measured and correlated to identify relationships with glycocalyx shedding. Future work should use either Hedin and Hahn [42] or Hasselgren [43] technique to measure capillary leakage.

5. Conclusions

Whilst the effects of surgery are known to cause shedding of the EG, clinical surgical research into protective treatments for this structure are still in the early stages. The aim of this systematic review was to investigate treatments listed on large clinical databases to protect against EG degradation in surgery. Reviewing the pooled data from 385 patients in 8 RCTs, interventions looked to target (1) dampening the inflammatory response or (2) fluid maintenance during surgery. From these initial studies, we have seen some promising results. Methylprednisolone is shown to target the transient increase of syndecan-1 levels immediately after surgery. There is evidence that UTI can reduce the degradation of the EG with a reduction of albumin extravasation. Building from the initial findings of the included studies in this systematic review, and following the proposed suggestions for study design, the hope is that future RCTs will improve uniformity and maximise confidence for finding protective treatments for the EG in surgery.

Author Contributions: Conceptualization, H.Q.R.B.K. and G.C.R.; methodology, H.Q.R.B.K.; software, H.Q.R.B.K.; writing—original draft preparation H.Q.R.B.K.; writing—review and editing H.Q.R.B.K. and G.C.R. Both authors have read and agreed to the published version of the manuscript.

Funding: This research received no external funding.

Institutional Review Board Statement: Not applicable.

Informed Consent Statement: Not applicable.

Data Availability Statement: The data presented in this study are available within the article.

Conflicts of Interest: The authors declare no conflict of interest.

References

1. Beyer, A.M.; Gutterman, D.D. Regulation of the human coronary microcirculation. *J. Mol. Cell. Cardiol.* **2012**, *52*, 814–821. [CrossRef]
2. Yau, J.W.; Teoh, H.T.; Verma, S. Endothelial cell control of thrombosis. *BMC Cardiovas. Disord.* **2015**, *15*, 1–11. [CrossRef] [PubMed]
3. Reitsma, S.; Slaaf, D.W.; Vink, H.; van Zandvoort, M.A.M.J.; oude Egbrink, M.G.A. The endothelial glycocalyx: Composition, functions, and visualization. *Eur. J. Physiol.* **2007**, *454*, 345–359. [CrossRef]
4. Gilepsie, G.P.; Muller, U. Mechanotransduction by hair Cells: Models, molecules, and mechanisms. *Cell* **2009**, *139*, 33–44. [CrossRef]
5. Pahakis, M.Y.; Kosky, J.R.; Dull, R.O.; Tarbell, J.M. The role of endothelial glycocalyx components in mechanotransduction of fluid shear stress. *Biochem. Biophys. Res. Commun.* **2007**, *355*, 228–233. [CrossRef]
6. Squire, J.M.; Chew, M.; Nneji, G.; Barry, J.; Michel, C. Quasi-periodic substructure in the microvessel endothelial glycocalyx: A possible explanation for molecular filtering. *J. Struct. Biol.* **2001**, *136*, 239–255. [CrossRef] [PubMed]

7. Zeng, Y.; Tarbell, J.M. The adaptive remodeling of endothelial glycocalyx in response to fluid shear stress. *PLoS ONE* **2014**, *9*, 1–15. [CrossRef]
8. Robich, M.; Ryzhov, S.; Kacer, D.; Palmeri, M.; Peterson, S.M.; Quinn, R.D.; Carter, D.; Shephard, F.; Hayes, T.; Sawyer, D.B.; et al. Prolonged cardiopulmonary bypass is associated with endothelial glycocalyx degradation. *J. Surg. Res.* **2020**, *251*, 287–295. [CrossRef]
9. Chappell, D.; Hofmann-Kiefer, K.; Jacob, M.; Rehm, M.; Briegel, J.; Welsch, U.; Conzen, P.; Becker, B.F. TNF-alpha induced shedding of the endothelial glycocalyx is prevented by hydrocortisone and antithrombin. *Basic Res. Cardiol.* **2009**, *104*, 78–89. [CrossRef]
10. Uchimido, R.; Schmidt, E.P.; Shapiro, N.I. The glycocalyx: A novel diagnostic and therapeutic target in sepsis. *Crit. Care* **2019**, *23*, 1–12. [CrossRef] [PubMed]
11. Triantafyllou, C.; Nikolao, M.; Ikonomidis, I.; Bamias, G.; Kouretas, D.; Andreadou, I.; Tsoumani, M.; Thymis, J.; Papaconstantinou, I. Effects of anti-inflammatory treatment and surgical intervention on endothelial glycocalyx, peripheral and coronary microcirculatory function and myocardial deformation in inflammatory bowel disease patients: A two-arms two-stage clinical trial. *Diagnostics* **2021**, *11*, 993. [CrossRef]
12. Johansson, P.; Stensballe, J.; Rasmussen, L.S.; Ostrowski, S.R. A high admission syndecan-1 level, a marker of endothelial glycocalyx degradation, is associated with inflammation, protein C depletion, fibrinolysis, and increased mortality in trauma patients. *Ann. Surg.* **2011**, *254*, 194–200. [CrossRef]
13. Yilmaz, O.; Afsar, B.; Ortiz, A.; Kanbay, M. The role of endothelial glycocalyx in health and disease. *Clin. Kidney J.* **2019**, *12*, 611–619. [CrossRef] [PubMed]
14. Rahbar, E.; Cardenas, J.C.; Baimukanova, G.; Usadi, B.; Bruhn, R.; Pati, S.; Ostrowski, S.R.; Johansson, P.I.; Holcomb, J.B.; Wade, C.E. Endothelial glycocalyx shedding and vascular permeability in severely injured trauma patients. *J. Transl. Med.* **2015**, *13*, 1–7. [CrossRef] [PubMed]
15. Devaraj, S.; Yun, J.; Adamson, G.; Galvez, J.; Jialal, I. C-reactive protein impairs the endothelial glycocalyx resulting in endothelial dysfunction. *Cardiovas. Res.* **2009**, *84*, 479–484. [CrossRef] [PubMed]
16. Chappell, D.; Bruegger, D.; Potzel, J.; Jacob, M.; Brettner, F.; Vogeser, M.; Conzen, P.; Becker, B.F.; Rehm, M. Hypervolemia increases release of atrial natriuretic peptide and shedding of the endothelial glycocalyx. *Crit. Care* **2014**, *18*, 538–546. [CrossRef] [PubMed]
17. Rehm, M.; Bruegger, D.; Christ, F.; Conzen, P.; Thiel, M.; Jacob, M.; Chappell, D.; Stoeckelhuber, M.; Welsch, U.; Reichart, B.; et al. Shedding of the endothelial glycocalyx in patients undergoing major vascular surgery with global and regional ischemia. *Circulation* **2007**, *116*, 1896–1906. [CrossRef]
18. Gybel-Brask, M.; Rasmussen, R.; Stensballe, J.; Johansson, P.A.; Ostrowski, S.R. Effect of delayed onset prostacyclin on markers of endothelial function and damage after subarachnoid haemorrhage. *Acta. Neurochir.* **2017**, *159*, 1073–1078. [CrossRef]
19. Halenarova, K. Effects of corticosteroids on endothelial dysfunction in cardiac surgery patients. *Crit. Care Med.* **2019**, *47*, 94. [CrossRef]
20. Tosif, S. Preoperative microvascular protection in patients undergoing major abdominal surgery. Unpublished work. 2018.
21. Yonsei University. The effect of balanced crystalloid versus 5% albumin on endothelial glycocalyx degradation in patients undergoing off-pump coronary artery bypass surgery. Unpublished work. 2019.
22. Brettner, F.; Chappell, D.; Nebelsiek, T.; Hauera, D.; Schellinga, G.; Becker, B.F.; Rehm, M.; Weis, F. Preinterventional hydrocortisone sustains the endothelial glycocalyx in cardiac surgery. *Clin. Hemorheol. Microcirc.* **2019**, *71*, 59–70. [CrossRef] [PubMed]
23. Kim, T.K.; Nam, K.; Cho, Y.J.; Min, J.J.; Hong, Y.J.; Park, K.U.; Hong, D.M.; Jeon, Y. Microvascular reactivity and endothelial glycocalyx degradation when administering hydroxyethyl starch or crystalloid during off-pump coronary artery bypass graft surgery: A randomised trial. *Anaesthesia* **2017**, *72*, 204–213. [CrossRef]
24. Lindberg-Larsen, V.; Ostrowski, S.R.; Lindberg-Larsen, M.; Rovsing, M.L.; Johansson, P.I.; Kehlet, H. The effect of pre-operative methylprednisolone on early endothelial damage after total knee arthroplasty: A randomised, double-blind, placebo-controlled trial. *Anaesthesia* **2017**, *72*, 1217–1224. [CrossRef]
25. Mennander, A.A.; Shalaby, A.; Oksala, N.; Leppanen, T.; Hamalainen, M.; Huovinen, S.; Zhao, F.; Moilanen, E.; Tarkka, M. Diazoxide may protect endothelial glycocalyx integrity during coronary artery bypass grafting. *Scand. Cardiovasc. J.* **2012**, *46*, 339–344. [CrossRef]
26. Nemme, J.; Krizhanovskii, C.; Ntika, S.; Sabelnikovs, O.; Vanags, I.; Hahn, R.G. Hypervolemia does not cause degradation of the endothelial glycocalyx layer during open hysterectomy performed under sevoflurane or propofol anesthesia. *Acta. Anaesthesiol. Scand.* **2019**, *64*, 538–545. [CrossRef]
27. Pesonen, E.; Keski-Nisula, J.; Andersson, S.; Palo, R.; Salminen, J.; Suominen, P.K. High-dose methylprednisolone and endothelial glycocalyx in paediatric heart surgery. *Acta. Anaesthesiol. Scand.* **2016**, *60*, 1386–1394. [CrossRef]
28. Wang, J.; Wu, A.; Wu, Y. Endothelial Glycocalyx Layer: A Possible Therapeutic Target for Acute Lung Injury during Lung Resection. *BioMed Res. Int.* **2017**, *1*, 1–8. [CrossRef]
29. Bruegger, D.; Rehm, M.; Abicht, J.; Paul, J.O.; Stoeckelhuber, M.; Pfirrmann, M.; Reichart, B.; Becker, B.F.; Christ, F. Shedding of the endothelial glycocalyx during cardiac surgery: On-pump versus off-pump coronary artery bypass graft surgery. *J. Thorac. Cardiovasc. Surg.* **2009**, *138*, 1445–1447. [CrossRef] [PubMed]

30. Willis, B.A.; Oragui, E.E.; Dung, N.M.; Loan, H.T.; Chau, N.V.; Farrar, J.J.; Levin, M. Size and charge characteristics of the protein leak in dengue shock syndrome. *J. Infect. Dis.* **2004**, *190*, 810–818. [CrossRef]
31. Mulivor, A.W.; Lipowsky, H.H. Role of glycocalyx in leukocyte-endothelial cell adhesion. *Am. J. Physiol. Heart Circ. Physiol.* **2002**, *283*, 1282–1291. [CrossRef] [PubMed]
32. Keski-Nisula, J.; Pesonen, E.; Olkkola, K.T.; Peltola, K.; Neuvonen, P.J.; Tuominen, N.; Sairanen, H.; Andersson, S.; Suominen, P.K. Methylprednisolone in neonatal cardiac surgery: Reduced inflammation without improved clinical outcome. *Ann. Thorac. Surg.* **2013**, *95*, 2126–2132. [CrossRef] [PubMed]
33. Chappell, D.; Jacob, M.; Hofmann-Kiefer, K.; Bruegger, D.; Rehm, M.; Conzen, P.; Welsch, U.; Becker, B.F. Hydrocortisone preserves the vascular barrier by protecting the endothelial glycocalyx. *Anesthesiology* **2007**, *107*, 776–784. [CrossRef]
34. Chappell, D.; Dorfler, N.; Jacob, M.; Rehm, M.; Welsch, U.; Conzen, P.; Becker, B.F. Glycocalyx protection reduces leukocyte adhesion after ischemia/reperfusion. *Shock* **2010**, *34*, 133–139. [CrossRef]
35. Kilger, E.; Weis, F.; Briegel, J.; Frey, L.; Goetz, A.E.; Reuter, D.; Nagy, A.; Schuetz, A.; Lamm, P.; Knoll, A.; et al. Stress doses of hydrocortisone reduce severe systemic inflammatory response syndrome and improve early outcome in a risk group of patients after cardiac surgery. *Crit. Care Med.* **2003**, *31*, 1068–1074. [CrossRef]
36. Checchia, P.A.; Bronicki, R.A.; Costello, J.M.; Costello, J.M.; Nelson, D.P. Steroid use before pediatric cardiac operations using cardiopulmonary bypass: An international survey of 36 centers. *Pediatr. Crit. Care Med.* **2005**, *6*, 441–444. [CrossRef] [PubMed]
37. Norberg, A.; Hahn, R.G.; Li, H.; Olsson, J.; Prough, D.S.; Børsheim, E.; Wolf, S.; Minton, R.K.; Svensén, C.H. Population volume kinetics predicts retention of 0.9% saline infused in awake and isoflurane-anesthetized volunteers. *Anesthesiology* **2007**, *107*, 24–32. [CrossRef] [PubMed]
38. Jacob, M.; Bruegger, D.; Rehm, M.; Welsch, U.; Conzen, P. Contrasting effects of colloid and crystalloid resuscitation fluids on cardiac vascular permeability. *Anesthesiology* **2006**, *104*, 1223–1231. [CrossRef] [PubMed]
39. Pepys, M.B.; Hirschfield, G.M. C-reactive protein: A critical update. *J. Clin. Investig.* **2013**, *111*, 1805–1812. [CrossRef]
40. Rehm, M.; Haller, M.; Orth, V.; Kreimeier, U.; Jacob, M.; Dressel, H.; Mayer, S.; Brechtelsbauer, H.; Finsterer, U. Changes in blood volume and hematocrit during acute perioperative volume loading with 5% albumin or 6% hetastarch solutions in patients before radical hysterectomy. *Anesthesiology* **2001**, *95*, 849–856. [CrossRef]
41. Hahn, R.G.; Hasselgren, E.; Håkan, B.; Markus, Z.; Zdolsek, J. Biomarkers of endothelial injury in plasma are dependent on kidney function. *Clin. Hemorheol. Microcirc.* **2019**, *72*, 161–168. [CrossRef] [PubMed]
42. Hedin, A.; Hahn, R.G. Volume expansion and plasma protein clearance during intravenous infusion of 5% albumin and autologous plasma. *Clin. Sci.* **2005**, *106*, 217–224. [CrossRef] [PubMed]
43. Hasselgren, E.; Zdolsek, M.; Zdolsek, J.; Björne, H.; Krizhanovskii, C.; Stelia, N.; Hahn, R.G. Long intravascular persistence of albumin 20% in postoperative patients. *Anesth. Analg.* **2019**, *129*, 1232–1239. [CrossRef] [PubMed]

MDPI
St. Alban-Anlage 66
4052 Basel
Switzerland
Tel. +41 61 683 77 34
Fax +41 61 302 89 18
www.mdpi.com

Applied Sciences Editorial Office
E-mail: applsci@mdpi.com
www.mdpi.com/journal/applsci

www.ingramcontent.com/pod-product-compliance
Lightning Source LLC
LaVergne TN
LVHW071258170726
843515LV00006BA/1431

* 9 7 8 3 0 3 6 5 4 4 2 7 4 *